Lecture Notes in Computer Science 782

Edited by G. Goos and J. Hartmanis

Advisory Board: W. Brauer   D. Gries   J. Stoer

Jürg Gutknecht (Ed.)

Programming Languages and System Architectures

International Conference
Zurich, Switzerland, March 2-4, 1994
Proceedings

Springer-Verlag

Berlin Heidelberg New York
London Paris Tokyo
Hong Kong Barcelona
Budapest

Series Editors

Gerhard Goos
Universität Karlsruhe
Postfach 69 80
Vincenz-Priessnitz-Straße 1
D-76131 Karlsruhe, Germany

Juris Hartmanis
Cornell University
Department of Computer Science
4130 Upson Hall
Ithaca, NY 14853, USA

Volume Editor

Jürg Gutknecht
Institut für Computersysteme, ETH Zentrum
CH-8092 Zürich, Switzerland

CR Subject Classification (1991): D, B.3, B.5, B.7, C.1-3, F.1, F.3

ISBN 3-540-57840-4 Springer-Verlag Berlin Heidelberg New York
ISBN 0-387-57840-4 Springer-Verlag New York Berlin Heidelberg

CIP data applied for

© Springer-Verlag Berlin Heidelberg 1994
Printed in Germany

Typesetting: Camera-ready by author
SPIN: 10132011   45/3140-543210 - Printed on acid-free paper

Frontierland

by Jürg Gutknecht (Program Chair)

Programming languages and system architectures –
two fundamental pillars of computer science,
rich in tradition and yet so modern,
mutually dependent and yet so separate,
tightly coupled and yet a big gap between,
stimulating as well as impeding one another,
close friends and yet bitter opponents.

Programming languages and system architectures –
two topics representing the frontiers of two different worlds,
the world of algorithms and the world of electronic circuits,
the world of formalists and the world of physicists.

This conference is an adventure in a field of high voltage,
in a land where two different cultures clash,
in a land that is both fertile and barren.

We are invited to explore this exciting land
under the guidance of some world-renowned researchers,
and we are well-advised to accept the offer gratefully,
because opportunities like this are rare.
There are not many experts in such a difficult terrain,
and only a few have ever celebrated bridging the gap.

One of the most outstanding among them
will have particularly good cause for a celebration:
By pure chance, the conference coincides
with Niklaus Wirth's 60th birthday.

Happy birthday!

Program Committee

J. Gutknecht, ETH Zürich, CH (Chairman)
R. P. Cook, Microsoft, USA
C. Coray, EPF Lausanne, CH
O. J. Dahl, University of Oslo, N
E. W. Dijkstra, University of Texas, USA
G. Goos, TU Karlsruhe, D
S. Graham, UC Berkeley, USA
D. Gries, Cornell University, USA
D. Hanson, Princeton University, USA
B. Kernighan, AT&T Bell Labs, USA
B. W. Lampson, DEC SRC, USA
J. Ludewig, TU Stuttgart, D
J. Misra, University of Texas, USA
H. Mössenböck, ETH Zürich, CH
R. Needham, Cambridge University, UK
S. Owicki, DEC SRC, USA
G. Pomberger, University of Linz, A
P. Rechenberg, University of Linz, A
M. Reiser, IBM Research, CH
B. A. Sanders, ETH Zürich, CH
P. Schulthess, University of Ulm, D
A. Shaw, University of Washington, USA
A. Tanenbaum, Vrije Universiteit Amsterdam, NL
P. D. Terry, Rhodes University, SA
L. Tesler, Apple Corp., USA
J. Welsh, University of Queensland, AUS
N. Wirth, ETH Zürich, CH

Organizing Committee

C. A. Zehnder, ETH Zürich (Chairman)
W. Gander, ETH Zürich
M. Franz, ETH Zürich

Contents

Interconnecting Computers:
Architecture, Technology, and Economics

Butler W. Lampson

Systems Research Center, Digital Equipment Corporation
One Kendall Sq., Bldg 700, Cambridge, MA 02138
lampson@src.dec.com

Abstract. Modern computer systems have a recursive structure of processing and storage
elements that are interconnected to make larger elements:
Functional units connected to registers and on-chip cache.
Multiple processors and caches connected to main memories.
Computing nodes connected by a message-passing local area network.
Local area networks bridged to form an extended LAN.
Networks connected in a wide-area internet.
All the computers in the world exchanging electronic mail.
Above the lowest level of transistors and gates, the essential character of these connec-
tions changes surprisingly little over about nine orders of magnitude in time and space.
Connections are made up of nodes and links; their important properties are bandwidth, la-
tency, connectivity, availability, and cost. Switching is the basic mechanism for connecting lots of things. There are many ways to implement it, all based on multiplexing and
demultiplexing. This paper describes some of them and gives many examples. It also
considers the interactions among the different levels of complex systems.

1 Introduction

A point of view is worth 80 points of IQ.
                                    Alan Kay

A computing system is part of a complex web of interconnections. We impose order
on this web by organizing it hierarchically: a system is made up of connected subsystems, and is itself connected to other parts of the larger system that contains it. Figure
1 shows some of this structure as it exists today, ranging from an individual processor
register to the world-wide Internet. It is striking that there is a range of at least seven
orders of magnitude in both the number of components (shown on the right) and the
minimum time for communication (shown on the left).

In spite of this enormous variation, the interconnections themselves are remarkably
similar in design. Both the interface that an interconnection offers to a system and the
structure of the interconnection itself are taken from a small set of design alternatives.
This paper describes many of these alternatives and illustrates them with examples
drawn from every level of the figure.

An interface deals either with messages or with storage, and is characterized by a
few performance parameters: bandwidth, latency, connectivity, and availability. Section 2 describes these variations.

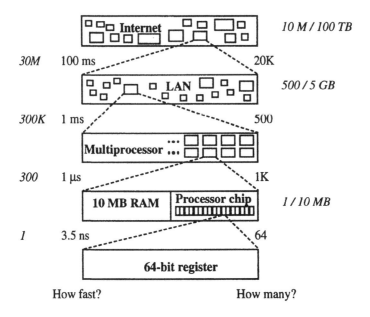

Fig. 1. Scales of interconnection. Relative speed and size are in italics.

An implementation of a connection is made up of links and nodes. We explain this classification in Section 3 and study links in Section 4. A node in turn can be classified as a converter, a multiplexer, or a switch, and the next three sections are devoted to these components.

A critical property of both links and nodes is that they can be built by composing lower-level links and nodes according to uniform principles. Because different levels in this recursion are so similar, a study of interconnections in general reveals much about the details of any particular one. In the simplest case the interfaces in the composite are the same as those of the whole. This is the subject of Section 8, and Section 9 then treats the general case.

Section 10 gives a brief treatment of fault tolerance, and we end with a conclusion.

Another unifying theme is how the evolution of silicon and fiber optic technology affects interconnection.

As more devices can fit on a single chip, it becomes feasible to use wide on-chip data paths, and to depend on control that is both complex and fast as long as the speed can be obtained by using more gates. The second fact tends to make designs similar at different scales, since it means that a fast, low-level implementation does not have to be extremely simple.

As the bandwidth available on a single fiber rises toward the tens of terabits/second that seems to be feasible in the long run, a big system can increasingly have the same interconnection bandwith as a small one. This too tends to make designs similar at different scales.

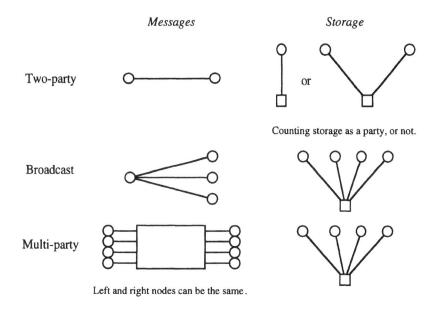

	Messages	*Storage*

Two-party

Counting storage as a party, or not.

Broadcast

Multi-party

Left and right nodes can be the same.

Fig. 2. Communication styles: messages and storage.
Legend: ○ = active node, □ = storage node

2 Interfaces for communication

In the Turing tarpit everything is possible, but nothing is easy.
Alan Perlis

The duality between states and events is a recurring theme in computer science. For us it is the choice between messages and storage as the style of communication. Both are universal: you can do anything with one that you can do with the other, and we will see how to implement each one using the other. But the differences are important.

Figure 2 shows the system structures for various communication patterns using messages and using storage. The pictures reflect the role of storage as a passive, possible shared repository for data.

Multi-party communication requires addresses, which can be flat or hierarchical. A flat address has no structure: the only meaningful operation (other than communication) is equality. A hierarchical address, sometimes called a path name, is a sequence of flat addresses or simple names, and if one address is a prefix of another, then in some sense the party with the shorter address contains, or is the parent of, the party with the longer one. Usually there is an operation to enumerate the children of an address. Flat addresses are usually fixed size and hierarchical ones variable, but there are exceptions. An address may be hierarchical in the implementation but flat at the interface, for instance an Internet address or a URL in the World Wide Web. The examples below should clarify these points.

2.1 Messages

The interface for messages is the familiar send and receive operations. The simplest form is a blocking receive, in which a process or thread of control doing a receive waits until a message arrives. If the receiver wants to go on computing it forks a separate thread to do the receive. The alternative is an interrupt when a message arrives; this is much more complicated to program but may be necessary in an old-fashioned system that has no threads or expensive ones.

A range of examples serves to illustrate the possibilities:

System	Address	Sample address	Delivery Ordered	Reliable
J-machine[4]	source route	4 north, 2 east	yes	yes
IEEE 802 LAN	6 byte flat	FF F3 6E 23 A1 92	no	no
IP	4 byte hierarchical	16.12.3.134	no	no
TCP	IP + port	16.12.3.134 / 3451	yes	yes
RPC	TCP + procedure	16.12.3.134 / 3451 / Open	yes	yes
E-mail	host name + user	lampson@src.dec.com	no	yes

Usually there is some buffering for messages that have been sent but not yet received, and the sender is blocked (or perhaps gets an error) when the buffering is exhausted. This kind of "back-pressure" is important in many implementations of links as well; see Section 4. If there is no buffering the system is said to be "synchronous" because the send and the receive must wait for each other; this scheme was introduced in CSP [7] but is unpopular because the implementation must do extra work on every message. The alternative to blocking the send is to discard extra messages; the 802 and IP interfaces do this. Except in real-time systems where late messages are useless, an interface that discards messages is usually papered over using backoff and retry; see Section 4.

Message delivery may be ordered, reliable, both, or neither. Messages which may be reordered or lost are often cheaper to implement in a large system, so much cheaper, in fact, that ordered reliable messages are best provided by end-to-end sequence numbering and retransmission on top of unordered unreliable messages [10]. TCP on IP is an example of this, and there are many others.

Often messages are used in an asymmetrical "request–response", "client–server", or "remote procedure call" pattern [2] in which the requester always follows a send immediately with a receive of a reply, and the responder always follows a receive with some computation and a send of a response. This pattern simplifies programming because it means that communicating parties don't automatically run in parallel; concurrency can be programmed explicitly as multiple threads if desired.

A message interface may allow broadcast or multicast to a set of receivers. This is useful for barrier synchronization in a multiprocessor; the bandwidth is negligible, but the low latency of the broadcast is valuable. At the opposite extreme in performance broadcast is also useful for publishing, where latency is unimportant but the bandwidth may be considerable. And broadcast is often used to discover the system configuration at startup; in this application performance is unimportant.

It's straightforward to simulate storage using messages by implementing a storage server that maintains the state of the storage and responds suitably to `load` and `store` messages. This is the simplest and most popular example of the client–server pattern.

2.2 Storage

The interface for storage is the familiar `load` and `store` operations. Like `receive` and `send`, these operations take an address, and they also return or take a data value. Normally they are blocking, but in a high-performance processor the operations the programmer sees are implemented by non-blocking `load` and write-buffered `store` together with extra bookkeeping.

Again, some examples show the range over which this interface is useful:

System	Address	Sample address	Data value
Main memory	32-bit flat	04E72A39	1, 2, 4, or 8 bytes
File system [20]	path name	/udir/bwl/Mail/inbox/214	0-4 Gbytes
World Wide Web	protocol + host name + path name	http://src.dec.com/ SRC/docs.html	typed, variable size

Storage has several nice properties as a communication interface. Like request–response, it introduces no extra concurrency. It provides lazy broadcast, since the contents of the storage is accessible to any active party. And most important, it allows caching as an optimization that reduces both the latency of the interface and the bandwidth consumed. Of course nothing is free, and caching introduces the problem of cache coherence [9, 11], the treatment of which is beyond the scope of this paper.

There are two ways to simulate messages using storage. One is to construct a queue of waiting messages in storage, which the receiver can poll. A common example is an input/output channel that takes its commands from memory and delivers status reports to memory. The other method is to intercept the `load` and `store` operations and interpret them as messages with some other meaning. This idea was first used in the Burroughs B-5000 [1], but became popular with the PDP-11 Unibus, in which an input/output device has an assigned range of memory addresses, sees `load` and `store` operations with those addresses, and interprets the combination of address and data as an arbitrary message, for instance as a tape rewind command. Most input/output systems today use this idea under the name "programmed I/O". An example at a different level is the Plan 9 operating system [16], the Unibus of the '90s, in which everything in the system appears in the file name space, and the display device interprets loads and stores as commands to copy regions of the bitmap or whatever. The World Wide Web does the same thing less consistently but on a much larger scale.

2.3 Performance

The performance parameters of a connection are:

— *Latency*: how long a minimum communication takes. We can measure the latency in bytes by multiplying the latency time by the bandwidth; this gives the capacity penalty for each separate operation. There are standard methods for minimizing the effects of latency:

Caching reduces latency when the cache hits.

Prefetching hides latency by the distance between the prefetch and the use.

Concurrency tolerates latency by giving something else to do while waiting.

— *Bandwidth*: how communication time grows with data size. Usually this is quoted for a two-party link. The "bisection bandwidth" is the minimum bandwidth across a set of links that partition the system if they are removed; it is a lower bound on the possible total rate of communication. There are standard methods for minimizing the cost of bandwidth:

Caching saves bandwidth when the cache hits.

More generally, locality saves bandwidth when cost increases with distance.

Combining networks reduce the bandwidth to a hot spot by combining several operations into one, several loads or increments for example [17].

— *Connectivity*: how many parties you can talk to. Sometimes this is a function of latency, as in the telephone system, which allows you to talk to millions of parties but only one at a time.

— *Predictability*: how much latency and bandwidth vary with time. Variation in latency is called "jitter"; variation in bandwidth is called "burstiness". The biggest difference between the computing and telecommunications cultures is that computer communication is basically unpredictable, while telecommunications service is traditionally highly predictable.

— *Availability*: the probability that an attempt to communicate will succeed.

Uniformity of performance at an interface is often as important as absolute performance, because dealing with non-uniformity complicates programming. Thus performance that depends on locality is troublesome, though often rewarding. Performance that depends on congestion is even worse, since congestion is usually much more difficult to predict than locality. By contrast, the Monarch multiprocessor [17] provides uniform, albeit slow, access to a shared memory from 64K processors, with a total bandwidth of 256 Gbytes/sec and a very simple programming model. Since all the processors make memory references synchronously, it can use a combining network to eliminate many hot spots.

3 Implementation components

An engineer can do for a dime what any fool can do for a dollar.
Anonymous

Communication systems are made up of links and nodes. Data flows over the links. The nodes connect and terminate the links. Of course, a link or a node can itself be a communication system made up of other links and nodes; we study this recursive structure in Sections 8 and 9.

There are two kinds of nodes: converters and switches. A converter connects two links of different types, or a terminal link and the client interface. The simplest switches connect one link to many; they are called multiplexers and demultiplexers depending on whether the one link is an output or an input. A general switch connects any one of a set of input links to any one of a set of output links.

The bandwidth of a connection is usually the minimum bandwidth of any link or node. The latency is the sum of several terms:

    link time, which consists of

        time of flight for one bit, usually at least half the speed of light, plus

        message size (number of bits) divided by bandwidth;

    switching time;

    buffer delays;

    conversion time (especially at the ends).

The cost of a connection is the total cost of its links and nodes. We study the cost of physical links in Section 4. Nodes are made of silicon and software, and software runs on silicon. Hence the cost of a node is governed by the cost of silicon, which is roughly proportional to area in a mature process. Since 1960 the width of a device (transistor or wire) on a silicon die has been cut in half every five years. The number of devices per unit area increases with the square of the width, and the speed increases linearly. Thus the total amount of computing per unit area, and hence per unit cost (measured in device-cycles), grows with the cube of the width and doubles three times every five years, or every 20 months [6, 13].

The cost of a node therefore tends to zero as long as it effectively uses the ever-increasing number of devices on a chip. There are two consequences:

— Concurrency on the chip is essential, in the form of wide busses and multiple function units.

— Complex control can be made fast as long as it can take advantage of lots of gates. Instead of a large number of sequential steps, each doing a small amount of work, it's possible to have lots of concurrent finite state machines, and to use lots of combinational logic to do more work in a single cycle.

We see typical results in Ethernet interface chips that cost $10, or in a high-bandwidth, low-latency, robust, reliable, low-cost switched network like Autonet [22].

In addition, dramatic improvements in fiber optics mean that almost as much bandwidth is available on long distance links as locally, and at much lower cost that in the past [5].

4   Links

There are many kinds of physical links, with cost and performance that vary based on length, number of drops, and bandwidth. Here are some current examples. Bandwidth is in bytes/second, and the "+" signs mean that software latency must be added.

Medium	Link	Bandwidth		Latency		Width
Alpha chip	on-chip bus	2.2	GB/s	3.6	ns	64
PC board	RAM bus	0.5	GB/s	150	ns	8
	PCI I/O bus	133.0	MB/s	250	ns	32
Wires	HIPPI	100	MB/s	100	ns	32
	SCSI	20	MB/s	500	ns	16
LAN	FDDI	12.5	MB/s	20 +	μs	1
	Ethernet	1.25	MB/s	100 +	μs	1
Wireless	WaveLAN	.25	MB/s	100 +	μs	1
Fiber	OC-48	300	MB/s	5	μs/km	1
Coax cable	T3	6	MB/s	5	μs/km	1
Copper pair	T1	0.2	MB/s	5	μs/km	1
Copper pair	IDSN	16	KB/s	5	μs/km	1
Broadcast	CAP 16	3	MB/s	3	μs/km	6 MHz

A physical link can be unidirectional ("simplex") or bidirectional ("duplex"). A duplex link may operate in both directions at the same time ("full-duplex"), or in one direction at a time ("half-duplex"). A pair of simplex links running in opposite directions form a full-duplex link, as does a half-duplex link in which the time to reverse direction is negligible.

To increase the bandwidth of a link, run several copies of it in parallel. This goes by different names in different branches of our subject; "space division multiplexing" and "striping" are two of them. Common examples are

Parallel busses, as in the first five lines of the table.

Switched networks: the telephone system and switched LANs (see Section 7).

Multiple disks, each holding part of a data block, that can transfer in parallel.

Cellular telephony, using spatial separation to reuse the same frequencies.

In the latter two cases there must be physical switches to connect the parallel links.

Another use for multiple links is fault tolerance, discussed in Section 10.

Many links do not have a fixed bandwidth that is known to the sender, because of multiplexing inside the link. Instead, some kind of *flow control* is necessary to match the flow of traffic to the link's capacity. A link can provide this in two ways:

— By dropping excess traffic and signaling "trouble" to the sender, either explicitly or by failing to return an acknowledgment. The sender responds by waiting for a while and then retransmitting. The sender increases the wait by some factor after every trouble signal and decreases it with each trouble-free send. In this "exponential back-off" scheme the sender is using the wait as an estimate of the link capacity. It is used in the Ethernet and in TCP [14, 8]

— By supplying "back-pressure" that tells the sender how much it can send without suffering losses. This can take the form of start and stop signals, or of "credits" that allow a certain amount of traffic to be sent. The number of unused credits the sender has is called its "window". Let b be the bandwidth at which the sender can send when it has permission and r be the time for the link to respond to new traffic from the sender. A start–stop scheme can allow rb units of traffic between a start and a stop; a link that has to buffer this traffic will overrun and lose traffic if r is too large. A credit scheme needs rb credits when the link is idle to keep running at full bandwidth; a link will underrun and waste bandwidth if r is too large. The failure mode of the credit scheme is usually less serious. Start–stop is used in the Autonet [22] and on RS-232 serial lines under the name XON-XOFF; credits are used in TCP [8].

Either of these schemes can be implemented on an individual link. An alternative is to let internal links simply drop excess traffic and to implement backoff end-to-end [19]. TCP does this, and it confusingly also uses credits to keep the receiver's buffers from overflowing.

5   Converter nodes

Many converters from one kind of link to another connect a fast link to a slow one and are therefore part of a multiplexer or demultiplexer. Most other converters are for backward compatibility. They are usually cheap, because by the time an old link must be connected to a new one, hardware technology has improved so much that the old link is simple to implement. A glance at the back of a Macintosh or a stereo receiver shows how many different connectors you can buy for a small amount of money.

The converters that terminate connections are another matter. For a simple, synchronous link that is designed along with its end nodes, like the bus between a register file and a functional unit on a processor chip, the converter is implemented in simple, fast hardware and presents a design problem only if a lot of switching is involved.

Terminating a network link is much more complicated because of requirements for standardization and fault-tolerance. Furthermore, the link is usually specified without much attention to the problem of terminating it. A network converter consists of an "adapter" or "controller" together with "driver" software. In computer applications the driver is usually the main source of latency and often a serious bandwidth bottleneck as well, especially when individual messages are small. To see why this is true, consider Amdahl's rule that one instruction of useful work needs one bit of I/O. If a message is 20 bytes (a common size for multiprocessors) and we want to keep the driver overhead below 10%, there are only 16 instructions available for handling each message. It takes a lot of care to handle a message this cheaply [3, 4, 21]. Certainly it cannot be done unless the controller and the driver are designed together.

6   Multiplexer nodes

A multiplexer combines traffic from several input links onto one output link, and a demultiplexer separates traffic from one input link onto several output links. The multiplexed links are called "sub-channels" of the one link, and each one has an address. Figure 3 shows various examples.

There are three main reasons for multiplexers:

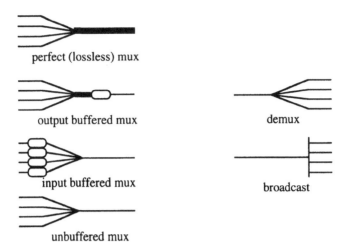

Fig. 3. Multiplexers and demultiplexers. Traffic flows from left to right.

— Traffic must flow between one node and many, for example when the one node is a busy server or the head end of a cable TV system.

— One wide wire may be cheaper than many narrow ones, because there is only one thing to install and maintain, or because there is only one connection at the other end. Of course the wide wire is more expensive than a single narrow one, and the multiplexers must also be paid for.

— Traffic aggregated from several links may be more predictable than traffic from a single one. This happens when traffic is bursty (varies in bandwidth) but uncorrelated on the input links. An extreme form of bursty traffic is either absent or present at full bandwidth. This is standard in telephony, where extensive measurements of line utilization have shown that it's very unlikely for more than 10% of the lines to be active at one time.

There are many techniques for multiplexing. In the analog domain:

— *Frequency division* (FDM) uses a separate frequency band for each sub-channel, taking advantage of the fact that e^{int} is a convenient basis set of orthogonal functions. The address is the frequency band of the sub-channel. FDM is used to subdivide the electromagnetic spectrum in free space, on cables, and on optical fibers.

— *Code division* multiplexing (CDM) uses a different coordinate system in which a basis vector is a time-dependent sequence of frequencies. This smears out the cross-talk between different sub-channels. The address is the "code", the sequence of frequencies. CDM is used for military communications and in a new variety of cellular telephony.

In the digital domain time-division multiplexing (TDM) is the standard method. It comes in two flavors:

— *Fixed* TDM, in which *n* sub-channels are multiplexed by dividing the data sequence on the main channel into fixed-size slots (single bits, bytes, or whatever) and assigning every *n*th slot to the same sub-channel. Usually all the slots are the same size, but it's sufficient for the sequence of slot sizes to be fixed. A 1.5 Mbit/sec T1 line, for example, has 24 sub-channels and "frames" of 193 bits. One bit marks the start of the frame, after which the first byte belongs to sub-channel 1, the second to sub-channel 2, and so forth. Slots are numbered from the start of the frame, and a sub channel's slot number is its address.

— *Variable* TDM, in which the data sequence on the main channel is divided into "packets". One packet carries data for one sub-channel, and the address of the sub-channel appears explicitly in the packet. If the packets are fixed size, they are often called "cells", as in the Asynchronous Transfer Mode (ATM) networking standard. Fixed-size packets are used in other contexts, however, for instance to carry load and store messages on a programmed I/O bus. Variable sized packets (up to some maximum which either is fixed or depends on the link) are usual in computer networking, for example on the Ethernet, token ring, FDDI, or Internet, as well as for DMA bursts on I/O busses.

All these methods fix the division of bandwidth among sub-channels except for variable TDM, which is thus better suited to handle the burstiness of computer traffic. This is the only architectural difference among them. But there are other architectural differences among multiplexers, resulting from the different ways of implementing the basic function of *arbitrating* among the input channels. The fixed schemes do this in a fixed way that is determined which the sub-channels are assigned. This is illustrated at the top of Figure 3, where the wide main channel has enough bandwidth to carry all the traffic the input channels can offer. Arbitration is still necessary when a sub-channel is assigned to an input channel; this operation is usually called "circuit setup".

With variable TDM there are many ways to arbitrate, but they fall into two main classes, which parallel the two methods of flow control described in Section 4.

— *Collision*: an input channel simply sends its traffic, but has some way to tell whether it was accepted. If not, it "backs off" by waiting for a while, and then retries. The input channel can get an explicit and immediate collision signal, as on the Ethernet [14], or it can infer a collision from the lack of an acknowledgment, as in TCP [8].

— *Scheduling*: an input channel makes a request for service and the multiplexer eventually grants it; I/O busses and token rings work this way. Granting can be centralized, as in many I/O busses, or distributed, as in a daisy-chained bus or a token ring [18].

Flow control means buffering, as we saw in Section 4, and there are several ways to arrange buffering around a multiplexer, shown on the left side of Figure 3. Having the buffers near the arbitration point is good because it reduces r and hence the size of the buffers. Output buffering is good because it tolerates a larger r across the multiplexer, but the buffer may cost more because it has to accept traffic at the total bandwidth of all the inputs.

A multiplexer can be centralized, like a T1 multiplexer or a crosspoint in a crossbar switch, or it can be distributed along a bus. It seems natural to use scheduling with a centralized multiplexer and collision with a distributed one, but the examples of the

Monarch memory switch [17] and the token ring [18] show that the other combinations are also possible.

Multiplexers can be cascaded to increase the fan-in. This structure is usually combined with a converter. For example, 24 voice lines, each with a bandwidth of 64 Kb/s, are multiplexed to one 1.5 Mb/s T1 line, 30 of these are multiplexed to one 45 Mb/s T3 line, and 50 of these are multiplexed to one 2.4 Gb/s OC-48 fiber which carries 40,000 voice sub-channels. In the Vax 8800 16 Unibuses are multiplexed to one BI bus, and 4 of these are multiplexed to one internal processor-memory bus.

Demultiplexing uses the same physical mechanisms as multiplexing, since one is not much use without the other. There is no arbitration, however; instead, there is *addressing*, since the input channel must select the proper output channel to receive each sub-channel. Again both centralized and distributed implementations are possible, as the right side of figure 3 shows. In a distributed implementation the input channel is broadcast to each output channel, and an address decoder picks off the sub-channel as its data fly past. Either way it's easy to broadcast a sub-channel to any number of output channels.

7   Switch nodes

A switch is a generalization of a multiplexer or demultiplexer. Instead of connecting one link to many, it connects many links to many. Figure 4(a) is the usual drawing for a switch, with the input links on the left and the output links on the right. We view the links as simplex, but usually they are paired to form full-duplex links so that every input link has a corresponding output link which sends data in the reverse direction.

A basic switch can be built out of multiplexers and demultiplexers in the two ways shown in Figure 4(b) and 4(c). The latter is sometimes called a "space-division" switch since there are separate multiplexers and demultiplexers for each link. Such a switch can accept traffic from every link provided each is connected to a different output link. With full-bandwidth multiplexers this restriction can be lifted, usually at a considerable cost. If it isn't, then the switch must arbitrate among the input links, generalizing the arbitration done by its component multiplexers, and if input traffic is not reordered the average switch bandwidth is limited to 58% of the maximum by "head-of-line blocking".

Some examples reveal the range of current technology. The range in latencies for the LAN switches is because they receive an entire packet before starting to send it on.

Medium	Link	Bandwidth		Latency		Links
Alpha chip	register file	13.2	GB/s	3.6	ns	6
Wires	Cray T3D	85	GB/s	1	μs	2K
	HIPPI	1.6	GB/s	1	μs	16
LAN	FDDI Gigaswitch	275	MB/s	10–400	μs	22
	Switched Ethernet	10	MB/s	100–1200	μs	8
Copper pair	Central office	80	MB/s	125	μs	50K

It is also possible to use storage as a switch of the kind shown in Figure 4(b). The storage device is the common channel, and queues keep track of the addresses that in-

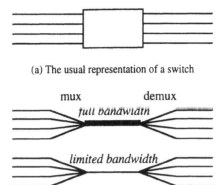

(a) The usual representation of a switch

(b) A mux–demux implementation

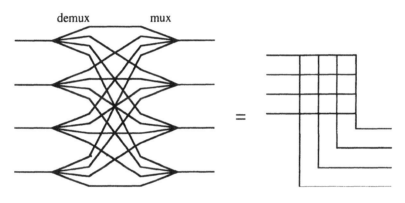

(c) A demux–mux implementation, often drawn as a crossbar

Fig. 4. Switches.

put and output links should use. If the switching is implemented in software the queues are kept in the same storage, but sometimes they are maintained separately. Bridges and routers usually implement their switches this way.

8 Composing switches

Any idea in computing is made better by being made recursive.
Brian Randell

Having studied the basic elements out of which interconnections are made, we can now look at how to compose them. We begin by looking at how to compose switches to make a larger switch with the same interface; the next section examines the effect of changing the interface.

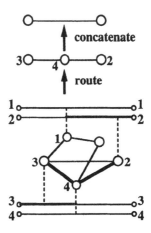

Fig. 5. Composing switches.

8.1 Concatenating links

First we observe that we can concatenate two links using a connector, as in the top half of Figure 5, to make a longer link. This structure is sometimes called a "pipeline". The only interesting thing about it is the rule for forwarding a single traffic unit: can the unit start to be forwarded before it is completely received ("wormholes") [15], and can parts of two units be intermixed on the same link ("interleaving")? As we shall see, wormholes give better performance when the time to send a unit is not small, and it is often not because a unit is often an entire packet. Furthermore, wormholes mean that a connector need not buffer an entire packet.

The latency of the composite link is the total delay of its component links (the time for a single bit to traverse the link) plus a term that reflects the time the unit spends on links. With no wormholes this term is the sum of the times the unit spends on each link (the size of the unit divided by the bandwidth of the link). With wormholes and interleaving, it is the time on the slowest link. With wormholes but without interleaving, if there are alternating slow and fast links $s_1\,f_1\,s_2\,f_2\,...\,s_n\,f_n$ on the path (with f_n perhaps null), it is the total time on slow links minus the total time on fast links. A sequence of links with increasing times is equivalent to the slowest, and a sequence with decreasing times to the fastest. We summarize these facts:

Wormhole	Interleaving	Time on links
No	—	$\Sigma\,t_i$
Yes	No	$\Sigma\,ts_i - \Sigma\,tf_i = \Sigma\,(ts_i - tf_i)$
Yes	Yes	$\max t_i$

The moral is to use either wormholes or small units. A unit shouldn't be too small on a variable TDM link because it must always carry the overhead of its address. Thus ATM cells, with 48 bytes of payload and 5 bytes of overhead, are about the smallest practical units (though the Cambridge slotted ring used cells with 2 bytes of payload). This is not an issue for fixed TDM, and indeed telephony uses 8 bit units.

There is no need to use wormholes for ATM cells, since the time to send 53 bytes is small in the intended applications. But Autonet [22], with packets that take milliseconds to transmit, uses wormholes, as do multiprocessors like the J-machine [4] which have short messages but care about every microsecond of latency and every byte of network buffering. The same considerations apply to pipelines.

8.2 Routing

If we replace the connectors with switch nodes, we can assemble a mesh like the one at the bottom of Figure 5. The mesh can implement the bigger switch that surrounds it and is connected to it by dashed lines. The path from node 3 to node 4 is shown by the heavy lines in both the mesh and the switch. The pattern of links between switches is called the "topology" of the mesh.

The new mechanism we need to make this work is *routing*, which converts an address into a "path", a sequence of decisions about what output link to use at each switch. Routing is done with a map from addresses to output links at each switch. In addition the address may change along the path; this is implemented with a second map, from input addresses to output addresses.

There are three kinds of addresses. In order of increasing cost to implement the maps, and increasing convenience to the end nodes, they are:

— *Source* routing: the address is just the sequence of output links to use; each switch strips off the one it uses. The IBM token ring and several multiprocessors [4, 23] use this. A variation distributes the source route across the path; the address (called a "virtual circuit") is local to a link, and each switch knows how to map the addresses on its incoming links. ATM uses this variation.

— *Hierarchical* routing: the address is hierarchical. Each switch corresponds to one node in the address tree and knows what links to use to get to its siblings, children, and parent. The Internet and cascaded I/O busses use this.

— *Flat* routing: the address is flat, and each switch knows what links to use for every address. Broadcast networks like Ethernet and FDDI use this; the implementation is easy since every receiver sees all the addresses and can just pick off those destined for it. Bridged LANs also use flat routing, falling back on broadcast when the map is inadequate. The mechanism for routing 800 numbers is mainly flat.

8.3 Deadlock

Traffic traversing a composite link needs a sequence of resources (most often buffer space) to reach the end, and usually it acquires a resource while holding on to existing ones. This means that deadlock is possible. The left side of Figure 6 shows the simplest case: two nodes with a single buffer pool in each, and links connecting them. If traffic must acquire a buffer at the destination before giving up its buffer at the source,

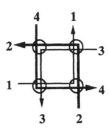

Fig. 6. Deadlock.

it is possible for all the messages to deadlock waiting for each other to release their buffers.

The simple rule for avoiding deadlock is well known: define a partial order on the resources, and require that a resource cannot be acquired unless it is greater in this order than all the resources already held. In our application it is usual to treat the links as resources and require paths to be increasing in the link order. Of course the ordering relation must be big enough to ensure that a path exists from every sender to every receiver.

The right side of Figure 6 shows what can happen even on a simple rectangular grid if this problem is ignored. The four paths use links as follows: 1—EN, 2—NW, 3—WS, 4—SE. There is no ordering that will allow all four paths, and if each path acquires its first link there is a deadlock.

The standard order on a grid is: $l_1 < l_2$ iff they are head to tail, and either they point in the same direction, or l_1 goes east or west and l_2 goes north or south [15]. So the rule is: "Go east or west first, then north or south." On a tree $l_1 < l_2$ iff they are head to tail, and either both go up toward the root, or l_2 goes down away from the root. The rule is thus "First up, then down." On a DAG impose a spanning tree and label all the other links up or down arbitrarily [22].

9   Layers

There are three rules for writing a novel.
Unfortunately, no one knows what they are.

Somerset Maugham

In the last section we studied systems composed by plugging together components with the same interface such as Internet routers, LAN bridges, and telephone switches and multiplexers. Here we look at systems in which the interface changes. When we implement an interface on top of a different one we call the implementation a "layer".

The simplest kind of layering is "encapsulation", in which we stick converters on the ends of a link that implements one interface to get a link that implements a different one; see Figure 7. Examples are transporting Internet packets over an 802 LAN, or over DECnet, or the reverse encapsulations of 802 or DECnet packets over the Internet. Another way to think of this is as multiplexing several protocols over a single link. As usual, multiplexing needs an address field in each message, so it is prudent and cus-

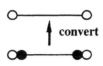

Fig. 7. Encapsulation.

tomary to provide a "protocol type" field in every link interface. A "version number" field plays a similar role on a smaller scale.

Here is encapsulation in the large. We can build

	What	*Why*
a)	a TCP reliable transport link	function: reliable stream
b)	on an Internet packet link	function: routing
c)	on the PPP header compression protocol	performance: space
d)	on the HDLC data link protocol	function: packet framing
e)	on a 14.4 Kbit/sec modem line	function: byte stream
f)	on an analog voice-grade telephone line	compatibility
g)	on a 64 Kbit/sec digital line multiplexed	function: bit stream
h)	on a T1 line multiplexed	performance: aggregation
i)	on a T3 line multiplexed	performance: aggregation
j)	on an OC-48 fiber.	performance: aggregation

This stack is ten layers deep. Each one serves some purpose, tabulated in the right column and classified as function, performance, or compatibility. Note that compatibility caused us to degrade a 64 Kbit/sec stream to a 14.4 Kbit/sec stream in layers (f) and (g) at considerable cost; the great achievement of ISDN is to get rid of those layers.

On top of TCP we can add four more layers, some of which don't look so much like encapsulation:

	What	*Why*
w)	mail folders	function: organization
x)	on a mail spooler	function: storage
y)	on SMTP mail transport	function: routing
z)	on FTP file transport	function: reliable char arrays

Now we have 14 layers with two kinds of routing, two kinds of reliable transport, three kinds of stream, and three kinds of aggregation. Each serves some purpose that isn't served by other, similar layers. Of course many other structures could underlie the filing of mail messages in folders.

Here is an entirely different example, an implementation of a machine's load instruction:

	What	*Why*
a)	load from cache	function: data access
b)	miss to second level cache	performance: space
c)	miss to RAM	performance: space
d)	page fault to disk	performance: space

Layer (d) could be replaced by a page fault to other machines on a LAN that are sharing the memory [12] (function: sharing), or layer (c) by access to a distributed cache over a multiprocessor's network (function: sharing). Layer (b) could be replaced by access to a PCI I/O bus (function: device access) which at layer (c) is bridged to an ISA bus (compatibility).

The standard picture for a communication system is the OSI reference model, which shows peer-to-peer communication at each of seven layers: physical, data link, network, transport, session, presentation, and application. The peer-to-peer aspect of this picture is not as useful as you might think, because peer-to-peer communication means that you are writing a concurrent program, something to be avoided if at all possible. At any layer peer-to-peer communication is usually replaced with client-server communication as soon as possible.

It should be clear from these examples that there is nothing magic about any particular arrangement of layers. The same load/store function is provided for the file data type by NFS and other distributed file systems [20], and for an assortment of viewable data types by the World Wide Web. What is underneath is both similar and totally different. Furthermore, it is possible to collapse layers in an implementation as well as add them; this improves efficiency at the expense of compatibility.

We have seen several communication interfaces and many designs for implementing them. The basic principle seems to be that any interface and any design can be useful anywhere, regardless of how lower layers are done. Something gets better each time we pile on another abstraction, but it's hard to predict the pattern beforehand.

10 Fault-tolerance

The simplest strategies for fault-tolerance are

>Duplicate components, detect errors, and ignore bad components.

>Detect errors and retry.

>Checkpoint, detect errors, crash, reconfigure without the bad components, and restart from the checkpoint.

Highly available systems use the first strategy, others use the second and third. The second strategy works very well for communications, since there is no permanent state to restore, retry is just resend, and many errors are transient.

A more complex approach is to fail over to an alternate component and retry; this requires a failover mechanism, which for communications takes the simple form of changes in the routing database. An often overlooked point is that unless the alternate component is only used as a spare, it carries more load after the failure than it did before, and hence the performance of the system will decrease.

In general fault tolerance requires timeouts, since otherwise you wait indefinitely for a response from a faulty component. Timeouts in turn require knowledge of how long things should take. When this knowledge is precise timeouts can be short and failure detection rapid, conditions that are usually met at low levels in a system. It's common to design a snoopy cache, for instance, on the assumption that every processor will respond in the same cycle so that the responses can be combined with an or gate. Higher up there is a need for compatibility with several implementations, and each lower level with caching adds uncertainty to the timing. It becomes more difficult to set timeouts appropriately; often this is the biggest problem in building a fault-tolerant system. Perhaps we should specify the real-time performance of systems more carefully, and give up the use of caches such as virtual memory that can cause large variations in response time.

All these methods have been used at every level from processor chips to distributed systems. In general, however, below the level of the LAN most systems are synchronous and not very fault-tolerant: any permanent failure causes a crash and restart. Above that level most systems make few assumptions about timing and are designed to keep working in spite of several failures. From this difference in requirements follow many differences in design.

11 Conclusion

We have seen that both interfaces and implementations for interconnecting computing elements are quite uniform at many different scales. The storage and message interfaces work both inside processor chips and in the World Wide Web. Links, converters, and switches can be composed by concatenation, routing, and layering to build communication systems over the same range. Bandwidth, latency, and connectivity are always the important performance parameters, and issues of congestion, flow control, and buffering arise again and again.

This uniformity arises partly because the ideas are powerful ones. The rapid improvement in silicon and fiber optics technology, which double in cost/performance every two years, also plays a major role in making similar designs appropriate across the board. Computer scientists and engineers should be grateful, because a little knowledge will go a long way.

References

1. R. Barton: A new approach to the functional design of a digital computer. *Proc . Western Joint Computer Conference* (1961)
2. A. Birrell and B. Nelson: Implementing remote procedure calls. *ACM Transactions on Computer Systems* 2, 39-59 (1984)
3. D. Culler et al.: Fine-grain parallelism with minimal hardware support: A compiler-controlled threaded abstract machine. *4th ACM Conference on Architectural Support for Programming Languages and Operating Systems*, 164-175 (1991)
4. W. Dally: A universal parallel computer architecture. *New Generation Computing* 11, 227-249 (1993)
5. P. Green: The future of fiber-optic computer networks. *IEEE Computer* 24, 78-87 (1991)
6. J. Hennessy and N. Jouppi: Computer technology and architecture: An evolving interaction. *IEEE Computer* 24, 18-29 (1991)
7. C. Hoare: Communicating sequential processes. *Communications of the ACM* 21, 666-677 (1978)
8. V. Jacobsen: Congestion avoidance and control. *ACM SigComm Conference*, 1988, 314-329
9. L. Lamport: How to make a multiprocessor computer that correctly executes multiprocess programs. *IEEE Trans. Computers* C-28, 241-248 (1979)
10. B. Lampson: Reliable messages and connection establishment. In S. Mullender (ed.) *Distributed Systems*, Addison-Wesley, 1993, 251-282
11. D. Lenkosi et al.: The Stanford Dash multiprocessor. *IEEE Computer* 25, 63-79 (1992)
12. K. Li and P. Hudak: Memory coherence in shared virtual memory systems. *ACM Transactions on Computer Systems* 7, 321-359 (1989)
13. C. Mead and L. Conway: *Introduction to VLSI Systems*. Addison-Wesley, 1980

14. R. Metcalfe and D. Boggs: Ethernet: Distributed packet switching for local computer networks. *Communications of the ACM* 19, 395-404 (1976)
15. L. Ni and P. McKinley: A survey of wormhole routing techniques in direct networks. *IEEE Computer* 26, 62-76 (1993)
16. R. Pike et al.: The use of name spaces in Plan 9. *ACM Operating Systems Review* 27, 72-76 (1993)
17. R. Rettberg et al.: The Monarch parallel processor hardware design. *IEEE Computer* 23, 18-30 (1990)
18. F. Ross: An overview of FDDI: The fiber distributed data interface. *IEEE Journal on Selected Areas in Communication* 7 (1989)
19. J. Saltzer, D. Reed, and D. Clark: End-to-end arguments in system design. *ACM Transactions on Computer Systems* 2, 277-288 (1984)
20. M. Satyanarayanan: Distributed file systems. In S. Mullender (ed.) *Distributed Systems*, Addison-Wesley, 1993, 353-384
21. M. Schroeder and M. Burrows: Performance of Firefly RPC. *ACM Transactions on Computer Systems* 8, 1-17 (1990)
22. M. Schroeder et al.: Autonet: A high-speed, self-configuring local area network using point-to-point links. *IEEE Journal on Selected Areas in Communication* 9, 1318-1335 (1991)
23. C. Seitz: The cosmic cube. *Communications of the ACM* 28, 22-33 (1985)

Languages and Interactive Software Development *

Susan L. Graham

Computer Science Division – EECS, University of California
Berkeley, CA 94720 USA

Abstract. Most software is developed using interactive computing systems and substantial compute-power. Considerable assistance can be given to the developer by providing language-based support that takes advantage of analysis of software artifacts and the languages in which they are written. In this paper, some of the technical challenges and new opportunities for realizing that support are discussed. Some language design issues that affect the implementation of language-based services are summarized. The paper concludes with some proposals for assisting user understanding of language documents.

1 Introduction

The development and maintenance of software has evolved from the days of punched cards, long turnaround, and — by today's standards — small, slow, computers, to a world in which developers work interactively, using powerful and visually rich workstations and having on-line access to information. As the technology has changed, languages, tools and development practices have followed an evolutionary path. Although researchers have proposed a variety of useful and visionary approaches to exploiting the benefits of interaction [5, 7, 13, 14, 15, 23, 27], common practice has lagged behind. One of the reasons is the effort and expense of building interactive services for each new language or environment.

Powerful interactive systems create both technical challenges and new opportunities for providing language-based services. In this paper we survey some of the technical issues that arise in providing language support in interactive environments. We also consider some of the ways in which the heritage of off-line batch processing still influences the design of languages and tools, and suggest some departures from past practice.

By the word *language*, we will mean formal languages, not natural, spoken languages. Developers use many such languages in their work. The most obvious

---

* This research was supported in part by the Advanced Research Projects Agency of the U.S. Department of Defense under Grant MDA972-92-J-1028, and by the National Science Foundation under Infrastructure Grant CDA-8722788. The content of the information does not necessarily reflect the position or the policy of the U.S. Government.

are programming languages, or possibly design or specification languages. In addition, most systems provide a variety of document description languages (for example, TeX), and a plethora of command languages for interactive shells, configuration management and build process specification languages, languages for information management and retrieval, languages for customizing tools, and so forth.

The discussion that follows draws most of its examples from programming languages and uses a program as an example of a linguistic entity. Much of the discussion pertains equally well to other kinds of languages, and to components as well as complete entities. Indeed, it is for many of the other languages that language-based support is absent. We will sometimes use the term *language documents* to suggest the more general domain. In this context, one can think of tools such as the formatter of a LaTeX document or the **make** program applied to a **Makefile** as a kind of compiler or interpreter.

The potential for language-based support has been demonstrated in many single-language environments. Notable among them are LISP and Smalltalk systems. Our interest is in making the benefits more easily available for other languages. One of the ways to make language-based services more widely available is to develop language-based tools that are *description-driven*, that is, instantiated by specifying a particular language. That approach is facilitated by the use of formal declarative specifications as much as possible, to minimize recoding.

The Ensemble project at Berkeley is investigating the technology to provide interactive language-based services for both formal languages and multi-media natural language documents. Many of the ideas and observations in this paper are the outgrowth of that research effort and its predecessors, the Pan project [4] and the VorTeX project [6].

In the remainder of the paper, we first sketch some of the language-based services we have in mind. Next we describe some of the technical issues that arise in interactive language processing, followed by a discussion of the effect of certain language design choices on that processing. We highlight two issues that demand special attention in an interactive environment – the consequences of incomplete information, and the nature of document presentation. Finally, we consider the use of language services for user understanding of language documents.

2 Interactive Language-based Services

One of the important characteristics of an interactive system is the ability of a user to engage in a dialogue with the system. That property is exploited in a modest way by contemporary text-editors, such as Emacs [22], that incorporate services that check well-formedness properties of the document being edited and assist the user in discovering and correcting mistakes. Two familiar examples are checking natural language spellings and detecting unmatched bracketing characters such as parentheses. Using the Emacs extension language, it is possible to define language modes that extend checking to other syntactic forms, and also provide some assistance in introducing those forms.

The kinds of language-based services we have in mind are in keeping with the examples just given. The system "understands" the languages in which a document is written and can respond accordingly. A rich dialogue can ensue if the system knows not only syntactic properties of the language but also some semantic properties[2]. Even greater benefits can be achieved if the system can extend its services to documents that are compound, both in the sense of containing multiple languages, and in the sense of having structure and relationships that transcend a single file or storage unit.

Our emphasis in this paper is on services that require fine-grained structure and analysis. "In-the-large" services at the granularity of modules, chapters, functions, etc. are equally valuable, but involve a different set of issues and engineering decisions.

Consider the services of an interactive editor. A text editor deals with the language of text, consisting of characters, words and lines. (In some direct manipulation systems, that language is enriched to include fonts, sizes, colors, and other format-related attributes.) A program with a textual representation can be written and modified as a textual document. Among the other services offered by a text editor are search and navigation operations, such as "move to the next line", or "replace all occurrences of foo by bar".

If the program is regarded as a document in the programming language rather than the language of text, then similar services can be provided in that language, for example "move to the next variable", or "rename (*definition and uses of*) type fooType as barType within the innermost scope in which fooType is defined". Thus textual navigation and search/replace commands are extended to linguistic forms of those operations. Another example is to extend textual notions of cut/paste to structural or semantic operations.

In order to support structural and semantic versions of text operations, the system must maintain information about the syntactic and semantic structure of the document. Once linguistic information is available, new services can be provided. Familiar examples include highlighting or elision of structural components, formatting, class hierarchy navigation, and call-graph browsing.

The opportunity also exists to provide new operations based on information derived from analysis or user annotation. Van De Vanter [25] gives the following example. Many programs contain both mainstream problem-solving code and a possibly substantial amount of code intended to handle special cases, errors, and other infrequently occurring situations. If a developer is attempting to understand an unfamiliar program, then it is the mainstream code that is of interest initially. Understanding is enhanced if the developer can visually identify the mainstream code and downplay the rest. Identifying which code is mainstream might require the combined efforts of automated analysis and annotation by the author of the program, probably at some earlier time. If the identification of

---

[2] In keeping with the use of the term in compilation, we use the term *semantic* to refer also to some properties that are syntactic in the traditional linguistic sense.

mainstream code is available, browsing and display techniques can be used to show it to the developer.

3 Interactive Language Processing

In order to provide interactive language-based services, a system builds a structural description of the program, augmented with annotations, links, and auxiliary information. Early systems, such as the Cornell Program Synthesizer [24], and Mentor [8] required that the system maintain a structurally well-formed program, by providing a *structure editor* interface that limited the user to structural modifications. Later *syntax understanding* systems support text editor interfaces[3]. Structure is inferred by analysis (namely, some form of parsing). Policies for analysis include analysis on user demand, periodic analysis (e.g. when the user pauses), continuous analysis, or analysis when some structural operation such as navigation is invoked. In many of these systems, structure also can be provided explicitly through the use of templates.

The reason that early systems require structural well-formedness is not only to eliminate the need for structural analysis, but, more importantly, to facilitate other structurally-based forms of analysis. For example, traditional kinds of static semantic checking might be carried out using some sort of attribute grammar technology on an abstract syntax tree. Even less formally specified analyses tend to be based on a structural representation.

Many interactive systems are based on the use of *incremental* analysis algorithms. An incremental algorithm is one that updates existing information by first determining what has been modified and then propagating the consequences of the modification. Historically the motivation for incremental algorithms as opposed to recomputation was performance – to reduce the number and extent of updates when local changes are made to large documents. As computing speeds have increased, the performance issue has declined in importance.

There are other important benefits of incremental algorithms. By determining what has changed, a system can provide a variety of change-related user feedback. (An example is highlighting on a screen.) More importantly, by determining what has *not* changed, a system can preserve information that cannot be reconstructed by analysis, such as user annotation, or other information provided by external agents. Incremental algorithms allow changes to be made in place, rather than constructing a new version and comparing it with the old one to detect changes.

Using incremental analysis instead of "batch" analysis has two important characteristics. First, the processing order is largely temporal rather than structural. In other words, change-based analysis or services propagate from the changes outward rather than from the beginning of the document forward. Many languages are designed not only with the assumption that the language document will be read in textual order (a cognitive issue), but that it will be processed in textual order (a translation and analysis issue). For example, some languages require that entities designated by identifiers be defined before they are used. Some

---

[3] Later versions of the Cornell system support some text editing as well.

languages define the scope of a definition to extend from the textual occurrence of the definition to the nearest following textual scope-closing delimiter (or to a corresponding "undefine" directive). Since an incremental algorithm does not process the document in textual order, special mechanisms are needed to handle textual ordering, particularly if the ordered occurrences are not adjacent. Second, it is commonplace for processing to be done with incomplete information and incomplete documents, either because the program is only partially developed, or because conceptually atomic changes are made in a sequence of steps. One consequence is that some of the information needed for analysis may be absent.

4 Language Issues

There are a variety of language design issues that complicate interactive language processing. Some of them are issues that cause difficulties for formal specification of languages. Often those same features complicate specifications used for compilation. An example is the need for unbounded lookahead to make certain parsing decisions. The more interesting issues in the context of this paper are the ones in which the temporal processing order comes into play.

In considering the material in this section, the knowledgeable reader might object that the language design issues are all problems that the language design community recognizes and can avoid. However, to be most useful, language-based services should be available for *all* languages. In fact, it is for badly designed languages that help is needed most! Additionally, designers of other kinds of languages seem to be making the familiar mistakes and then some.

4.1 Information Feedback

Conventional language analysis is decomposed into three stages

1. lexical transformation of a sequence of characters – screened or filtered to remove formatting information and commentary – into a sequence of tokens,
2. syntactic transformation of a token sequence into phrase structure,
3. "semantic" analysis to achieve bindings, name resolution, and attributions.

Although these stages can be composed into a single syntax-directed analysis in which lexical analysis and semantic analysis are embedded, the architecture used by many optimizing or retargetable compilers and by most interactive language processors is to construct an explicit structural representation first, and then to determine bindings and attributions by traversal of the structure, perhaps using some sort of attribute evaluation formalism. If the stages are independent, then there are known techniques to carry out each stage incrementally [2, 3, 18, 21]. However, if earlier analyses require feedback from later ones, then a multi-phase analysis becomes awkward.

Example: C/C++ Type Names

Consider the fragment

$$a(*b);$$

If **a** names a function, then the fragment denotes a call of function **a** with argument *b. However, if **a** denotes a C type, for instance in a context containing

typedef int a;

or a C++ class, then the fragment defines **b** to be a variable whose values are pointers to entities of the type or class denoted by **a**.

It is normally desirable to use different structural representations of the call **a(*b)** and the variable definition **a(*b)**. Yet the structural analysis of the fragment needs information about the binding of **a** that is determined by the later semantic analysis phase, which in turn requires a structural traversal. The need to know **typedef** bindings in order to determine structure also exists in C and C++ compilers. Since the textual occurrence of the **typedef** precedes uses of the defined identifier, maintaining binding information during parsing solves the problem, although at the possible cost of more complicated specification and additional mechanism.

In an interactive environment, in which attribution follows structural analysis in order to support incrementality appropriately, **typedef** bindings are necessarily a special case with a separate mechanism. Furthermore, if the contextual information is changed (for instance, the **typedef** is removed or its identifier spelling is changed) then not only must all fragments containing uses of **a** bound to the **typedef** be reparsed, but all attributions using attributes of **a** must be reevaluated. The reevaluation is initiated as a consequence of dependency information maintained by the incremental semantic evaluator. One way in which the reparsing can be triggered is by using a special token, say **TYPE-ID**, for identifiers designating types, and by replacing appropriately-bound occurrences of **ID** by **TYPE-ID** or vice-versa when type definitions are added or removed, in order to force a syntactic change event. Those replacements, in turn, require the existence of semantically analyzed bindings to reflect the scope rules of the language.

Example: User Operator Priority Settings

In several languages, notably PROLOG, ML, FIDIL [16] and Haskell [17] users can define a new infix operator and specify a parsing priority for it. The priority determines the structure of each expression in which the operator is used. The priority is normally associated with the operator during the attribution phase, and the operator definition usually depends on scope analysis. Thus many of the same issues arise as in the previous example. If there are a small number of possible priorities, as there are in FIDIL, then the token-based solution outlined in the earlier example can be used, by introducing a separate user-defined-operator token for each priority. However if, as is the case for PROLOG, and for early versions of ML, there are a large (effectively unbounded) number of levels, then a complicated reparsing strategy may be required.

4.2   Contextually-determined Information

In some older languages, notably FORTRAN, PL/I, and some dialects of BASIC, keywords are not reserved, and the use of an identifier as a keyword is inferred from context. Although methods exist to do the appropriate analysis, they are not necessarily supported by description-driven tools.

The use of embedded sublanguages is also determined by context. One example is formatting specifications in output directives. Another is math mode in TeX. Typically, these sublanguages have their own syntactic and semantic rules. Syntax analyzers in translators often handle sublanguages by maintaining a global state variable and switching among analyzers as the sublanguage boundary is crossed. Incremental analyzers must also be able to determine that state, but not by use of a global variable. Instead the sublanguage boundaries are part of the structure, either in the form of attributions, or in the form of links to separate structures. The treatment of sublanguages must be robust if the determining contextual cues are unavailable.

4.3   Formatting-related Syntax Rules

Some language designers introduce syntax rules – intended to provide readability or typing convenience for the user – that may complicate language specification or processing. The problems with FORTRAN whitespace insensitivity are well known to compiler writers.

Ends of lines are often used as delimiters, sometimes in complicated ways. FORTRAN, COBOL, and UNIX shell languages are examples of line-oriented languages in which escape symbols are required to prevent textual line breaks from being treated as logical end-of-statement delimiters. In Icon [12], a line break sometimes serves as a statement delimiter; but only if the last token before the line break and the first token after it are not part of the same construct.

In Haskell, indentation can be used to indicate nesting, in place of explicit bracketing tokens, thereby making some line breaks, white space, and column positions significant. In **make** [9], commands must be indented by at least one tab character (and not by the visually indistinguishable sequence of blank characters). In the former case, the reader is helped at the expense of the language tools. In the latter case, the opposite is true.

4.4   Preprocessors and Macros

Macros are often used for abbreviation or language extension. If a macro language and the language in which it is used have the same lexical and syntactic rules, as is the case in many LISP dialects, then the major complication in incremental analysis comes from the treatment of errors. Macro expansion and analysis of the expanded language can be intermixed, as long as the system retains enough information to replace the expansions when a macro definition is changed. That information is also needed to support selective viewing of macro expansions. Since the expansions are structurally well-formed, both replacement and selective viewing reuse mechanisms available for other purposes.

If the macro language transforms the text to which it is applied at the character level, or even at the token level, as is the case with macro processing embedded in preprocessors such as the C preprocessor (cpp), then the situation is more difficult. The nature of such a preprocessor is intrinsically a sequential rewriting of the text. Since expansion in different contexts may create different structure, an unexpanded macro call can cause syntactically invalid text in the language to which the macros are applied. Also, since expansion causes changes in structure, selective viewing of expansions and macro redefinition become more complicated for the system to support.

Finally, if arbitrary file inclusion is incorporated, as it is in cpp, then the benefits of sharing copies of expansions, which are an important engineering consideration, may be difficult or impossible to achieve. Since multiple inclusions of the same file are used in the absence of linguistic support for interface modules, and those files are often large, the space required for multiple copies and the time for multiple analyses can be significant.

5 Incomplete Documents

The language design discussion illustrates the fact that language analysis often uses information that is available in a complete, well-formed document, but may be missing in a document that is being developed or modified. In a translator, if a document is incomplete or not well-formed then it is in error. Syntactic and semantic error processing mechanisms are invoked, and appropriate diagnostic information is generated for the developer.

However, from the developer's point of view, incompleteness or partial modification are different from errors. Suppose we refer to all these situations by the less-loaded term *anomalies*. Since anomalies are a normal occurrence, system services should be maintained in their presence. That has the following consequences for the design of an incremental system.

Partial structure and analysis must be available. Structure and attribution-based services such as navigation, display, and querying should continue to be available. Change-based analysis provides considerable help in this regard, since the structure and properties of unchanged components can be retained even if they are part of anomalous constructs.

Knowledge about anomalies is important to the user. Since the anomalies are part of the linguistic information about the document, linguistic services such as navigation and highlighting should apply to them as well.

The user should decide when to resolve anomalies. If an interactive system is to assist a developer, then the developer must be able to choose the order in which to work. That means that the services must not degrade if anomalies are unresolved. The requirement in early structure editors that syntax anomalies be absent was an important factor in their lack of wide-spread use.

It is because of the need to support anomalies, that some of the language design issues we have summarized are particularly problematical. If one is to provide language-based services using description-driven tools, then language features requiring special handling impose an additional barrier. Additionally, a system that is robust in the presence of anomalies, must provide linguistic services in the absence of linguistic information. For instance, the system must choose some way to format a(*b) for display even if its structure is unknown, and must have some reasonable policy about communicating type anomalies that stem from a lack of binding information rather than a mistake on the part of the developer. Van De Vanter, Ballance, and Graham [26] provide further discussion of these issues.

6   Presentation and User Comprehension

An important part of the task of constructing and modifying language documents is human comprehension of both their form (i.e. the use of the language) and their content. The previous discussion has suggested some ways in which interactive tools can assist developers in using a language. Interactive systems can also assist developers in understanding the content of a language document. We will touch on two kinds of assistance – better readability and better association and preservation of auxiliary information. Space does not permit discussion of an important third kind of assistance, namely better understanding of dynamic behavior or of the output of language tools.

6.1   Presentation

The programming language research community has long realized that the same language can be represented in more than one form. The notion of abstract syntax goes back to the 1960's, along with the pejorative reference to concrete syntax as "syntactic sugar". Nevertheless, one of the characteristics of most language definitions is the inclusion of rather specific rules for their concrete syntax. The reason for the precision of the definitions, of course, is that they serve as input specifications for language-processing tools, notably compilers and interpreters. That precision was essential in the days of punched cards and paper tape.

The progression from keypunches to text editors to *language-sensitive* editors has done little to loosen the rules of concrete syntax. Syntax is usually expressed as a sequence of ASCII characters. Font shifts (for example, emboldened keywords) and colors are sometimes used, and layout styles are sometimes incorporated automatically. It is an easy matter to map those enhancements back to an ASCII character sequence.

In an interactive computing environment, the user prepares and modifies a language document on-line and also invokes a compiler, interpreter, or other language processor on-line. There is no reason that the form of the language read and written by the user need be the same as the input representation to a language tool, as long as an appropriate input representation can be generated automatically from the human-created version. The user should be able to use a representation that facilitates comprehension; not one that has been designed to ease translation. On the other hand, the input to the language translator need not be burdened with formatting-motivated syntax rules.

An interesting study by Baecker and Marcus [1] suggests some ways in which appearance can affect human understanding of language documents. They devised a variety of ways of presenting C programs to improve their readability. An example appears in Figure 1. Many other researchers have proposed partic-

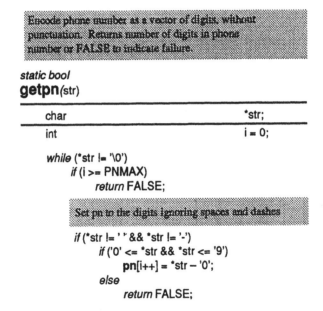

Fig. 1. A program presentation example from Baecker/Marcus [1, pg. 61]

ular stylistic choices of concrete syntax, formatting, and appearance to enhance understanding. Studies such as that of Oman and Cook [20] demonstrate that careful use of typographic effects can strongly influence how well programmers, either novice or expert, understand a program.

The work of Baecker and Marcus is intended primarily for display, not for interaction. By building interactive tools that treat appearance separately from structure and content, styles of presentation can be designed for comprehension and can be customized for user preferences. Concrete syntax can be defined to facilitate tool building and not be cluttered with readability considerations.

To support presentations such as the one in Figure 1, the system must provide high-quality typography and formatting, reordering of abstract syntax components, and the ability to map back and forth between the presentation and the components of the abstract syntax representation. In addition, the notion of ordering used for navigation must be carefully considered to avoid user confusion.

The Ensemble project at Berkeley is developing the technology to support such separation of appearance. The structure, the semantics, and one or more styles of appearance are formally specified for a language[4]. Appearance is specified by a *presentation schema*, which provide the rules that are applied incrementally as the document changes. The first version of the presentation system is summarized in a conference paper [11]; a later version is described in Munson's dissertation [19].

6.2 Annotation

One common language feature intended for human comprehension is the comment. In most text-based languages, comments are text as well, and are separated from other constructs in the language by delimiters. Comments are normally ignored by language translators or interpreters. Some researchers have proposed systems of formal annotation as well, that are, in effect, an embedded sublanguage with formal properties.

If an interactive system supports separate mechanisms for presentation of language documents, and if it supports structural representations of language documents, then additional kinds of annotation become possible. A simple example is to associate comments, whether formal or in natural language, with semantically attributed structural components of the document, rather than with textual positions. These annotations can remain bound to the structural components (which might be important user abstractions) even if the textual document changes. Furthermore, it is possible to use the richness of multimedia to provide non-textual static or dynamic, or even audible comments, such as an animation of the use of a data abstraction. By associating computable predicates with the comments, mechanisms can be provided to discover and signal their possible lack of validity as the document changes. It is also possible to use increasingly common notions of hyperlinks to associate electronic information outside the document with its entities and structures.

In this scenario, a program in its entirety consists of much more than the instructions for an abstract execution engine. The input to a compiler or interpreter is obtained by extracting the relevant view in the appropriate representation. Interactive execution or debugging are easily incorporated in this point of view.

Annotations can be provided by tools as well as by people. The semantic attributes calculated by an incremental analyzer, or the attributes derived from data flow analysis constitute annotations. As another example, execution profiling data can be associated with components of a program and used both for improved compilation and for focusing the developer's attention.

---

[4] The conventional concrete syntax must also be specified, in order to pass language documents into and out of the Ensemble system.

An important property for a system that supports rich forms of annotation is flexible access to the information. It is useful to regard an annotated document as a semantically rich database, in which the information can be queried or viewed as desired. The technical challenge is to support that behavior even though the system representation of the information is not at all that of a database.

6.3 Information Retention

The final step in regarding a language document as a human-centered artifact in which comprehension is an important property is the retention of information over time. One of the ways change-based analysis can be used is to preserve a language-based modification history, as opposed to the textual versioning systems in use today, and to retain the associated annotations. Some of the technology needed for that kind of history is a consequence of the mechanisms used for language-based *undo* services. Another often-suggested idea is to record, along with changes to a document, cognitive information such as a design rationale for the change. The impediments to realizing that idea are primarily nontechnical. The advantages of using a language-based approach lie in associating the rationale with the change itself and not just its natural-language description.

Acknowledgements

Many of the ideas expressed in this paper are the outgrowth of collaborations with research students and colleagues in the Pan project and the Ensemble project at Berkeley. Bruce Forstall's M.S. report [10] contains an extensive discussion of language specification issues from which some of the material in Section 4 is drawn. Michael Van De Vanter's dissertation [25] provides a wealth of insights about the ways in which an interactive system can assist user understanding of language documents, some of which are summarized in Section 2 and Section 6.

References

1. Ronald M. Baecker and Aaron Marcus. *Human Factors and Typography for More Readable Programs.* Addison-Wesley, Reading, Massachusetts, 1990.
2. Robert A. Ballance, Jacob Butcher, and Susan L. Graham. Grammatical abstraction and incremental syntax analysis in a language-based editor. In *Proc. SIGPLAN '88 Conf. on Programming Language Design and Implementation*, pages 185–198, Atlanta, Georgia, June 22–24, 1988. Appeared as SIGPLAN Notices, 23(7), July 1988.
3. Robert A. Ballance and Susan L. Graham. Incremental consistency maintenance for interactive applications. In K. Furukawa, editor, *Proc. Eighth International Conf. on Logic Programming*, pages 895–909. The MIT Press, Cambridge, Massachusetts and London, England, June 1991.

4. Robert A. Ballance, Susan L. Graham, and Michael L. Van De Vanter. The Pan language-based editing system. *ACM Transactions on Software Engineering and Methodology*, 1(1):95–127, January 1992.

5. David R. Barstow, Howard E. Shrobe, and Erik Sandewall, editors. *Interactive Programming Environments*. McGraw-Hill, New York, 1984.

6. Pehong Chen, John L. Coker, Michael A. Harrison, Jeffrey W. McCarrell, and Steven J. Procter. The VorTeX document preparation environment. In *Proc. Second European Conf. on TeX for Scientific Documentation, Lecture Notes in Computer Science No. 236*, pages 32–54, Strasbourg, France, June 1986. Springer-Verlag.

7. Reidar Conradi, Tor M. Didriksen, and Dag Wanvik, editors. *Advanced Programming Environments*. Number 244 in Lecture Notes in Computer Science. Springer-Verlag, Berlin, Heidelberg, New York, 1986.

8. Véronique Donzeau-Gouge, Gérard Huet, Gilles Kahn, and Bernard Lang. Programming environments based on structured editors: The MENTOR experience. In David R. Barstow et al., editor, *Interactive Programming Environments*, pages 128–140. McGraw-Hill, New York, 1984.

9. S. I. Feldman. *Make—A Program for Maintaining Computer Programs*. Bell Laboratories, Murray Hill, NJ, 1978. In the Unix programmer's manual, vol. 2.

10. Bruce T. Forstall. Programming language specification for editors. Master's report, Computer Science Division—EECS, University of California, Berkeley, November 1991.

11. Susan L. Graham, Michael A. Harrison, and Ethan V. Munson. The Proteus presentation system. In *SIGSOFT '92: Proceedings of the Fifth ACM SIGSOFT Symposium on Software Development Environments*, pages 130–138. ACM Press, December 1992. *ACM Software Engineering News 17* (5), December 1992.

12. R. E. Griswold and M. T. Griswold. *The Icon Programming Language*. Prentice-Hall, Englewood Cliffs, N.J., 1983.

13. Peter Henderson, editor. *Proceedings of the ACM SIGSOFT/SIGPLAN Software Engineering Symposium on Practical Software Development Environments*, 1984. *ACM SIGPLAN Notices, 19* (5), and Software Engineering Notes 9 (3), May 1984.

14. Peter Henderson, editor. *Proceedings of the ACM SIGSOFT/SIGPLAN Software Engineering Symposium on Practical Software Development Environments*, 1986. *ACM SIGPLAN Notices, 22* (1), January 1987.

15. Peter Henderson, editor. *ACM SIGSOFT '88: Third Symposium on Software Development Environments*, 1988. *ACM SIGPLAN Notices, 24* (2), Feb. 1989 and *Software Engineering Notes 13* (5), Nov. 1988.

16. Paul N. Hilfinger and Phillip Colella. FIDIL: A language for scientific programming. Technical report, Lawrence Livermore National Lab., Livermore, CA, January 1988.

17. Paul Hudak and Philip Wadler. *Report on the Functional Programming Language Haskell*, 1990.

18. Fahimeh Jalili and Jean H. Gallier. Building friendly parsers. In *Conf. Record Ninth ACM Symposium on Principles of Programming Languages*, pages 196–206, 1982.

19. Ethan V. Munson. *Proteus: An Adaptable Presentation System for a Software Development and Multimedia Document Environment*. Ph.d. dissertation, Computer Science Division – EECS, University of California, Berkeley, 1994. To appear.

20. Paul Oman and Curtis R. Cook. Typographic style is more than cosmetic. *Communications of the ACM*, 33(5):506–520, May 1990.

21. Thomas Reps. *Generating Language-Based Environments*. The MIT Press, Cambridge, Massachusetts and London, England, 1984.
22. Richard M. Stallman. Emacs: The extensible, customizable, self-documenting display editor. In *Proceedings, ACM SIGPLAN/SIGOA Symposium on Text Manipulation*, pages 147–156, Portland, Oregon, June 8-10, 1981. Published as SIGPLAN Notices 16(6), June 1981.
23. Richard N. Taylor, editor. *SIGSOFT '90 Proceedings of the Fourth Symposium on Software Development Environments*, Irvine, CA, December 3–5 1990. ACM SIGSOFT Software Engineering Notes, 15(6), December 1990.
24. Tim Teitelbaum and Thomas W. Reps. The Cornell Program Synthesizer: A syntax-directed programming environment. *Communications of the ACM*, 24(9):563–573, September 1981.
25. Michael L. Van De Vanter. *User Interaction in Language-Based Editing Systems*. Ph.d. dissertation, Computer Science Division – EECS, University of California, Berkeley, December 1992. Available as Technical Report No. UCB/CSD-93-726.
26. Michael L. Van De Vanter, Robert A. Ballance, and Susan L. Graham. Coherent user interfaces for language-based editing systems. *International Journal of Man-Machine Studies*, 37(4):431–466, 1992.
27. Herbert Weber, editor. *SIGSOFT '92 Proceedings of the Fifth ACM Symposium on Software Development Environments*, Tyson's Corner, VA, December 9–11 1992. ACM SIGSOFT Software Engineering Notes, 17(5), December 1992.

Mechanized Support for Stepwise Refinement

Jan L.A. van de Snepscheut
California Institute of Technology

This note describes a notation for formula manipulation and an editor that provides support for the production of programs through the process of stepwise refinement.

1 Introduction

Stepwise refinement is the method of gradually developing programs from their specification through a number of steps. This method was first proposed by E.W. Dijkstra [2], [3], [4] and N. Wirth [8], [9]. As [8] puts it

> In each step, one or several instructions of the given program are decomposed into more detailed instructions. This successive decomposition or refinement of specifications terminates when all instructions are expressed in terms of an underlying computer or programming language, and must therefore be guided by the facilities available on that computer or language. ...
>
> Every refinement step implies some design decisions. It is important that these decisions be made explicit, and that the programmer be aware of the underlying criteria and of the existence of alternative solutions.

Both authors give elegant and convincing examples of the application of this method. In both cases, however, the process is an informal one. In [1], R.J. Back lays a mathematical foundation under this process by viewing refinement as a partial order on state transformers.

Although stepwise refinement is a simple method, it is not widely used in practice because it is often tedious and, as a result, error-prone. In this note, we describe an editor that is geared to the production of programs via stepwise refinement by automating the tedious parts and by making explicit the transformations carried out in each step as well as the conditions under which they apply. Numerous systems support program transformation or theorem proving but (almost) none of them reduce the amount of labor required by the practicing programmer who uses the system. There is (almost) always some aspect of the mechanization that forces the programmer to pay attention to details that are only tangential to the program development itself. The driving force behind our design is to compete with paper and pencil, so to speak, by actually reducing the amount of work done by the programmer. The editor is called *proxac* for program and proof transformation and calculation.

2 Overview

The editor presents a number of windows, including a window that contains the text being edited and a window that contains the transformation rules that can be applied. For example, if the edit window contains the text

$$s = mss.(n+1) \ \land \ 0 \leq n < N$$

then application of rule

$$mss.n = \mathbf{MAX}(j \mid 0 \leq j \leq n \rhd mes.j)$$

transforms the text into

$$
\begin{aligned}
& s = mss.(n+1) \ \land \ 0 \leq n < N \\
= \ & \{ \ mss[n := n+1] \ \} \\
& s = \mathbf{MAX}(j \mid 0 \leq j \leq n+1 \rhd mes.j) \ \land \ 0 \leq n < N
\end{aligned}
$$

(We will turn to the interpretation of these formulae later on.) In the current version of the proxac system, a rule is selected by clicking with the mouse on the rule (see [7] for details). The editor supports the tedious part of this rewriting in the sense that it matches the given text to the selected rule; it determines the "longest" subformula that matches one side of the rule (namely, $mss.(n+1)$ if variable n in the rule is replaced by $n+1$); it then carries out this substitution in the right-other side of the rule to produce the rewrite. The old and new lines are connected by the hint $mss[n := n+1]$ to indicate which rule was applied and which substitution was carried out. Including this information in the text helps in making the transformations explicit. The author of the text is the one who selects the rule that is being applied, the edit program carries out the other actions. Notice that the text being produced is in the format suggested by W.H.J. Feijen.

We have cheated a little bit in the example since we did not indicate in the rule that n is a variable and all the other quantities are constants. Also, transformation rules are applicable only under certain conditions; in this case the condition is $0 \leq n \leq N$. The full version of the transformation rule is, therefore, as follows.

$$\textbf{rule } mss : (n \mid 0 \leq n \leq N \, \triangleright \, mss.n = \textbf{MAX}(j \mid 0 \leq j \leq n \, \triangleright \, mes.j))$$

In addition to the actions described earlier, the editor checks that the applicability condition is met. Since the transformation is applied in a conjunction where $0 \leq n < N$ is one of the terms, the condition holds and the rule applies.

We continue the example with one more rule.

$$\textbf{rule } split : (x, y, z \, \triangleright \, (x \leq y \leq z + 1) = (x \leq y \leq z \ \lor \ x \leq y = z + 1))$$

Rules can be viewed in different ways. The *split* rule is an algebraic identity, not a definition. But a rule like *mss* can be viewed as an explicit definition of function *mss*. The second view is a special case of the first view. We prefer the first view since it provides a great economy in formal labor, even though it has the danger of leading to inconsistencies (since the algebraic properties are postulated instead of proved).

Application of these rules leads to the following text.

$$
\begin{aligned}
& s = mss.(n + 1) \ \ \land \ \ 0 \leq n < N \\
= \ & \{ \ mss[n := n + 1] \ \} \\
& s = \textbf{MAX}(j \mid 0 \leq j \leq n + 1 \, \triangleright \, mes.j) \ \ \land \ \ 0 \leq n < N \\
= \ & \{ \ split[x := 0, y := j, z := n] \ \} \\
& s = \textbf{MAX}(j \mid 0 \leq j \leq n \, \triangleright \, mes.j) \uparrow mes.(n + 1) \ \ \land \ \ 0 \leq n < N \\
= \ & \{ \ mss \ \} \\
& s = mss.n \uparrow mes.(n + 1) \ \ \land \ \ 0 \leq n < N
\end{aligned}
$$

Notice that the last step is the *mss* rule applied in the opposite direction. Also notice that the second transformation step produces term

$$s = \textbf{MAX}(j \mid 0 \leq j \leq n \ \lor \ j = n + 1 \, \triangleright \, mes.j)$$

but the editor reduces this further to

$$s = \textbf{MAX}(j \mid 0 \leq j \leq n \, \triangleright \, mes.j) \uparrow mes.(n + 1)$$

through an application of the range disjunction and one-point rules for quantification. It shows that $\uparrow$ is the infix operator that corresponds to quantifier **MAX** just like $\lor$ corresponds to $\exists$ and $+$ corresponds to $\sum$. These correspondences are not built into the editor; they are specified through the following statements.

declare *INFIX* 20 ↑

property *ASSOCIATIVE*(↑) ∧ *DUAL*(↑) = ↑

declare *QUANTIFIER* **MAX**

property *INFIXOPERATOR*(**MAX**) = ↑

The first line declares ↑ to be an infix operator with precedence level 20. The second line states that it is associative and commutative (that is, it is its own dual). The third line declares quantifier **MAX** and the fourth line gives the correspondence between the two new operators. The associativity and commutativity of ↑ are necessary to make **MAX** a well-defined quantifier. They also enable a lot of simplifications that are automatically applied by the system. By writing the rules and properties in a small but rather general language instead of a richer language with more built-in facts, we gain the ability to extend the application domain of our editor to algebraic manipulations that were not necessarily foreseen. In particular, we show how it can be used to set up a calculus of stepwise refinement.

3 Refinement calculus

In this section, we develop a formalization of the refinement calculus within the framework of our transformation method. The refinement calculus introduced by R.J. Back in [1] is based on the weakest preconditions introduced in [4]. It is based on an ordering relation on programs, written as $s0 \sqsubseteq s1$ for programs $s0$ and $s1$ to denote that $s0$ can be refined by $s1$. Two properties are essential for stepwise refinement. The first is that $\sqsubseteq$ be reflexive and transitive because this justifies the fact that a sequence of steps can be used to refine a specification into an executable program. The second is monotonicity of the program constructs because this justifies that refining one subprogram by another refines the whole program. Notice that this view of refinement requires that programs and specifications be treated on equal footing. Hence, specifications are treated as programs, but we continue refining a program until it contains no specifications. (See the quote in section 1.)

As a first attempt, we may introduce some program constructs. For example, sequential composition will be denoted by semicolon and the empty statement by *skip*.

declare *INFIX* 0 ⊑

property *TRANSITIVE*(⊑) ∧ *REFLEXIVE*(⊑)

declare *INFIX* 1 ;

declare *skip*

property $UNIT(;) = skip$

property $ASSOCIATIVE(;)$

property $\forall(s0, s1, t0, t1 \mid s0 \sqsubseteq s1 \wedge t0 \sqsubseteq t1 \triangleright s0; t0 \sqsubseteq s1; t1)$

Notice that this does not provide a definition of ; even though it is claimed to be an associative operator. A definition-based style would have to prove this result from the definition, which would depend on the associativity of function composition. The last line states the monotonicity of sequential composition.

Though mathematically elegant, formalization of weakest preconditions leads to a complication in their practical use. The complication is due to the difference between program variables and mathematical variables. J.J. Lukkien provided the following example to illustrate the confusion that may arise. Suppose we want to prove the correctness of program

$$i := 100; \ DO \ i \neq 0 \rightarrow i := i - 1 \ OD$$

with respect to precondition *true* and postcondition *true*. All we need to do is to prove termination. Using invariant $i \geq 0$ and bound function i, our proof obligation is to show that the conjunction of the invariant and the guard implies a decrease of the bound function, that is,

$$i \geq 0 \wedge i \neq 0 \wedge i = C \ \Rightarrow \ wp.(i := i - 1).(i < C)$$

for all constants C. Using a naive formalization, we may proceed as follows

$$i \geq 0 \wedge i \neq 0 \wedge i = C \ \Rightarrow \ wp.(i := i - 1).(i < C)$$
$$= \ \{ \ i = C \ \}$$
$$i \geq 0 \wedge i \neq 0 \wedge i = C \ \Rightarrow \ wp.(i := i - 1).(C < C)$$
$$= \ \{ \ \text{algebra} \ \}$$
$$i \geq 0 \wedge i \neq 0 \wedge i = C \ \Rightarrow \ wp.(i := i - 1).false$$
$$= \ \{ \ \text{law of excluded miracle} \ \}$$
$$i \geq 0 \wedge i \neq 0 \wedge i = C \ \Rightarrow \ false$$
$$= \ \{ \ \text{algebra} \ \}$$
$$\neg(i = C > 0)$$

and we are stuck. The problem, of course, is that one should not allow the substitution of C for i in the argument of wp. The solution is to distinguish between i on the left-hand side and i on the right-hand side by making both sides boolean functions instead of boolean scalars. In particular, the second argument of wp becomes a boolean function that maps argument i to the boolean value $i < C$. We write this function as $(i \triangleright i < C)$. In this way the problem disappears. Unfortunately, so does the practicality of the *wp* calculus. For example, the weakest precondition of statement $i := i - 1$ with respect to postcondition $i < C$ is written as $wp.(i := i - 1).(i \triangleright i < C)$. It becomes even worse when the statement is to be understood in a state where i is not the only program variable. If the program has variables i, j, and k, then the aforementioned precondition becomes $wp.(i := i - 1).(i, j, k \triangleright i < C)$. The size of the formula grows with the number of program variables and this greatly impacts its practical use.

In [6], C. Morgan provides an alternative formalization of the refinement calculus. It is based on the specification statement, written as $v : [pre, post]$, in which v is called the frame, and *pre* and *post* are the precondition and postcondition. Its effect is given as (see [6])

> If the initial state satisfies the precondition *then* change only the variables listed in the frame so that the resulting final state satisfies the postcondition.

The rules for calculating with specification statements do not involve *wp*'s and thereby avoid the problem mentioned above.

The notation used for a specification statement is not that of an infix operator. It is a notation involving three arguments; the first is a list of variables, and the other two are single expressions. In our formalism, we write

notation $!LIST : [!, !]$

and, presently, we cannot express the restriction that the elements if the first list are variables. We have no need for expressing the semantics of the specification statement other than how it can be refined by other programs, as discussed below.

Using the specification statement, we postulate the following property of sequential composition.

$$v : [P, R] \sqsubseteq v : [P, Q]; v : [Q, R]$$

for all predicates P, Q, and R. Next, we introduce the assignment statement. Assignment statement $x := E$ is a refinement of any specification statement that contains x in its frame (in addition to a possibly empty list of variables v) and such that the postcondition in which x is replaced by E is implied by the precondition. In our formalism, we write

$$v, x : [P, Q] \sqsubseteq x := E$$

for all v, x, E, P, and Q provided $P \Rightarrow Q[x := E]$. Finally, we introduce the loop construct. We write

notation $DO ! \rightarrow ! OD$

for a loop with one guarded command and we postulate

$$v : [P, P \ \wedge \ \neg b]$$

$$\sqsubseteq$$

$$DO \ b \rightarrow v : [P \ \wedge \ b \ \wedge \ bf = BF, P \ \wedge \ bf < BF] \ OD$$

provided $P \wedge b \ \Rightarrow \ bf > 0$. For our example, we can get away with a simpler form of the loop in which there is an integer variable that is increased in steps of one from one given value to another given value. Using a more specific refinement rule implies less work upon application since part of the proof obligations can be taken care of when constructing (or postulating) this rule. The rule we will use is given in the text below. For the sake of completeness we also list a rule for strengthening the postcondition and a rule for introducing local variables. The latter notation is a bit more complicated because it restricts the scope of local variables, an issue that we are not concerned with here. This text is the entire refinement calculus as far as we need it for the example. In other cases we need more rules and the full version is about four times the size of the short version listed here. Whenever we develop a new program, we want to use these refinement rules and definitions in the same way we want to use a module of procedures and definitions in a program. We use the same mechanism: rules and definitions can be collected in a module, and the module can be imported by another text.

module *refine*
notation $!LIST \ : \ [\,!\,,\,!\,]$
notation $DO\,!\ \rightarrow\,!\ OD$
notation $VAR \ ?LIST(v) \ BEGIN \ !(v) \ END$
declare $INFIX \ 2 \ :=$
declare $INFIX \ 1 \ ;$
property $ASSOCIATIVE(;)$
rule $StrengthenPost : (v, P, Q, R \ \triangleright$
$\qquad v : [P, Q] \sqsubseteq v : [P, Q \wedge R])$
rule $Block : (v, w, P, Q \ \triangleright$
$\qquad v : [P, Q] \sqsubseteq VAR \ w \ BEGIN \ w, v : [P, Q] \ END \,)$
rule $Assignment : (v, x, E, P, Q \mid P \ \Rightarrow \ Q[x := E] \ \triangleright$
$\qquad v, x : [P, Q] \ \sqsubseteq \ x := E)$
rule $Semicolon : (v, P, Q, R \ \triangleright$
$\qquad v : [P, R] \ \sqsubseteq \ v : [P, Q]; v : [Q, R])$
rule $SemicolonAssignment : (v, x, E, P, Q \ \triangleright$
$\qquad v, x : [P, Q] \ \sqsubseteq \ v, x : [P, Q[x := E]]; x := E)$
rule $UpLoop : (v, i, pre, P, from, to \mid P \ \Rightarrow \ from \leq to \ \triangleright$
$\qquad v, i : [pre, \ P \wedge i = to]$
$\qquad \sqsubseteq$
$\qquad v : [pre, \ P[i := from]]; \ i := from;$
$\qquad DO \ i \neq to \rightarrow v : [P \wedge from \leq i < to, \ P[i := i+1] \wedge from \leq i < to];$
$\qquad\qquad i := i + 1$
$\qquad OD)$

property $\forall(s0, s1, t0, t1 \mid s0 \sqsubseteq s1 \wedge t0 \sqsubseteq t1 \triangleright s0; t0 \sqsubseteq s1; t1)$
property $\forall(b, s0, s1 \mid s0 \sqsubseteq s1 \triangleright DO\ b \to s0\ OD \sqsubseteq DO\ b \to s1\ OD\ )$
property $\forall(w, s0, s1 \mid s0 \sqsubseteq s1 \triangleright$
$\quad VAR\ w\ BEGIN\ s0\ END \sqsubseteq VAR\ w\ BEGIN\ s1\ END\ )$

Notice that we have included a rule, viz. *SemicolonAssignment*, that is strictly superfluous because it follows from the two rules that precede it. However, we often have a situation in which we know that a specification statement $v, x :$ $[P, Q]$ will include an assignment $x := E$. By letting it be the last statement in a sequential composition, we compute specification $v, x : [P, Q[x := E]]$ preceding it so that the combination is a proper refinement. By writing the combination as a single rule, the proxac system will compute and simplify predicate $Q[x := E]$. If we use the *Semicolon* rule instead, the author has to postulate this predicate and the system will verify its use in the subsequent refinement steps. The additional rule reduces the author's work.

In the module that contains the definition, we might want to prove that the more specific loop rule follows from the general *Loop* rule. Such a proof is given here.

rule $UpLoop : (v, i, pre, P, from, to \mid P \Rightarrow from \leq to \triangleright$

$v, i : [pre, P \wedge i = to]$
$\sqsubseteq \quad \{ \ Semicolon[P := pre, Q := P \wedge from \leq i \leq to, R := P \wedge i = to, v := (v, i)] \ \}$
$v, i : [pre, P \wedge from \leq i \leq to]; v, i : [P \wedge from \leq i \leq to, P \wedge i = to]$
$\sqsubseteq \quad \{ \ SemicolonAssignment[x := i, E := from, P := pre,$
$\qquad\qquad\qquad Q := P \wedge from \leq i \leq to] \ \}$
$v : [pre, P[i := from]]; i := from; v, i : [P \wedge from \leq i \leq to, P \wedge i = to]$
$\sqsubseteq \quad \{ \ Loop[P := P \wedge from \leq i \leq to, bf := to - i, v := (v, i), b := i \neq to] \ \}$
$v : [pre, P[i := from]]; i := from;$
$DO\ i \neq to \to v, i : [\ P \wedge from \leq i < to \wedge to - i = BF,$
$\qquad\qquad\qquad P \wedge from \leq i \leq to \wedge to - i < BF]$
OD
$\sqsubseteq \quad \{ \ SemicolonAssignment[\ P := P \wedge from \leq i < to, x := i, E := i + 1,$
$\qquad\qquad\qquad Q := P \wedge from \leq i \leq to \wedge to - i < BF] \ \}$
$v : [pre, P[i := from]]; i := from;$
$DO\ i \neq to \to v : [P \wedge from \leq i < to, P[i := i + 1] \wedge from \leq i < to];$
$\qquad i := i + 1$
$OD)$

Usage of this long version of the *UpLoop* rule is identical to usage of the version listed in the module text. The external view of a rule with a calculational body is that of a rule with the body reduced to its first and last line with a connective deduced from the sequence of connectives. In this reduction, transitivity of $\sqsubseteq$ is essential. After a rule has been written it is shown in abbreviated form in the rules window so that it can be applied by a mouse click.

Notice that we have now given a proof of the correctness of the *UpLoop* rule. The mechanism for developing the proof is identical to the mechanism for refining a program.

4   An example of a program derivation

In this section we illustrate the use of the refinement rules to derive a program from its specification. The program is well-known and so is its derivation. Our focus of attention is the support given by the proxac system.

In some steps of the proof above (and in some steps of program derivations below, but not in any other earlier step), some variables of rules cannot be determined by pattern matching. As a result, the author of the text will need to give the proxac system hints regarding these unresolved variables. In this section, we indicate hints by underlining them.

The programming problem is known as the *maximum segment sum* problem (see [5]). Given is an array a of $N \geq 0$ integers. A segment of the array is a contiguous subsequence of the array. A segment has a segment sum, viz. the sum of all its array elements. The problem is to write a program to determine the maximum segment sum. We formalize the problem as

module *mss*
import *refine*
declare $QUANTIFIER$ $\sum$
property $INFIXOPERATOR(\sum) = +$
declare $INFIX$ 20 $\uparrow$
property $ASSOCIATIVE(\uparrow)$ $\wedge$ $DUAL(\uparrow) = \uparrow$ $\wedge$ $IDEMPOTENT(\uparrow)$
declare $QUANTIFIER$ **MAX**
property $INFIXOPERATOR(\textbf{MAX}) = \uparrow$
property $0 \leq N$
rule $mss : (n \mid 0 \leq n \leq N \triangleright mss.n = \textbf{MAX}(j \mid 0 \leq j \leq n \triangleright mes.j))$
rule $mes : (j \mid 0 \leq j \leq N \triangleright mes.j = \textbf{MAX}(i \mid 0 \leq i \leq j \triangleright sum.i.j))$
rule $sum : (i,j \mid 0 \leq i \leq j \leq N \triangleright sum.i.j = \sum(h \mid i \leq h < j \triangleright a.h))$
edit $s : [true, s = mss.N]$

We recognize the rules that we had in section 2. We use $\uparrow$ for the infix maximum operator. The problem is to write a program for computing $mss.N$, that is, a program that refines $s : [true, s = mss.N]$. We will need a loop, and this will lead to a specification statement in the loop body that contains $mss.n$ in the precondition and $mss.(n + 1)$ in the postcondition. Given the calculation in section 2, we know that the latter can be rewritten as $mss.n \uparrow mes.(n + 1)$ which means that we are tempted to introduce $mes.(n+1)$ in the loop invariant. However, upon termination of the loop, $n = N$, and $mes.(N + 1)$ is undefined. We must, therefore, decrease by one the argument of mes and calculate $mes.(n+1)$ from $mes.n$ when needed. The programming problem can now be formalized as finding a refinement for $n, r, s : [true, s = mss.n \ \wedge \ r = mes.n \ \wedge \ n = N]$.

If we had not noticed the problem with undefinedness of $mes.(N+1)$, we would have proceeded with $mes.(n+1)$ in the invariant. We would get stuck later on where a step cannot be justified because

$$0 \le n < N \ \Rightarrow\ 0 \le n+1 < N$$

cannot be established. We would not have been led into undefined results!

$$
\begin{aligned}
&s : [true, s = mss.N] \\
\sqsubseteq\ &\{\ Block[v := s, w := (n,r), P := true, Q := s = mss.N]\ \} \\
&VAR\ n, r\ BEGIN\ \underline{n, r, s : [true, s = mss.N]}\ END \\
\sqsubseteq\ &\{\ StrengthenPost[v := (n, r, s), P := true, Q := s = mss.N, \\
&\qquad\qquad\quad \underline{R := r = mes.N \wedge n = N}]\ \} \\
&VAR\ n, r\ BEGIN\ n, r, s : [true, s = mss.N \wedge r = mes.N \wedge n = N]\ END \\
=\ &\{\ n = N\ \} \\
&VAR\ n, r\ BEGIN\ n, r, s : [true, s = mss.n \wedge r = mes.n \wedge n = N]\ END
\end{aligned}
$$

Notice how the second step introduces $Q \wedge R$ in which R in turn is a conjunction. Since conjunction is associative, no parentheses surround R. We have found these kind of aspects instrumental in keeping down the amount of detail that the author has to deal with and, hence, the number of steps needed to complete a program derivation. We now focus attention on the latter specification statement and ignore the surrounding block. When doing so, the system keeps track of the context in which this narrowing of attention occurs.

$$
\begin{aligned}
&n, r, s : [true, s = mss.n \wedge r = mes.n \wedge n = N] \\
\sqsubseteq\ &\{\ UpLoop[v := (r, s), i := n, pre := true, P := s = mss.n\ \wedge\ r = mes.n, \\
&\qquad\qquad from := 0, to := N]\ \} \\
&r, s : [true, s = mss.0\ \wedge\ r = mes.0];\ n := 0; \\
&DO\ n \ne N \to r, s : [s = mss.n\ \wedge\ r = mes.n\ \wedge\ 0 \le n < N, \\
&\qquad\qquad\qquad s = mss.(n+1)\ \wedge\ r = mes.(n+1)\ \wedge\ 0 \le n < N]; \\
&\qquad\qquad n := n + 1 \\
&OD \\
\sqsubseteq\ &\{\quad r, s : [true, s = mss.0\ \wedge\ r = mes.0] \\
&\ =\quad \{\ mss[n := 0]\ \} \\
&\qquad r, s : [true, s = mes.0\ \wedge\ r = mes.0] \\
&\ \sqsubseteq\quad \{\ SemicolonAssignment[v := r, \underline{x := s}, E := r, P := true, \\
&\qquad\qquad\qquad\qquad\qquad Q := s = \underline{mes.0}\ \wedge\ r = mes.0]\ \} \\
&\qquad r, s : [true, r = mes.0];\ s := r \\
&\ =\quad \{\ mes[j := 0]\ \} \\
&\qquad r, s : [true, r = sum.0.0];\ s := r \\
&\ =\quad \{\ sum[i := 0, j := 0]\ \} \\
&\qquad r, s : [true, r = 0];\ s := r \\
&\ \sqsubseteq\quad \{\ Assignment[v := s, x := r, E := 0, P := true, Q := r = 0]\ \} \\
&\qquad r := 0;\ s := r \\
&\quad\}
\end{aligned}
$$

$r := 0; \ s := r; \ n := 0;$
$DO \ n \neq N \ \to \ r, s : [s = mss.n \ \wedge \ r = mes.n \ \wedge \ 0 \leq n < N,$
$\qquad\qquad\qquad s = mss.(n+1) \ \wedge \ r = mes.(n+1) \ \wedge \ 0 \leq n < N];$
$\qquad\qquad n := n+1$
OD

Notice that the above calculation contains a nested calculation. The step to replace $r, s : [true, s = mss.0 \ \wedge \ r = mes.0]$ by $r := 0; \ s := r$ consists of five steps by itself. Replacing them in the context where that specification statement occurs is justified by the monotonicity of sequential composition. We continue by narrowing attention to the specification statement in the loop body. We will need two more rules; they are not related to refinement but to ranges in quantifications. Since we have both ranges of the form $0 \leq i \leq j$ and of the form $i \leq h < j$, we have two split rules.

rule $split : (x, y, z \ \triangleright \ (x \leq y < z+1) = (x \leq y < z \ \vee \ x \leq y = z))$
rule $split : (x, y, z \ \triangleright \ (x \leq y \leq z+1) = (x \leq y \leq z \ \vee \ x \leq y = z+1))$

We continue the refinement.

$r, s : [s = mss.n \ \wedge \ r = mes.n \ \wedge \ 0 \leq n < N,$
$\qquad\quad s = mss.(n+1) \ \wedge \ r = mes.(n+1) \ \wedge \ 0 \leq n < N]$
$= \quad \{ \ mss[n := n+1] \ \}$
$r, s : [s = mss.n \ \wedge \ r = mes.n \ \wedge \ 0 \leq n < N,$
$\qquad\quad s = \mathbf{MAX}(j \mid 0 \leq j \leq n+1 \triangleright mes.j) \ \wedge$
$\qquad\quad r = mes.(n+1) \ \wedge \ 0 \leq n < N]$
$= \quad \{ \ split[x := 0, y := j, z := n] \ \}$

$r, s : [s = mss.n \ \wedge \ r = mes.n \ \wedge \ 0 \leq n < N,$
$\qquad\quad s = \mathbf{MAX}(j \mid 0 \leq j \leq n \triangleright mes.j) \uparrow mes.(n+1) \ \wedge$
$\qquad\quad r = mes.(n+1) \ \wedge \ 0 \leq n < N]$
$= \quad \{ \ mss \ \}$
$r, s : [s = mss.n \ \wedge \ r = mes.n \ \wedge \ 0 \leq n < N,$
$\qquad\quad s = mss.n \uparrow mes.(n+1) \ \wedge \ r = mes.(n+1) \ \wedge \ 0 \leq n < N]$
$= \quad \{ \ r = mes.(n+1) \ \}$
$r, s : [s = mss.n \ \wedge \ r = mes.n \ \wedge \ 0 \leq n < N,$
$\qquad\quad s = mss.n \uparrow r \ \wedge \ r = mes.(n+1) \ \wedge \ 0 \leq n < N]$
$\sqsubseteq \quad \{ \ SemicolonAssignment[v := r, \underline{x := s}, E := s \uparrow r,$
$\qquad\qquad\qquad P := s = mss.n \ \wedge \ r = mes.n \ \wedge \ 0 \leq n < N,$
$\qquad\qquad\qquad Q := s = (mss.n) \uparrow r \ \wedge \ r = mes.(n+1) \ \wedge \ 0 \leq n < N] \ \}$
$r, s : [s = mss.n \ \wedge \ r = mes.n \ \wedge \ 0 \leq n < N,$
$\qquad\quad s \uparrow r = mss.n \uparrow r \ \wedge \ r = mes.(n+1) \ \wedge \ 0 \leq n < N];$
$s := s \uparrow r$
$= \quad \{ \quad r = mes.(n+1)$
$\qquad\quad = \quad \{ \ mes[j := n+1] \ \}$
$\qquad\qquad r = \mathbf{MAX}(i \mid 0 \leq i \leq n+1 \triangleright sum.i.(n+1))$
$\qquad\quad = \quad \{ \ split[x := 0, y := i, z := n] \ \}$

$$r = \mathbf{MAX}(i \mid 0 \le i \le n \triangleright sum.i.(n+1)) \uparrow (sum.(n+1).(n+1)))$$
$$= \quad \{ sum[i := n+1, j := n+1] \}$$
$$r = \mathbf{MAX}(i \mid 0 \le i \le n \triangleright sum.i.(n+1)) \uparrow 0$$
$$= \quad \{ sum[j := n+1] \}$$
$$r = \mathbf{MAX}(i \mid 0 \le i \le n \triangleright \sum(h \mid i \le h < n+1 \triangleright a.h)) \uparrow 0$$
$$= \quad \{ split[x := i, y := h, z := n] \}$$
$$r = \mathbf{MAX}(i \mid 0 \le i \le n \triangleright \sum(h \mid i \le h < n \triangleright a.h) + a.n) \uparrow 0$$
$$= \quad \{ sum[j := n] \}$$
$$r = \mathbf{MAX}(i \mid 0 \le i \le n \triangleright sum.i.n + a.n) \uparrow 0$$
$$= \quad \{ factor \}$$
$$r = (\mathbf{MAX}(i \mid 0 \le i \le n \triangleright sum.i.n) + a.n) \uparrow 0$$
$$= \quad \{ mes[j := n] \}$$
$$r = (mes.n + a.n) \uparrow 0$$
$$\}$$
$$r, s : [s = mss.n \ \wedge \ r = mes.n \ \wedge \ 0 \le n < N,$$
$$\qquad s \uparrow r = mss.n \uparrow r \ \wedge \ r = mes.n + a.n \ \wedge \ 0 \le n < N];$$
$$s := s \uparrow r$$
$$\sqsubseteq \quad \{ Assignment[v := s, x := r, E := (r + a.n) \uparrow 0,$$
$$\qquad P := s = mss.n \ \wedge \ r = mes.n \ \wedge \ 0 \le n < N,$$
$$\qquad Q := s \uparrow r = mss.n \uparrow r \ \wedge \ r = mes.n + a.n \uparrow 0 \ \wedge \ 0 \le n < N] \}$$
$$r := (r + a.n) \uparrow 0; s := s \uparrow r$$

The last-but-one step in the subcalculation is a step labeled *factor* and this is one of the built-in transformations. However, since we did not specify that addition distributes over maximum, the proxac system is unable to verify the correctness of this transformation and will print a question asking

$$\text{Context implies: } \mathbf{MAX}(i \mid 0 \le i \le n \triangleright sum.i.n + a.n) =$$
$$\mathbf{MAX}(i \mid 0 \le i \le n \triangleright sum.i.n) + a.n \qquad ?$$

The author of the text can decide to add the distribution property, to prove it, or to ignore the question.

When we widen the focus again from the specification statement in the loop body, we end up with the text

> *VAR* n, r
>
> *BEGIN*
>
> $\quad r := 0; s := r; n := 0;$
>
> $\quad DO \ n \ne N \rightarrow r := (r + a.n) \uparrow 0; s := s \uparrow r; n := n + 1 \ OD$
>
> *END*

and this program solves the problem at hand.

5 Conclusion

The total text of the program derivation is quite long, much longer than the program text itself. This observation is often used as an argument against the use of stepwise refinement or against formal methods. The derivation consists of a total of 31 steps, 8 of them being narrowing and widening the focus of attention. Of the remaining 23 steps, 16 steps require no hint at all. The 7 hints that had to be given have been underlined. These hints are the only input given to the system in addition to each mouse click that selects a rule and triggers its application. As a result, the total input is comparable in size to the resulting program and not to the derivation. One major benefit of using this system is that design decisions have been made explicit. Another major benefit is that all steps have been mechanically verified. We feel that this derivation shows that the use of a formal system for stepwise refinement of programs puts no extra burden on the programmer, and competes well with paper and pencil. Of course, we have used a set of rules that constitute the refinement calculus, but this is an investment that is amortized over the development of many programs. We have also written explicitly what the specification of the problem is. We don't think that a responsible programmer delivers a program without a specification, so this does not constitute extra work.

The transformation rules we have used are rather elementary. One can come up with more complicated rules that correspond to many steps in our present repertoire. This reduces the number of steps to complete a program derivation; however, the increase in the number of rules may make it harder to use them.

One can view the transformations as the commands of a programming language for formula manipulation. The transformations that we have described here, correspond to the elementary commands. We have used the notation of functions for describing those rules. By extending the notation with function composition, we construct composite transformation commands. By extending the notation with conditionals and a fixpoint operator, we obtain a complete programming language. These extensions allow us to construct what are sometimes called tactics. Tactics and their semantics are beyond the scope of the present paper.

6 Acknowledgement

The proxac editor can be found in ftp directory jan/proxac on cs.caltech.edu. A more detailed description can be found in [7] and an up-to-date version thereof can be found in the same ftp directory. Writing this editor and developing the notations used was, and is, a challenging undertaking. Diana Finley was instrumental in getting this project underway. Greg Davis contributed many ideas and helped get the program to the point where it actually became usable. My thanks go to both of them.

References

[1] R.J.R. Back. *On the Correctness of Refinement Steps in Program Development.* PhD thesis, University of Helsinki, 1978. Report A-1978-4.

[2] E.W. Dijkstra. A Constructive Approach to the Problem of Program Correctness. *BIT*, 8:174–186, 1968.

[3] E.W. Dijkstra. Notes on Structured Programming. In O.J. Dahl, E.W. Dijkstra, and C.A.R. Hoare, editors, *Structured Programming.* Academic Press, 1971.

[4] E.W. Dijkstra. *A Discipline of Programming.* Prentice-Hall, 1976.

[5] D. Gries. A Note on the Standard Strategy for Developing Loop Invariants and Loops. *Science of Computer Programming*, 12:207–214, 1982.

[6] C. Morgan. *Programming from Specifications.* Series in Computer Science (C.A.R. Hoare, ed.). Prentice-Hall International, 1990.

[7] J.L.A. van de Snepscheut. JAN 183. Proxac: an Editor for Program Transformation. Technical Report CS 93-33, California Institute of Technology, 1993.

[8] N. Wirth. Program Development by Stepwise Refinement. *Communications of the ACM*, 14:221–227, 1971.

[9] N. Wirth. *Systematic Programming.* Prentice-Hall, 1973.

Hardware and Software : The Closing Gap

C. A. R. Hoare and I. Page

Computing Laboratory, University of Oxford,
Oxford OX1 3QD, U.K.

Abstract

The study of computing science is split at an early stage between the branches dealing separately with hardware and software; and there is a corresponding split in later professional specialisation. This paper explores the essential unity and overlap of the two branches. The basic concepts are those of occam, taken as a simple example of a high-level programming language; its notations may be translated by the laws of programming to the machine code of a conventional machine. Almost identical transformations can produce the networks of gates and flip-flops which constitute a hardware design. These insights are being exploited in hybrid systems, implemented partly in hardware and partly in software. A TRAM-standard printed circuit board called HARP has been constructed for such applications. It links a transputer by shared memory with a commercial Field Programmable Gate Array. Prospects for application are discussed.

1   Correctness of Design

The design of a complex engineering product like a real time process control system is ideally decomposed into a progression of related phases. It starts with an investigation of the properties and behaviour of the process evolving within its environment, and an analysis of requirements for its optimal or satisfactory performance, or at least for its safety. From these is derived a specification of the electronic or program-controlled components of the system. The project then may pass through an appropriate series of design phases, culminating in a program expressed in a high level language. After translation into the machine code of the chosen computer, it is loaded into memory and executed at high speed by electronic circuitry. Additional application-specific hardware may be needed to embed the computer into the system which it controls. Each of these phases presents a conceptual gap, as wide and challenging as that between hardware and software. Reliability of the delivered system requires that all the gaps be closed. It is achieved not just by testing, but by the quality of thought and meticulous care exercised by analysts, designers, programmers and engineers in all phases of the design.

This has been a description of an ideal that is rarely achieved in any field of engineering practice. Nevertheless, an ideal forms the best basis for long-term research into engineering method. The goal of this research is to discover and formalise methods which reduce the risks and simplify the routines of the design task, and give fuller scope for the exercise of human skill and invention in meeting product requirements at low cost and in good time. The goal of this paper is to convey an impression of the methods and intermediate results of the research. It illustrates them by the techniques of provably correct compilation, either to machine code or to hardware. Finally, it describes a project for mixing hardware and software implementation of programs, particularly for embedded applications.

In principle, the transition between one design phase and the next is marked by delivery of a document, expressed in some more or less formal notation. Each phase starts with study and acceptance of the document produced by the previous phase; and ends with the delivery of another document, usually formulated at a lower level of abstraction, closer to the details of the eventual implementation. Each designer seeks high efficiency at low cost; but is constrained by an absolute obligation that the final document must be totally correct with respect to the initial document for this design phase. Thus the requirements must be faithfully reflected in the specification, the specification must be fully achieved by the design, the design must be correctly implemented by the program, the program must be accurately translated to machine code, which must be reliably executed by the hardware. Although we have used different words in English to describe the correctness relation at each different level of design, we shall show that conceptually it is the same relation in all cases, namely logical implication, denoted by $\Leftarrow$.

When the system is eventually delivered and put into service, all that really matters is that the actual hardware delivered should meet the overall requirements of the system. This is guaranteed by a simple mathematical property of the implementation relation: it is transitive. If P is implemented by Q and Q is implemented by R then P is implemented by R:

$$\text{If } P \Leftarrow Q \text{ and } Q \Leftarrow R \text{ then } P \Leftarrow R.$$

However long the chain of intermediate documents, if each document correctly implements the previous one, the overall requirements will be correctly implemented by the delivered hardware.

We have given a very simple account of the design process, and the reason why it can validly be split into any number of phases. The account is highly abstract: in concrete reality, complications arise from the fact that each of the design documents is written in a different notation, adapted to a different conceptual framework at a different level of abstraction. For example, a requirements document for a real time system may use timing diagrams or temporal logic, a specification may use set theory (Z or VDM), a design may use flow charts or SSADM, a program may use ADA or C, the machine code may be INTEL 8080 and the hardware may be described in pictures or as a netlist of components and wires. How can we be certain that a document serving as an interface between one design phase and the next has been correctly understood (i.e. with the same meaning) by the specialists who produced it as a design and the different specialists who accepted it as a specification for the next phase? The utmost care and competence in each individual phase of design will be frustrated if bugs are allowed to congregate and breed in the interfaces between one phase and the next.

The solution is to interpret every one of the documents in the chain as a direct or indirect description at an appropriate level of abstraction of the observable properties and behaviour of some system or class of system or component that exists (or could be made to exist) in the real world. These descriptions can be expressed most precisely in the language which science has already shown to be most effective in describing and reasoning about the real world, namely the language of mathematics. Such descriptions use identifiers as free variables to stand for observations or measurements that could in principle be made of the real world system, for example the position and momentum of a physical point, plotter pen or projectile, or the initial and final values of a global variable of a computer program [1].

A simple example of a mathematically expressed requirement is that for a straight line constant speed trajectory of our point of interest. Let x_t be its displacement on the x axis at time t, and let a be the desired velocity. Then within some desired interval the difference between the actual and desired position should, within the relevant period, always be less than some permitted tolerance:

$$| x_t - at - x_0 | \leq 0.4$$

for all $t \in (3 \ldots 5)$. If we also want steady motion on the y-axis, this is stated separately:

$$| y_t - bt - y_0 | \leq 0.4,$$

for all $t \in (3 \ldots 5)$. The additional requirement is just conjoined by "and" to the original requirement. The use of conjunction to compose complex requirements from simple descriptions is a crucial advantage of the direct use of logical notations at the earliest stage of a design project.

In this example, the requirements formalise the permitted tolerances on the accuracy of implementation. Of course an implementation is permitted to achieve even greater accuracy. For example, suppose the behaviour of a particular implementation is described by

$$(x_t - at - x_0)^2 + (y_t - bt - y_0)^2 < 0.1,$$

for all $t \leq 5$. This implies both of the requirements displayed above; consequently the implementation correctly fulfills its specification.

This notion of correctness is perfectly general. Suppose a design document P and a specification S use consistent naming conventions for variables to describe observations of the same class of system; and suppose that P logically implies S. This means that every observation of any system described by P is also described by S, and therefore satisfies S as a specification. Certainly, no observation of the system described by P can violate the specification S. That is the basis of our claim that the relationship of correct implementation is nothing other than simple logical implication, the fundamental and familiar transitive relation that governs every single step of all valid scientific and mathematical reasoning. It should therefore be no surprise that it is also fundamental to all stages and phases of sound engineering design.

2 Hardware

The example we have just described might have been part of the requirement on a control system, formulated at the start of a design project. Our next example describes the actual behaviour of the ultimate components available for its implementation, right at the final phase of electronic circuit design and assembly. As in all descriptions of the real world, we choose to model it at a certain level of abstraction, which is only an approximation of reality. We have chosen a level which (subject to reasonable constraints) is known to be generally implementable in a wide range of technologies.

Let the variables x_t, y_t and w_t stand for voltages observable at time t on three distinct wires connected to an OR-gate (Fig. 1). The voltage takes one of two values, 0 standing for connection to ground and 1 standing for presence of electrical potential. The specification of the OR-gate is that the value of the output wire w is the greater of the values of the input wires x and y. This relationship cannot be guaranteed at all times, but only at regular intervals, at the end of each operational cycle of the circuit. For convenience, the unspecified duration of each cycle is taken to be the unit of time. The behaviour of the OR-gate is described as an equation:

$$w_t = x_t \vee y_t,$$

for all $t \in (0, 1, 2, \ldots)$, where the range of t is here and later restricted to the natural numbers. This means that observations can be made only at discrete intervals, understood to be on the rise of the relevant clock signal.

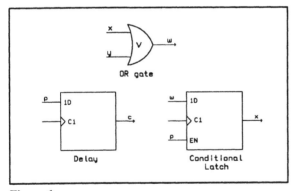

Figure 1.

Another example of a hardware circuit is the Delay element. On each cycle of operation, the voltage at its output c is the same as the voltage at its input p on the previous cycle of operation:

$$c_{t+1} = p_t.$$

The clock event which advances t is communicated to the Delay element by the global clock input signal (marked C1 in Fig. 1). Note that, we are unable to predict the initial value c_0, obtained when the hardware is first switched on. The correctness of any circuit using this component must not depend on the initial value. The description reflects a certain physical non-determinism, which cannot be controlled at this level of abstraction. As in the case of engineering tolerance, the specification must also allow for a range of possible outcomes; otherwise correctness will not be provable.

A useful variation of the Delay element is the Conditional Latch element (in engineering terms this is an edge-triggered flip-flop with a clock enable input). Here, the value of the output is changed only when a clock event occurs and the input control wire p is high; and then the new value of x is taken from the other input wire w. Otherwise the value of x remains unchanged. This behaviour is formally described:

$$x_{t+1} = (w_t \lhd p_t \rhd x_t),$$

where $a \lhd b \rhd c$ is read as "a if b else c".

A pair of hardware components is assembled by connecting output wires of each of them to like-named input wires of the other, making sure that any cycle of connections is cut by a Delay. Electrical conduction ensures that the value observed at the input ends of each wire will be the same as that produced at the output end. As a result, the combined behaviour of an assembly of hardware components is described surprisingly but exactly by a conjunction of the descriptions of their separate behaviours.

For example (Fig. 2), an assembly consisting of the OR-gate, the Delay element and the Conditional Latch is described by

$$w_t = x_t \vee y_t \quad \text{and} \quad c_{t+1} = p_t \quad \text{and} \quad x_{t+1} = (w_t \lhd p_t \rhd x_t).$$

The purpose of the internal wire w is solely to carry information from the OR-gate to the Conditional Latch. Its existence can therefore be concealed by existential quantification, with beneficial simplification of the behavioural description:

$$x_{t+1} = ((x_t \vee y_t) \lhd p_t \rhd x_t) \quad \text{and} \quad c_{t+1} = p_t.$$

The descriptive and deductive power of logical conjunction has been illustrated by two examples, one at the highest level of system requirement capture, and one at the lowest level of hardware implementation. In principle, the conjunction of the descriptions of all the hardware components of the system should imply the conjunction of all the requirements originally placed upon the system as a whole. In a simple system, this implication may be proved directly; otherwise it is proved through a series of intermediate design documents, perhaps including a program expressed in a high level language. A program also must be interpreted as an indirect description of its own behaviour when executed. We therefore need names to describe its observable features, and for reasons of our own we have chosen to reuse the names of the hardware wires.

Let p_t be an assertion true at just those times t when execution of the program starts, and let $c_{t'}$ be true at just those times t' at which it terminates. Let x_t be the value of a program variable x at time t, so $x_{t'}$ is the final value. With a slight simplification, the assignment statement

```
x := x OR y
```

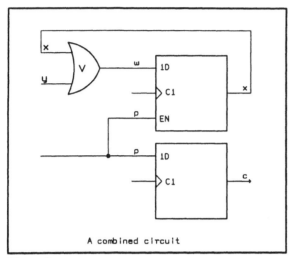

A combined circuit

Figure 2.

can now be defined as an abbreviation for

$$p_t \Rightarrow \exists t' \geq t. \ (x_{t'} = x_t \vee y_t) \text{ and } c_{t'}.$$

If the program starts at time t, then it stops at some later time t'; and at that time the final value of x is the disjunction of the initial values of x and y. Note that the execution delay $(t' - t)$ has been left deliberately unspecified. This gives design freedom for implementation in a variety of technologies, from instantaneous execution (achieved by compile time optimisation) to the arbitrary finite delays that may be interposed by an operating system in execution of a timeshared program.

Now we confess why we have chosen the same names for the software variables as the hardware wires. It demonstrates immediately that our example hardware assembly is also a valid implementation of the software assignment: the description of one of them implies the description of the other. The proof is equally simple: take the software termination time as exactly one hardware cycle after the start. It is the translation of both hardware and software notations into a common framework of timed observations that permits a proof of the correctness of this design step as a simple logical implication, thereby closing the intellectual and notational gap between the levels of hardware and software. Our example has been artificially simplified by use of exactly the same observation names at both levels. In general, it may be necessary to introduce a coordinate transformation to establish the link between them.

In principle, proof of correctness of a design step can always be achieved, as we have shown, by expanding the abbreviation of the relevant notations. But for a large system this would be impossibly laborious. What we need is a useful collection of proven equations and other theorems expressed wholly in the abbreviated notations; it is these that should be used to calculate, manipulate, and transform the abbreviated formulae, without any temptation or need to expand them. For example, the power of matrix algebra lies in the collection of equational laws which express associative and distributive properties of the operators:

$$A \times (B \times C) = (A \times B) \times C$$
$$(A + B) \times C = (A \times C) + (B \times C).$$

The mathematician who proves these laws has to expand the abbreviations, into a confusing clutter of subscripts and sigmas:

$$\Sigma_j A_{ij}(\Sigma_k B_{jk} C_{kl}) = \Sigma_k(\Sigma_j A_{ij} B_{jk})C_{kl}.$$

But the engineer who uses the laws just does not want to know.

Fortunately, programming notations have been proved by mathematicians [2, 3] to enjoy algebraic properties just as simple and useful as those of matrix algebra. For example, sequential composition of program statements is associative like multiplication; and it distributes leftwards into conditionals. Even without mathematical proof, an engineer can check quite easily that the two sides of these equations describe exactly the same possible sequences of execution of the commands P, Q and R.

3 Software

We have already given an example to show the power of observational reasoning on a small scale to prove the correctness of a simple circuit as a hardware implementation of a simple assignment statement of a programming language. To illustrate the power of algebraic reasoning, we will use it to derive a much more elaborate software implementation of the same assignment statement

```
x:= x OR y,
```

translating it step-by-step into the machine code of a simple computer.

A high-level language uses symbolic names (eg. x and y) for variables ranging over a certain finite range of integers. An actual computer stores these values in a random access memory (ram) at distinct locations, say 11 for x and 9 for y. Since we want to use the same high level language throughout our calculations, we represent the random access memory as an integer array, with (say) 8192 locations :

```
var ram :   array [0 .. 8191] of word;
```

An actual computer can access and change locations in this memory, for example by an assignment:

```
ram [11]  := ram [11] OR ram [9];
```

but it cannot access the high level variables x and y. We therefore need an abstraction function (data refinement), showing how the abstract variables x and y are represented concretely in ram. Fortunately this too can be expressed as an initial assignment which copies the values of x and y into the allocated locations before the program starts:

```
ram [11], ram [9] := x,y.
```

Similarly, the values x and y need to be copied back after the program finishes

```
x,y := ram [11], ram [9].
```

Although expressed as part of the program, these initial and final assignments are purely conceptual, and will not be executed in practice. In summary, the simple assignment can be validly replaced by a considerably longer one, whose only merit is that it is a slightly more faithful description of the way in which the eventual machine code will be executed.

```
x := (x OR y)
⇐      var ram :  array [0 .. 8191] of word;
       ram [11], ram [9]:= x, y;
       ram [11] := ram [11] OR ram [9];
       x, y := ram [11], ram [9]
```

This is the first of a series of transformations conceptually or actually made by a compiler. It is formulated as an implication whose proof is simultaneously a proof of the correctness of the transformation itself.

The next phase of translation concentrates on the essential assignment statement taken from the middle of the code

```
ram [11] := ram [11] OR ram [9].
```

In a conventional single-address machine code, this has to be split into three instructions, each of which refers to a general purpose register called A. The sole purpose of register A is to carry information between the instructions. It is therefore declared as a local variable, whose existence can be concealed by existential quantification in the same way as local wire w in the previous hardware example.

```
ram [11] := ram [11] OR ram [9]
⇐   var A : word;
    A := ram [11];
    A := A OR ram [9];
    ram [11] := A
```

Again, an algebraic calculation similar to symbolic execution shows that the mathematical meaning of the assignment is unchanged, so that this translation is also proved to be correct.

The next task is to implement the sequencing construction, denoted by semicolon, ensuring that the three assignments are executed properly in the right order. In a conventional computer, sequencing is implemented by a program pointer (called p), which in this case steps through four values, say 20, 21, 22 and 23. The value of p selects which of the three instructions is to be executed, and at the same time p itself is incremented by one

```
⇐   var p : integer;
    p:= 20;
    while (20 <= p) and (p < 23) do
      case p of
        20:  A, p := ram [11], p+1;
        21:  A, p := A OR ram [9], p+1;
        22:  ram [11], p := A, p+1;
      end;
    assert p=23
```

The last line is an assertion, ensuring that the code ends decently by falling off the end, rather than by a wild jump. Such assertions do not have to be executed; we will see that they are an essential aid to reasoning about designs and their correctness.

The purpose of the design so far is to ensure that each of the clauses of the case statement describes exactly the effect of execution of a single machine code instruction available on the target computer. Of course, in practice each instruction is represented as a bit-pattern, which packs together the operation code for the instruction and the address. Fortunately the execution of the instruction can again be accurately and conveniently described in the programming language itself, by a conventional interpreter or simulator for the machine code:

```
procedure interpret (instruction, address,: integer);
  case instruction of
    0:  A, p := ram [address], p+1;
    1:  ram [address], p := A, p+1;
    7:  A, p := A OR ram [address], p+1;
    ...
    ...
```

An interpreter like this is often decreed as a definition of the architecture of a computer, and is accepted by hardware designers as a formal specification document for their design. We will use it for the same purpose later.

The final transformation implements the basic idea of the stored program computer. The program store is represented by an array, say with 64 instructions

```
var code:  array [0 .. 63] of (instruction, address);
```

The action of the program loader is represented by an assignment of initial values to this array:

```
code[20], code[21], code[22] := (0, 11), (7, 9), (1, 11).
```

The program starts by initialising the sequence register, which is then used repeatedly as an index to fetch the successive instructions from the code array and execute them. The combined process of loading and running is described by

```
var code:  array [0 ..  63] of (instruction, address);
code [20], code [21], code [22]  := (0, 11), (7, 9), (1, 11);

p:= 20;
while (20 <= p) and (p < 23) do
   interpret (code [p]);
assert p < 23
```

Thus we reach the end of a long and complex series of transformations of a single simple assignment statement of a high level language into a complex structure of statements expressed in the same high level language. The only consolation for the length of the transformation process is its modularity. Each transformation of the series has separately introduced a separate architectural feature

1. random access memory and symbol tables

2. registers and single address instructions

3. sequence control

4. numeric encoding of instructions

5. the stored program concept.

This kind of unravelling and separation of concerns is an essential goal in engineering method. Careful modular structuring of the theory is essential to allow the same theorems to be used on a variety of source languages and different target architectures, including direct implementation in hardware.

The same principle of abstraction can be maintained in translation of the structural features of the language, for example sequential composition and conditionals. These are translated (or rather eliminated) by the algebraic technique of reduction to normal form. We define a normal form as an iterated conditional, where all the conditions are disjoint; when they are all false, the iteration terminates:

```
p:= k;
   while (k <= p) and (p < l) do
      case p of
         k:...    ;  ⎫
         ...      ;  ⎬ X
         l - 1:...;  ⎭
      end;
   assert p = l
```

If all the cases in the collection X can be represented as binary machine code instructions, it is fairly obvious that any complete program in normal form can be further translated into code for execution in a stored-program computer, using a loader and interpreter. Let us introduce the abbreviation [k, X, l] to stand for this normal form.

The remaining task is to show how to reduce every program of the source language into normal form. First, all the primitive expressions of the language are reduced (or more usually expanded) into the desired normal form; we have already shown how to do this for the assignments which are the ultimate components of a conventional program.

```
[k, QQQ, 1]  ;  [1, RRR, m]
           ⇐ [k, (QQQ; RRR), m]
```

Figure 3.

The next step is to eliminate all the composition operators of the language, one by one in a bottom-up fashion. The operands can therefore be assumed already to be in normal form; what is needed is that the result of elimination should also be in normal form. The algebraic laws used for this kind of elimination usually have the shape

$$NF1 \odot NF2 \Leftarrow NF3$$

where $\odot$ is the operator to be eliminated, and all three operands are in normal form, which does not contain $\odot$. Repeated application of such laws from left to right will eventually eliminate all occurrences of the given operator, leaving just a single normal form, which by transitivity of $\Leftarrow$ implements the original program.

An example of a reduction law for our compiler is one that eliminates sequential composition (Fig. 3). On the left hand side, the final assertion (-- p = 1) guarantees that the first operand of the sequence leaves the sequence register at exactly the value 1 at which the second operand expects to start. The second initialisation can therefore be omitted, and so can the assertion; the resulting code is just the concatenation (Q R) into a single case statement of the codes Q and R contributed by the two operands. This is shown in the case statement in Fig. 4, where Q and R stand for the list of cases contributed by Q and R respectively. The proof of this particular normal form theorem is quite elegant, and revealed some surprising mathematical insights.

A law for the elimination of a conditional is given in Fig. 4. It assumes availability in the machine code of a conditional jump

$$p := (p+1 \lhd A \rhd 1+1)$$

and an unconditional jump

$$p := m.$$

The first of these is planted at location k in the code memory, and the second at location 1.

All the laws that we have illustrated on a particular program can be generalised to apply to all programs expressed in the language. Each law can be proved algebraically from simpler laws of programming, and each proof can be checked individually by a

```
if A  then  [k + 1,  QQQ,  1]
       else  [l + 1,  RRR,  m]
   ⇐ [k,   (k:  p := (p if A else l) +1;
                QQQ;
            l:  p := m;
                RRR),
        m]
```

Figure 4.

computer algebraic system like OBJ3. Even better, each theorem is itself an algebraic transformation that can be directly executed by OBJ3. The result is an automatic general-purpose compiler that has been constructed as a byproduct of its own proof of correctness, thereby achieving the goal set by Dijkstra [4] many years ago. A prototype implementation of this philosophy has been explored by Augusto Sampaio in his recently completed Doctoral project at Oxford [5]. The technology is being further developed for a significantly larger subset of occam by a team of researchers in Kiel, under the leadership of Hans Langmaack.

4 Hardware from Software

The same philosophy and very similar techniques can close the gap which remains between the machine code produced by the compiler and the actual hardware circuits of the machine which executes the code. The intended behaviour of the machine has already been conveniently specified by an interpreter, written in the same high level language in which the source code was expressed. An obvious solution is to translate this interpreter into circuit notations that can be implemented directly by gates and flip-flops printed onto the surface of a silicon chip. Such a translator for the programming language occam [6] has been constructed by one of the authors (IP). For reasons of efficiency, it slightly generalises the parallel constructs of occam, and uses a global and discrete model of time [7]. It has been used successfully [8] for automatic synthesis of microprocessors. Some of these are application-specific, with an instruction set tailored to the job in hand. The method of compilation by correctness preserving transformations ensures that the microprocessor meets its specification; and may provide sufficient confidence for use in safety critical systems.

In many ways the translation to hardware is much more direct than translation to machine code. The global variables of the program denote registers that are implemented directly as groups of flip-flops, whose number and size are determined by the needs of each individual program. Thus there is no need for a symbol table of numbered locations in an external ram. Similarly, there is no fixed, centralised, function unit or instruction code: hardware is allocated to implement exactly the assignments and expressions of the source program. Neither is there any stored program nor interpreter. All these phases of conventional compilation can be omitted as can most of the run-time structures produced by a conventional compiler, leaving as the main task just the reduction to normal form, and the new and difficult task of component layout and routing of wires, which is not the subject of this paper.

The normal form for hardware is structurally and conceptually the same as for software. But since there is no code storage, the conditional testing of the value of the p-register has to be distributed through to the expressions on the right hand sides of the assignments, which must then be merged. The relevant transformations are demonstrated by the example:

```
IF
    p = 20
        A, p := e, f
    ELSE
        A, p := g, h

= A, p := (e, f) ◁ (p=20) ▷ (g, h).
```

After applying this transformation exhaustively, the body of the loop becomes a single parallel assignment with a rather complicated expression on the right hand side. But in principle it can be implemented in the manner illustrated in section 2.

In practice, a final optimisation must be made. The test (p=20) and the addition (p:=p+1) are quite expensive to implement directly in hardware. An attractive solution is to use only powers of two as values of p. Thus each test only has to test a single bit of the register, and incrementation is replaced by left shift, using a Delay element exactly as illustrated in section 2. This simple change of representation removes a major disadvantage of hardware over software implementation of normal sequential programs.

But the main advantage of hardware is that the components operate naturally in parallel with each other. This is achieved automatically, and without overhead; indeed, it is sequential execution in hardware that requires careful organisation of control signals and other overhead. Fortunately, the parallel constructions of occam are designed at a sufficiently high level of abstraction to be implemented efficiently in both hardware and software. The advantage of this abstraction is not just conceptual: the real practical pay-off is that the whole of an embedded system can be designed and implemented in the same programming language. Later, certain parts of it can be designated for translation into special purpose hardware, and other parts into machine code for execution on an attached general-purpose computer. In this way the gap between hardware and software has been safely closed, and there is no longer any possibility of errors congregating in the traditionally error-prone interfaces between them.

As a simple example we consider part of an interface from a robot position-measuring system [9]. The code shown here is extracted from an executable program which simulates the behaviour of a hardware interface to link a set of shaft encoder devices to a controlling computer. This fragment reads a two-bit code from a shaft encoder and alters a variable depending on whether the shaft has rotated by one unit.

```
SEQ
  encoder ?  current
  IF current  ≠ previous
    THEN
      IF current BIT 0  ≠ previous BIT 1
        THEN
          angle := angle - 1
        ELSE
          angle := angle + 1
  previous := current
```

This program could be compiled and executed on a conventional microprocessor, but it would be perhaps a hundred times slower than was required for our application. It is thus necessary to have an implementation in special-purpose hardware whose behaviour has been captured by our program. We therefore proceed by transforming the program into normal form. This transformation is done automatically by a hardware compiler, based on transformations similar to those of section 3.

```
SEQ
  start,c0,c1,c2,c3,c4,finished := 1,0,0,0,0,0,0
  WHILE ¬ finished
  PAR
    previous := current ◁ c2 ∨ c3 ∨ c4 ▷ previous
    current  := encoder ◁ (start ∨ c0) ∧ encoder_rdy ▷ current
    angle    := (angle+1 ◁ incr ▷ angle-1) ◁ decr ∨ incr ▷ angle
    c0       := (start ∨ c0) ∧ ¬ encoder_rdy
    c1       := (start ∨ c0) ∧ encoder_rdy
    c2       := decr
    c3       := incr
    c4       := c1 ∧ ¬ changed
    finished := c2 ∨ c3 ∨ c4
    start    := 0
  WHERE
    bitdiff  = current BIT 1 ≠ previous BIT 0
    changed = current ≠ previous
    incr     = (c1 ∧ changed) ∧ bitdiff
    decr     = (c1 ∧ changed) ∧ ¬ bitdiff
```

Our normal form program consists of a single parallel assignment (written here with PAR) embedded in a WHILE loop. This program captures the parallelism available in the original program, and was generated automatically from the original. It uses the control state variables (c0..c4, start, and finished) to implement the 'distributed program counter' optimisation mentioned previously; this is more commonly known by hardware engineers as 'one-hot' control encoding. To aid readability, no optimisations have been applied to this normal form program although a number are possible.

A normal form program such as this one can now be interpreted directly as a hardware circuit. Everything on the left hand side of the parallel assignment can be interpreted as hardware storage devices (flip-flops), and everything on the right hand side can be interpreted as a set of logic gates. In all but syntax, this program is the circuit diagram of hardware we can implement. The interpretation of this program as a circuit is sketched in Fig. 5 for comparison.

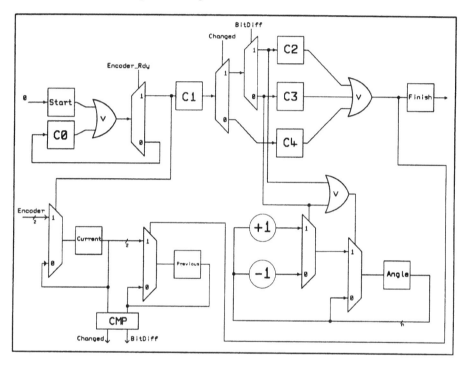

Figure 5.

The normal form program, when interpreted as hardware, proceeds with the parallel assignment being executed once each clock cycle. The number of clock cycles that the program takes to terminate represents the sequentiality of the program; and the complexity of the assignment statement represents its parallelism. Using the algebraic laws, we can balance the sequential and parallel aspects in different ways to achieve different cost/performance ratios for the hardware. This is typically accomplished by transformation of the original source program before conversion to normal form.

The validity of our hardware interpretation of a normal form program can be demonstrated by modelling each hardware element as a fragment of occam program and then showing that the parallel composition of these fragments is a refinement of the normal form program.

5   Hardware/Software Co-Design

Since we have encompassed software and hardware in the same theoretical framework, it is possible to develop implementations of programs that are realised partly in hardware and partly in software. We can also use this framework to trade off the hardware and software components against each other to achieve desirable cost/performance measures. We have developed a number of small-scale systems using our hardware compiler to produce the hardware components, and the Inmos occam2 compiler to produce the software components. We hope at some future date we will be able to compile complete hardware/software applications from a single source text, when we have solved the remaining problems of bridging the gap between these two closely-related languages.

Making it easier for programmers to produce circuits corresponding to their programs would be of little use unless there were some way of building these hardware circuits easily. Fortunately, there is a newly-available technology which does exactly this. The Field Programmable Gate Array, or FPGA, is a relatively new form of integrated circuit which can be programmed to act as almost any digital circuit, subject to the physical limitations of its size. The type of FPGA we use is a standard commercial chip consisting, essentially, of an array of gates and flip-flops that can be connected together via programmable switches, where the programmable elements are all implemented using static RAM. Changing the hardware configuration is accomplished by writing an appropriate set of bits to the RAM and can be repeated as often as required.

The major difference between this and previous hardware implementation technologies, is that the hardware can be reprogrammed rapidly by a purely software process. Today's FPGA chips can each implement a circuit with an equivalent complexity of some 20,000 gates in a matter of milliseconds. Even one of these chips is capable of hosting a small application program kernel and the size and speed of FPGAs is increasingly rapidly as a result of progress in VLSI technology. Their use allows design decisions to be deferred, or implementations to be upgraded while in service. These advantages and others have already made FPGA sales the fastest-growing segment of the digital integrated circuit market. It was the emergence of this technology which directly stimulated our interest in hardware compilation as a research topic.

However, many current uses of FPGAs ignore one of their most interesting characteristics. In a typical application, one or more FPGA chips will be configured on system power-up and will remain near-permanently in that state. The ability to reconfigure FPGAs during system operation holds out the possibility that tomorrow's computing systems can use reprogrammable hardware as a flexible resource which can be deployed to support whatever computation is running at the time. Thus, reconfigurable hardware may bring benefits similar to those offered by 'reconfigurable software' in its usual guise of the stored-program computer.

In order to explore hardware/software co-designs, we have designed and built the HARP board. This is a small, 17 x 9 cm, printed circuit board which conforms to the Inmos TRAM standard (Fig. 6). It is a daughter-board which can very simply be integrated with commercially available host boards for many computers, and with a wide variety of other parallel computing and input/output modules. These modules can readily be joined together to make larger systems.

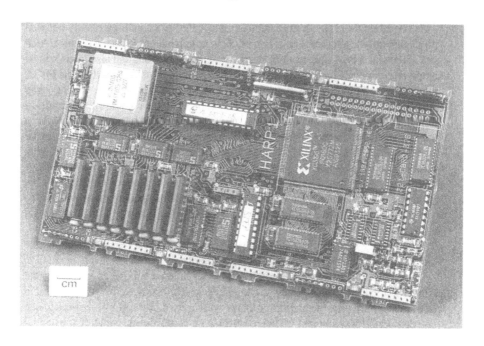

Figure 6.

Our HARP module consists of two computing systems intimately-connected with each other; one is based on a transputer, and the other on a large Xilinx FPGA [10]. The transputer could be replaced by any other fast, 32-bit microprocessor, but it is convenient for us because it supports parallelism in general, and the occam language in particular. Both systems have their own local memory and can run completely independently if required. However, they also share the same 32-bit bus, so that data exchange between them is fast and efficient. Some algorithms may be implemented with deep combinational hardware, such as exhibited by a parallel multiplier; others may be heavily pipelined with only tiny amounts of logic between the registers, such as in a systolic signal-processor. Consequently, it is necessary to operate the FPGA at different speeds, depending on the architectural style of our algorithm implementation. Thus the HARP board also has a 100MHz frequency synthesiser, controlled by the transputer, which can generate an arbitrary clock signal. The FPGA can be reconfigured by the transputer so that at one instant it might be supporting graphics or image-processing algorithms, and at the next it might be supporting data compression, spell-checking, or pattern matching.

As one example of the system implications of FPGA technology closely coupled with a conventional microprocessor, we briefly mention memory sharing on our HARP board. Since the FPGA chip has full access to the transputer bus and also its own local memory, it is simple to load the FPGA with a configuration which makes the local FPGA memory appear in the address space of the transputer. By loading this 'memory-map configuration', a transputer-based application can process its data, leaving the results in FPGA memory. A computing configuration can then be loaded into the FPGA, leaving it to process the data further. When the FPGA program has terminated, it can inform the transputer which then reloads the memory-map configuration to gain access to the data. We had originally planned that communications between the transputer and the FPGA would always be mediated by occam-style channels implemented by transfers over the bus. It was a pleasant discovery that this additional mode of operation was available to us. We take it as indicative of the greatly increased flexibility that reconfigurable logic imparts to systems which use it.

In our HARP system, the FPGA subsystem is perhaps best regarded as a flexible co-processor. Once the transputer has loaded a computing configuration into the FPGA, they may be equal partners in the computation, but unlike a normal co-processor, the hardware in the FPGA can be changed completely at any instant. If this style of flexible hardware/software implementation becomes widely used, we can also expect to see a corresponding development of microprocessor architecture in which a substantial amount of reconfigurable hardware is embedded in the microprocessor core. Indeed we regard the whole HARP board itself as an experimental prototype of a future generation of microprocessors. We are currently investigating the implications of FPGA and hardware compilation technologies on computer and system architectures in a joint project with Inmos.

In coupling FPGA technology with hardware compilation, we are trying to provide environments where programmers, as well as hardware engineers, can implement their programs to take advantage of the greater speed offered by hardware implementations. In a recent example, a programmer in our group having little understanding of hardware, has produced a small, elegant, spelling-checker in hardware that runs more than 25 times faster than a similar implementation of the same algorithm on a transputer. We could not pretend that the development of this application was nearly as easy or straightforward as we would like, but we have demonstrated feasibility, and it gives us hope that our main goals ultimately will be achieved.

One use of FPGA chips which we are beginning to investigate is the use of on-the-fly reconfiguration, for which we might use the term 'virtual hardware' by analogy with virtual memory. In this scenario, an application runs as far as it can with whatever hardware is loaded. When it can run no further, the environment swaps in further hardware on demand to enable processing to continue. We also expect our theoretical models and proof techniques to extend to such systems.

A particular area which is attracting our interest is that of multi-media applications. These are characterised by (i) a wide variety of information representations, many of them needing large amounts of data storage, with consequent emphasis on compression and efficient data transfer, (ii) computationally-intensive conversion programs which can extract high-level information from one domain of data and inject it into another, and

(iii) interfaces with a wide variety of input/output devices. The reconfigurable FPGA can lend support in each of these areas. As one example, the accuracy and detail of graphics rendering algorithms embedded in FPGAs could be traded off against throughput if the system became overloaded by a request for a high-speed animation sequence. Similarly, the same physical FPGA hardware could be supporting speech synthesis at one moment and video compression at the next. The occam CASE constructor provides a natural way of exploiting this style of reconfiguration.

We have looked at a number of applications of hardware compilation coupled with implementation via FPGA technology. Some of these have been speculative; others are a practical reality already in our laboratory and are being actively transferred into industrial practice. Our goal is to use these devices and our hardware/software compilation techniques to build flexible and dependable computing systems. We look forward to building a system in the future where users write programs or invoke applications, and literally have no idea whether the implementation generated is in software, in hardware, or in some mixture of the two; the compilation system will have ensured that the user's high-level constraints on cost and performance are met as closely as possible by the implementation (although we need further work to know how to express these constraints).

6 Conclusions

This paper is about closing a gap, not only between hardware and software but also between abstract theory and concrete practice. It has given two examples, drawn from either side of the gap — a correctness-driven approach to compiler design and an actual hardware board built to execute the target code. Clearly, this gap is still too wide to cover comfortably in one paper, and further work is in progress.

The gap between hardware and software is not the only gap that needs to be closed by further theoretical and practical investigations. We began with a summary of a number of the phases in the design of a complex engineering product, like a real-time process control system. It is our hope that the earlier stages of design, including specification and analysis of requirements, can be assisted by a systematic transformational approach that precludes the possibility of error; and that the gaps between all the phases may be securely closed in the same way as we have shown for the gap between hardware and software.

7 Acknowledgements

The work reported in this paper has been carried out by many colleagues supported by many research grants. The chief of these are the Esprit contracts 7071 (PROCOS), 7166 (CONCUR), and 7249 (OMI/HORN). We also wish to thank Mike Spivey for his helpful comments on this paper.

References

[1] C.A.R. Hoare, *Mathematical Logic and Programming Languages*, chapter 'Programs are Predicates', 141–154, Prentice-Hall, 1985.

[2] C.A.R. Hoare et al., 'Laws of programming', *Comm. ACM*, **30**(8), 672–686, (1987).

[3] A.W. Roscoe and C.A.R. Hoare, 'Laws of occam programming', P.R.G. Monograph, Oxford University Computing Laboratory, (1986).

[4] E.W. Dijkstra, 'A constructive approach to the problem of program correctness', *BIT*, **8**, 174–186, (1968).

[5] He Jifeng, C.A.R. Hoare, and A. Sampaio, 'Normal form approach to compiling specifications', *Acta Informatica*, (1994). to appear.

[6] Ian Page and Wayne Luk, 'Compiling occam into FPGAs', in *FPGAs*, 271–283, Abingdon EE&CS Books, (1991).

[7] Michael Spivey and Ian Page, 'How to Program in Handel', Technical Report, Oxford University Computing Laboratory, (1993).

[8] Ian Page, 'Parametrised Processor Generation', in *FPGAs 93*, to be published by Abingdon EE&CS Books (probably), (1993).

[9] Ian Page and Wayne Luk and Henry Lau, 'Hardware Compilation for FPGAs: Imperative and Declarative Approaches for a Robotics Interface', in *Proc. IEE Colloquium on Field-Programmable Gate Arrays – Technology and Applications, Ref. 1993/037*, pp. 9.1–9.4. IEE, (1993).

[10] Xilinx, San Jose, CA 95124, *The Programmable Gate Array Data Book (1993)*.

On Computing Power

Jean E. Vuillemin

Digital Equipment Corporation, Paris Research Laboratory
85 Av. Victor Hugo, 92563 Rueil Malmaison, Cedex France.

December 21, 1993

Abstract

We analyze in details some implementations of a challenging, yet simple application: CERN's *calorimeter*. We try both *general* purpose computer architectures (single and multi processors, Simd and Mimd), and *special* purpose electronics (full-custom, gate-array, FPGA) on the problem.

All measures are expressed in a single common *unit* for computing power: the Gbops[1]. It applies to all forms of digital processors, and across technologies. What's more, Noyce's thesis provides a reliable way to extrapolate Gbops benchmarks through *future time*, say up to year 2001.

The quantitative result of our analysis shows that *special* purpose processing is *an order of magnitude* more efficient than *general* purpose processing, on our specific problem. We show how to map the calorimeter on a *programmable active memory* PAM[2], at performance and cost comparable to those of fully dedicated implementations: orders of magnitude better than any general purpose implementation, in 1992. We argue that this current computational power advantage for PAM technology will *increase* with time.

Finally, we discuss how to program such novel *virtual* PAM computers in the **2Z** language, for very large synchronous designs.

---

[1] 10^9 binary operations per second
[2] large array of configurable logic

1 Noyce's Thesis

Since the advent of modern digital computers, we have lived through 40 years of an exponential growth which is unique in the technical history of mankind.

Thesis 1 (Noyce) *The computing power per unit doubles every year.*

This was first pointed out, in an equivalent form, by R. Noyce in the early sixties. Circuit technology integrates many contributions: they arise from almost every area of modern science and technology, spanning the range from solid state physics to digital computer architecture and programming languages.

Yet, as Noyce observed, the complex combined cumulative effect of all these punctual and discrete advances is *as if*, the feature size of our circuit manufacturing technology was simply shrinking *linearly* with time. As far as experimental evidence goes, and it is plentiful, the average shrink factor per year is about $\alpha \approx 1.25$. As documented further by C. Thacker, the shrink factor has remained statistically steady for over 30 years, and we have every reason to believe that it will keep doing so in the near future, say up to year 2001 (see [14]).

Let G (a natural number) be the number of logic *gates* which one can effectively fit within a unit area by the end of year y, and F (in Hertz) be the *frequency* at which one can reliably operate such gates. The *computing power* is the product of these two figures:

$$P = G \times F. \tag{1}$$

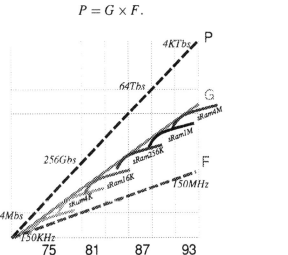

Fig. 1. Plots of the growth rates of G (number of storage bits), F (operating frequency) and P = GF for *static RAM* technology.

By the end of the next technology year $y' = y + 1$, the corresponding figures in (1) are $G' = \alpha^2 G$ and $F' = \alpha F$. Choosing $\alpha = \sqrt[3]{2}$, a good approximation, we find that $P' = 2P$, which is how we stated Noyce's thesis.

Keep in mind that it is just an observation about human science; it is neither a law from physics, nor a theorem from mathematics. It has to do with economics and technology questions, such as: Is GaAs faster than ECL? Will BiCMOS take over CMOS? At what cost?

2 Summary

Noyce's thesis provides a nice and simple model against which we attempt to analyze the impact of time, i.e. technology, on computer architecture. For this purpose, we start from a single application: CERN's *calorimeter*.

- The problem is simple enough to be fully stated in section 3. Its computing requirement is analyzed, step by step in section 5.

- It is part of a series of benchmarks put forward by CERN[3] [3]. The goal is to measure the performance of various computer architectures, in order to build the electronics required for the *Large Hadron Collider* LHC, before the turn of thé millennium.

- It is challenging, and well documented: [3] benchmarks a dozen electronic boxes, including some of the fastest current computers, on the calorimeter and other problems.

Our object is to complement CERN's *experimental* calorimeter benchmarks by a convergent, more *theoretical* analysis; and to use it in order to make some predictions regarding the future.

We try two types of solutions.

The first are representative of the computing power achieved by *general purpose* computer architectures on the calorimeter: this year's fastest computer on a chip; compared to both *massively* and *moderately* parallel implementations. We analyze the cycle time required for such machines, and predict a year when each should become technologically feasible.

The last is representative of the computing power delivered by *special purpose* digital architectures, specifically designed to perform the calorimeter's computation. An implementation in PAM technology is presented in section 7.2. It is

---

[3]Centre pour l'Etude des Réactions Nucléaires, in Geneva, Switzerland

representative of three related *design methodologies*: full-custom, gate-array and *field programmable gate array* FPGA.

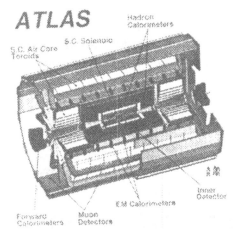

Fig. 2. CERN's view of the LHC.

We apply the same evaluation method to all cases:

1. First assess the *theoretical* computing power of the machine; all are expressed in a common unit, the Gbops.

2. Next, we analyze the *actual* computing power, as measured by running the machine on the calorimeter.

3 CERN's Calorimeter

The function of the calorimeter is to identify the position and most likely nature of a particle that traverses a digitized region of interest, which is a square $S = \{i,j : 0 \leq i,j < 20\}$. Within the LHC, such traversals occur at the rate of 100KHz, i.e. each $10\mu sec$.

The input is a pair of energy maps $(E', E''[i,j]$ for $i,j \in S)$ providing the line-by-line responses from two analog detectors, digitized down to 16 bits. The average input rate is 160MB/s, presented on two channels (32b, 20MHz each). The analysis of event (E', E'') is done by computing:

E The pixel-wise sum: $\qquad\qquad\qquad\qquad E \;=\; E' + E''.$

S The total energy: $\qquad\qquad\qquad\qquad S \;=\; \sum_S E[i,j].$

M The maximum energy: $\qquad\qquad\qquad\qquad M = E[i_m, j_m] = \max_{\mathcal{S}} E[i,j].$

O The first statistical moment: $\qquad\qquad\qquad O = \sum_{\mathcal{S}} r_m[i,j] \times E[i,j],$

centered at E's maximum. Here, $r_m[i,j]$ is a tabulated 8b integer approximation of the distance $\sqrt{(i - i_m)^2 + (j - j_m)^2}$.

P The peak energy: $\qquad\qquad\qquad\qquad P - \sum_{\mathcal{S}} p_m[i,j] \times E[i,j],$

here $p_m[i,j]$ is one if $|i - i_m| + |j - j_m| < 2$, zero otherwise.

D The final discriminant: $\qquad\qquad\qquad D = \operatorname{sign}(\alpha \frac{O}{S-M} - \beta \frac{P}{S}).$

The decision between an *electron* and a *hadronic jet* is based on computing the sign of D, for some (experimentally determined) suitable 16b integers α and β.

The whole computation may be carried out with 16-bit integers. The output rate is only 100 kb/s, one bit per energy pair (E', E''). The computation of the *maximum* implies that we must buffer a full energy map, between steps $\mathbf{E}, \mathbf{M}$ and steps $\mathbf{O}, \mathbf{P}, \mathbf{D}$.

4 The Gbops

Our only analytical tool so far is definition (1) of the computing power, which is a strictly *digital* measure. The exact *analog* process through which our *mathematical* computing power gets physically delivered does not matter here. What exactly is a *gate* is not important either: it only affects our measure by a constant multiplicative factor (provided that we keep *bounded fan-in*).

Our favorite accounting unit calls *one* any operation which is not more complex than a single bit-serial binary addition. Or subtraction, for that matter; or any gate with at most *three* inputs, and *one* bit of internal state.

Let **1 Gbops** be 10^9 *binary operations per second*, our unit for computing power. It is delivered by any *Bop* circuit, *operating* at 1 GHz.

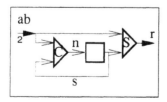

Each Bop circuit is made of two boolean functions $S, C \in \mathbf{B}^3 \mapsto \mathbf{B}$ and a synchronous register; they are connected as shown in the schema above, or the **2Z** code which follows.

```
Bop (a, b) = (s, r)
where
    n = C(a, b, s); // Next state
    s = reg(n);      // Flip flop
    r = S(a, b, s); // Result
end where;
```

Bop circuits include the *bit-serial binary adder*, which is obtained by choosing:

$$S(a,b,s) = (a+b+s) \cdot | \cdot 2,$$
$$C(a,b,s) = (a+b+s) \div 2.$$

The accounting rules which follow, for arithmetic and logic operations over n bit word inputs, are straightforward:

+ One $n+n \mapsto n+1$ bits addition each nano second is worth n Gbops. Subtraction, integer comparison and logical operations are bit-wise equivalent to addition.

× One $n \times m \mapsto n + m$ bits multiplication each nano second is worth nm Gbops. Division, integer shifts and transitive (see [15]) bit permutations are bit-wise equivalent to multiplication; consequently, so is a $n \mapsto m$ look-up table LUT, or RAM access.

5 Calorimeter Analysis

We now count the number of Gbops required by each step of the calorimeter.

E The input is composed of four digital flows: $4 \times 16b \times 20MHz$. We must add together the first and last two flows:
$$E_t[0] = E'_t[0] + E''_t[0];$$
$$E_t[1] = E'_t[1] + E''_t[1].$$
Each addition requires 16 binary operations per cycle: $P_e = 16 \times 2 \times 20M = 0.64$ Gbops.

S The input is : $2 \times 16b \times 20MHz$. We sum all energies from the same map: $P_s = 0.64$ Gbops.

M The maximum can be computed by using the sign of the subtraction to select the proper argument, at a cost of 640 Mbops; together with keeping up to date the 10b index (i_m, j_m): 0.4 Gbops. In addition, the maximum requires to *store a complete map*, in a $400 \times 16b$ double access RAM. So we charge $2 \times 10 \times 16 \times 40M = 12.8$ Gbops. Total: $P_m = 13.84$ Gbops.

O The first statistical moment is the most complex operation. It requires a 20b↦16b look-up table LUT for finding the distance $r_m[i,j]$: 12.8 Gbops. The 8b multiplication accounts for 5.12 Gbops. Total: $P_o = 17.92$ Gbops.

P The peak is the cheapest operation: at 5 additions 16b per *map*, it requires a negligible: $P_p = 8$ Mbops.

D The discriminant is expensive, $2 \times 16 \times 48$ *active bits*; it is executed once per map, so the computing power required here is only: $P_d = 154$ Mbops.

The computing power required by the complete calorimeter computation is the sum $\mathcal{P} = P_e + P_s + P_m + P_o + P_p + P_d$, namely $\boxed{33 \text{ Gbops}}$.

Note that our accounting does not take into consideration any of the required data movements: from input to processing unit; from processing unit to output. Such transport operations do not transform values; they do not directly contribute to the final decision D; they get charged here as overhead, exclusively accounted for in the *virtual* computing power of the implementation technology.

6 General Purpose Architectures

To make the analysis simple, we give *general purpose* technology *all benefits from the doubt*: caches are all assumed to be wide enough and fast enough, in order to provide each Cpu with data and valid instructions, at no latency but the minimum feasible.

Ignore the fact that both data and instructions caches would have to be *huge*, by 1992's standards. This permits to perfectly *streamline* the calorimeter's computation: unroll all loops, and take one cycle per fetch or store, on every memory access.

Ignore the fact that, in 1992, none of the general purpose machines by CERN could cope with the calorimeter's external input bandwidth of 160MB/s. So, the input had to be *faked* in the experiments.

6.1 Single Processor

In 1992, the highest performance microprocessor has 64b of data, clocked at 200MHz. The *virtual* computing power of this Cpu64b200MHz is $64 \times 0.2 = 12.8$ Gbops. We know from the calorimeter analysis that this processor is not fast enough.

Let us analyze the number of clock cycles required for running the calorimeter on a *reduced instruction set* Risc processor. In a streamlined code, the number of *cycles* required to process the calorimeter, for each $16b \times 20 \times 20$ energy map, is:

$$
\begin{aligned}
S &= C_e + C_s + C_m + C_o \\
&= 5 + 2 + 3 + (4 + 16), \\
C &= 400S + C_p + \dot{C_a} \\
&= 12.1 \text{ K cycles.}
\end{aligned}
$$

The calorimeter operates at 100KHz; so, the minimum cycle time at which we can expect to run this program is $\boxed{1.2 \text{ GHz}}$ Noyce's thesis predicts that this Cpu64b at 1.2GHz will be technologically feasible around year 2000.

The moment **M** gets computed in 4 cycles for the look-up table LUT, and 16 cycles for the actual multiplication. On a machine with a hardware multiplier, C_o gets reduced to 8 cycles. So the clock at which we need operate the calorimeter is only 720 MHz. A Fpu64b720MHz should be feasible by year 1998.

The virtual computing power of Cpu 64b at 1.2GHz is 87 Gbops. The virtual computing power of Fpu64b at 720MHz [4] is 1656 Gbops.

Fig. 3. In this popular micro processor, the area of the actual Cpu is only 1/200-th of the whole.

In both cases, the computing power actually expanded on the calorimeter is only 33 Gbops. The respective virtual to actual power ratios are about 3 and 47. Observe that Fpu delivers at most 4.8 Gbops when it is only computing additions or equivalently cheap operations: a very low utilization of the computing power virtually available in the multiplier.

---

[4] A 64b floating point unit, with 48b mantissa and 16b exponent, which operates at 100MHz delivers 230 Gbops.

Note that large data paths (32b or 64b) past 16b do not help the calorimeter: the whole computation can be performed on 16b integers, except for the final decision where 48b are convenient. To conclude on single processors, our best fit to the calorimeter are

Cpu16b1.2GHz The ratio between the 43 Gbops (equals 19 for Cpu plus 24 for the LUT and RAM) virtual power and the 35 Gbops actual power *is near one*. We achieve an optimal fit where the 16b computed in each cycle all contribute to the calorimeter's decision.

Fpu16b720MHz The ratio between the 189 Gbops virtual power and the 35 Gbops actual power is near five. The multiplier is used at less than one fifth of its capacity.

Although Cpu16b1.2GHz makes the single processor RISC software solution *optimal* for the data path of the calorimeter, it is hiding a *large* structure (with high Gbops virtual cost) for handling its hierarchical data and instruction memories. There is a *lot more* to Cpu16b1.2GHz than just its Alu16b. If you care to look at Figure 3, the part of interest here is only a *tiny* fraction of the silicon area in the full microprocessor.

6.2   Multi Processors

The class of massively parallel processors fares poorly on CERN's calorimeter. The strong experimental evidence provided in [6] can probably be explained by observing that any attempt to process the calorimeter on a pool of *slow* processors implies a *large* cost: the amount of *memory* required is proportional to the number k of processors used; so is the bandwidth required for communicating the proper data to the proper processing units.

Massively parallel solutions to the calorimeter are ruled out by economics and engineering problems. A 4k=4096 parallel processors 4b Simd operating at 12MHz machine 4kPP4b12MHz has a virtual computing power of 200 Gbops. Yet, the implied cost in memory and communication makes it incapable, in 1992, to compute the calorimeter near real time.

Processing independently the six steps of the algorithm is the best room available for parallelism in the calorimeter. Each processor performs some of the steps (E,S,M,O,P,D).

This multiple instruction, multiple data parallel machine operates at the speed of its slowest component, namely the moment unit **M**. Using here both a LUT and a multiplier 16b, we reduce C_o to eight cycles. Each 16b processor is now only required to operate at $40 \times 8 = 320$ MHz in order to process the calorimeter. Such a parallel Mimd processor 6PP16b320MHz should be feasible before 1996.

Note that the bandwidth required between the 6 processing units is $8 \times 80 = 640$ MB/s, a taxing requirement for all general purpose architectures.

Past such a simple six long assembly line organization, there is little more to be gained through parallel processing: the increase in storage and communication is not worth the benefit in effective operations.

7   Special Purpose Architectures

From the nature of the *physical* interface of the calorimeter (input on two HIPPI channels 32b20MHz, output on the host's TURBOchannel 32b25MHz), we know that a minimal size electronics has four printed circuit boards (say 6cm×8cm in 1992): two for input, one for output, and one board for the calorimeter algorithm per se, and connecting the other boards.

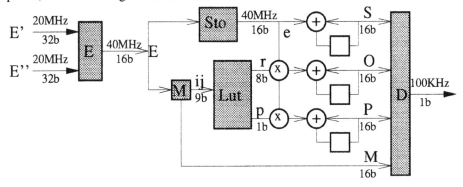

The calorimeter algorithm maps directly into eight functional units: schemas above, **2Z** code below.

```
Calorimeter({E', E''} : [32]) = D
where
  E = AddUnit(E', E''); // 100MBs peak output
  // 159 zeroes and a one, period 200 bits
  c159 = Sdd(2**159/(1-2**200));
  (ij, M) = MaxUnit(E, c159);
  (r, p) = Lut(ij, c159);
  e = Sto(E, c159);   // Delay 10mus
  reset c159 do
    S = SumUnit(e);
```

```
      P = PeakUnit(e, p);
      O = MomentUnit(e, r);
    end reset;
    D = OutputUnit(P, S, O, M, c159);
end where;
```

7.1 Hardware Blocks

We present the function of each atomic unit; when it is relevant, we provide the **2Z** code from which the PAM configuration for the corresponding unit can be derived. We also analyze the virtual computing power required by each step of the PAM implementation.

The whole design is synchronized to the 160MB/s input, by a 40MHz clock. Both the period ($10\mu s$) and the delay ($20\mu s$) are kept at their absolute minimal values. Each arithmetic unit is precisely tailored to its function: 16 bits parallel operators for steps 1 through 5, connected according to the schematics below. Step 6 is implemented in a fully bit-serial manner, to take best advantage of the low bandwidth requirement on this final electron/hadronjet decision.

E The input is composed of four parallel digital flows: $4 \times 16b \times 20MHz$. We first interleave $E'_t[0]$ and $E'_t[1]$ in time, so that $E'_{2t} = E'_t[0]$ and $E'_{2t+1} = E'_t[1]$; similarly, interleave $E''_t[0]$ and $E''_t[1]$ so as to produce E'' at 16b40MHz. In PAM technology, each interleave is realized by a specific column of 16 Pabs[5], at cost: $2 \times 16b40MHz = 1.28$ Gbops. We add together the two flows through a 16b adder at 40MHz. The required computing power is 3×16 Pabs: $P_e = 1.92$ Gbops.

M Computes the maximum M of the current map E at 16b100KHz and its index (i_m, j_m) at 10b100KHz. In PAM technology, the maximum is best implemented from *high to low* bits (see [7]). The computing power of this unit is $2 \times (16b + 10b) \times 40$ MHz $= 2.08$ Gbops.

Sto Double buffers the current $400 \times 16b$ energy map E while reading the previous one e. Both flows are 16b40MHz. In our PAM implementation, we use a $2 \times 400 \times 16b40MHz$ double access RAM: 12.8 Gbops, and 400 Mbops to control the addresses. Total: $P_{sto} = 13.2$ Gbops.

S Sum all energies from the same map, with a 16b accumulator: $P_s = 640$ Mbops.

---

[5]Programmable Active Bit

```
SumUnit (E:[16])=S:[16]
where
   R = Add(16)(E, S, 0);
   S = Reg(16)(R);
end where;
```

P The peak is the computed at full cost: $P_p = 640$ Mbops. Bit $p = p_m(i,j)$ is produced by the LUT.

```
PeakUnit (E:[16],d)=P:[16]
where
   for k<16 do // sum when d=1
      F[k] = E[k] & d
   end for;
   P = SumUnit(F);
end where;
```

O The first statistical moment is implemented by a 20b↦9b LUT for finding the 8b distance $r_m[i,j]$ and the 1b peak $p = p_m(i,j)$. The address control for this RAM uses 800 Mbops. Including the 8b multiplication, we find: $P_o = 18.72$ Gbops.

```
MomentUnit (E:[16],D:[8])  = O:[16]
where
   P = Mul(16, 8)(E, D);
   O = SumUnit(P[0..15]);
end where;
```

D Each of the M,P,S,O units takes inputs at 16b40MHz and produces 16b of output at 100KHz. The four outputs (M,P,S,O) are consumed by the decision unit, 4×16b at 100KHz, in order to produce the final decision D at 1b per 100KHz. The PAM discriminant is detailed in section 8.

Note that the virtual computing power rquired for our PAM calorimeter is only $\mathcal{P} = \boxed{39 \text{ Gbops}}$. The ratio between actual and virtual power is very near one, as for Cpu16b at 1.2GHz. The difference is that here, the whole chip area is devoted to the calorimeter computation. There is no hidden virtual cost for managing PAM data.

From this level of description, we can design and implement a *full-custom* solution in one chip. That makes up for a relatively empty calorimeter board: a single chip and lots of connectors.

An easier solution is to realize all but the LUT stage in a calorimeter *gate-array*; implement the LUT by a RAM; this is a two chips implementation of the calorimeter board.

We can implement the calorimeter on a generic PAM board (same size as all others). It is composed of two RAM banks, one FPGA, two input connectors and one output connector. It can be ready made from off the shelf components.

The only difference between our three boards is their cost per unit. All are functionally equivalent calorimeter implementations. They also have equal performance.

7.2 Programmable Active Memories

As our reader is not assumed to be familiar with this technology, let us survey some of the concepts in this new emerging field. The following is from [5]:

Definition 1 (PAM) *A PAM is a uniform array of identical cells all connected in the same repetitive fashion. Each cell, called a Pab (for Programmable Active Bit) is* configurable *enough so that the following holds true:* **any** *synchronous digital circuit can be realized, through suitable configuration of each Pab, on a large enough grid and for a slow enough clock.*

A Pab is the basic building block out of which FPGAs are built. There are many ways to construct a Pab which has the required generality. The FPGAs from [2], [9], [10] and [17] present four rather *different* implementations of the concept. The Bop circuit from section 4 provides another example of universal Pab.

It should be pointed out that the five Pab structures mentioned so far do not exactly have the same computing power: while it only takes one of either [17] programmable active bit to implement a serial adder, it takes two of [9] and four of [2] or [10] to realize the same function. Such factors must be accounted for in the detailed analysis of their virtual computing power. With our Pab=Bop choice, we simply count one binary operation per Pab.

It takes quite a bit more than our PAM definition to obtain a workable and powerful general purpose digital engine. The most important designs issues involved are thoroughly discussed by P. Bertin in [4]. Besides ours, which were built at INRIA and DEC-PRL, other successful PAM implementations have been reported, in particular at the Universities of Edinburgh [12], Zurich [11], and at Maryland's SRC [1]. Let us also mention [13] which is a large PAM, dedicated to hardware emulation.

The ratio between the theoretically available computing power, and that practically usable for the calorimeter is *much* lower for *dedicated* hardware than for general purpose solutions. PAM technology combines the best from both:

- being a universal virtual machine, the PAM can be configured to a wide class of computing units. As *software*, it is by no means limited to processing a single application.

- being configurable at the *gate and wire* level, a properly dimensioned PAM can emulate efficiently each special purpose hardware. A *fixed* size PAM, say 16×20 Pabs at 40MHz, has some well defined virtual computing power: 12.8 Gbops. With some design effort, it was found on a large number of test cases that such PAM can simulate *in real time*, any specific dedicated synchronous *hardware* whose computing power is less than 12.8 Gbops.

- the benefits derived from processing the calorimeter through special purpose hardware are large; they are representative of a wide class of applications, for which PAM technology provides today an optimal implementation medium.

We demonstrate in [6], through 10 benchmarks which cover a wide range of applications, drawn from arithmetics, algebra, geometry, physics, biology, audio, video and data compression that, our implementation vehicle DECPeRLe$_1$ consistently performs in the 100Gbops range.[6]

8 PAM Programming

By its nature, the PAM has lead us to implement hardware algorithms which are substantially bigger than anything yet attempted on single silicon chips: on the order of 250K gates (or 2M transistors), excluding RAMs. The sheer size of such designs has forced us to aggressively pursue the strictly synchronous design paradigm, throughout the PAM implementations reported in [6].

It has quickly become clear that arithmetic circuits are the key to success in this area. Obviously, each of our implementations only requires a finite arithmetic precision. However, any design system which claims to cover the whole spectrum (from 1b to 4Kb!) requires the ability to handle truly arbitrary precision arithmetic.

The natural mathematical domain into which this leads us is that of the 2adic numbers, both discovered and created by K. Hensel around 1900. In [16], we uncover some of the intimate relationships which exist between digital synchronous

---

[6]Note that the present definition of the Gbops is only one **half** of that used in [5]; the aim is to simplify out useless constant factors, as one serial bit of addition now amounts to one bop, no longer two as in [5].

circuits and 2adic numbers. Capitalizing on these results, [8] attempts to introduce a new programming, named **2Z** , whose main function is to help *concisely* define synchronous circuits.

The most classical features of **2Z** have already been illustrated through examples, since the beginning of this paper. Let us complement them by the source **2Z** code for the decision unit **D**. This fully exploits the facilities for bit serial arithmetic synthesis, which are quite unique to the **2Z** language.

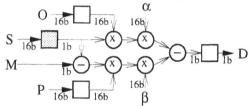

The **2Z** code corresponding to the above schema is:

```
OutputUnit ({P, S, O} : [16], M, c159) = D
where
   s = ParSer(16)(S, c159);
   enable c159 do
      O' = Reg(16)(O);
      P' = Reg(16)(P);
   end enable;
   // serial logic synthesis
   u = sMul(16)(s, O');
   l = sMul(16)(s - m, P');
   d' = (u * a) - (l * b);
   // Cern's constants
   a = 134535;
   b = 767665;
   // controls
   enable Fin do
      D = reg d'
   end enable;
   reset c159 do
      Fin = Sdd(2**48);
   end reset;
end where;
```

See [8] and [16] for more details.

9 Conclusion

Under Noyce's thesis, we have established the following.

I The computing power of a single fast Cpu 16b or 64b will grow by a factor *two* each three years. It should reach the 1.2GHz calorimeter frequency before year 2001. By then, the computing power actually delivered on the calorimeter will be 33 Gbops.

II The computing power available in a FPGA will grow by a factor *eight* each three years. Let us pick a for starting point 400 Pabs at 40MHz in 1992. By Noyce's thesis, the corresponding figures by year 2001 should be 25.6K Pabs and 320MHz: 8 Tbops per cm^2!

So, by year 2001, a single chip FPGA will have 200 times the computing power of the fastest sequential Cpu. A small PAM will be three orders of magnitude more powerfull than a single Cpu64b at 1.2GHz. An equivalent way to look at this: in 92, a 16×20 FPGA at 40MHz has the computing power of a vintage 2001 Cpu64b, at 1.2GHz. Two things are clear.

1. PAM technology will inevitably become an important contributor to the high power scientific computations, before the turn of the millennium.

2. General purpose computers will have to become multi-processors, with a relatively large number of processors, in order to sustain the competition.

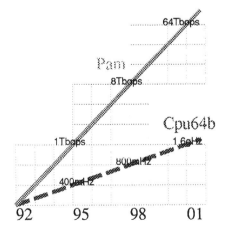

Fig. 4. The future of two technologies?

From our experience, it is clear that the main obstacles to the development of PAM technology come from the current state of computer aided design Cad system: it is much harder to program a PAM that a serial Cpu.

The Cad tools all run on sequential Cpus. The computing power available to run the Cad tools does not scale along with PAM technology.

The 2Z language is a small step towards meeting such Cad challenges. Orders of magnitude must be gained over the current design techniques, in order to implement the truly huge PAM designs for year 2001. We now deal with 10K Pabs; by then, we shall have 1M gates to design, place and route.

Based on our observations, it is tempting to venture the question:

What computing power will be available in a shoe size box by year 2001?

According to the theories exposed here, and taking 256 Gbops as a reference point for 1993, we predict:

> 68 Tbops!

By then, Niklaus Wirth will also be 68, if God permits.

References

[1] J. Arnold, D. Buell and E. Davis, *Splash II*, in *4th ACM Symposium on Parallel Algorithms and Architectures*, San Diego, California, USA (1992).

[2] Algotronix Ltd., *The Configurable Logic Data Book*, Edinburgh, UK (1990).

[3] J. Badier, R. Bock et al., *Evaluating Parallel Architectures for two Real-Time Applications with 100kHz repetition rate*, EAST note, CERN.

[4] P. Bertin, *Mémoires actives programmables: conception, réalisation et programmation*, Thèse, Université Paris 7, 1993.

[5] P. Bertin, D. Roncin, and J. Vuillemin, *Introduction to Programmable Active Memories*, in *Systolic Array Processors*, J. McCanny, J. McWhirter, E. Swartzlander Jr. editors, pp 301–309, Prentice-Hall (1989). Also as PRL report 3, Digital Equipment Corporation, Paris Research Laboratory, 85, av. Victor-Hugo, 92563 Rueil-Malmaison Cedex, France (1989).

[6] P. Bertin, D. Roncin, and J. Vuillemin, *Programmable Active Memories: a Performance Assessment*, in *Symposium on Integrated Systems*, Seattle, WA, USA, March 1993, MIT Press (1993). Also as PRL report 24, Digital Equipment Corporation, Paris Research Laboratory, 85, av. Victor-Hugo, 92563 Rueil-Malmaison Cedex, France (1993).

:

[7] P. Boucard, J. Vuillemin, and M. Shand, *Calorimeter Collision Detector on DECPeRLe₁*, PRL report 40, Digital Equipment Corporation, Paris Research Laboratory, 85, av. Victor-Hugo, 92563 Rueil-Malmaison Cedex, France (1994).

[8] F. Bourdoncle, J. Vuillemin, and G. Berry, *The 2Z Report*, PRL report 36, Digital Equipment Corporation, Paris Research Laboratory, 85, av. Victor-Hugo, 92563 Rueil-Malmaison Cedex, France (1994).

[9] W. S. Carter, K. Duong, R. H. Freeman, H. C. Hsieh, J. Y. Ja, J. E. Mahoney, L. T. Ngo, and S. L. Sze, *A User Programmable Reconfigurable Logic Array*, in *Proc. IEEE 1986 Custom Integrated Circuits Conference*, 233–235 (1986).

[10] Concurrent Logic, Inc., *Cli6000 Series Field-Programmable Gate Arrays*, Concurrent Logic Inc., 1270 Oakmead Parkway, Sunnyvale, CA94086, USA (1992).

[11] B. Heeb and C. Pfister, *Chameleon, a Workstation of a Different Colour*, in *2nd International Workshop on Field-Programmable Logic and Applications*, paper 5.6, Vienna, Austria (1992).

[12] T. A. Kean and J. P. Gray, *Configurable Hardware: two Case Studies of Micro-Grain Computation*, in *Systolic Array Processors*, J. McCanny, J. McWhirter, E. Swartzlander Jr. editors, pp. 310–319, Prentice-Hall (1989).

[13] Quickturn Systems, Inc., *RPM Emulation System Data Sheet*, Quickturn Systems, Inc., 325 East Middlefield Road, Mountain View, CA 94043, USA (1991).

[14] C. P. Thacker, *Computing in 2001*, Digital Equipment Corporation, Systems Research Center, 130 Lytton, Palo Alto, CA94301, U.S.A.

[15] J. Vuillemin, *A Combinatorial Limit to the Computing Power of VLSI Circuits*, IEEE trans. on Computers, Avril 1983.

[16] J. Vuillemin, *On Circuits and Numbers*, PRL report 25, Digital Equipment Corporation, Paris Research Laboratory, 85, av. Victor-Hugo, 92563 Rueil-Malmaison Cedex, France (1993).

[17] Xilinx, Inc., *The Programmable Gate Array Data Book*, Xilinx, 2100 Logic Drive, San Jose, CA 95124, USA (1991).

Increasing Memory Bandwidth for Vector Computations

Sally A. McKee, Steven A. Moyer[1], Wm. A. Wulf

Department of Computer Science, Thornton Hall
University of Virginia, Charlottesville, VA 22903

Charles Hitchcock[2]

Thayer School of Engineering
Dartmouth College, Hanover NH 03755

Abstract. Memory bandwidth is rapidly becoming the performance bottleneck in the application of high performance microprocessors to vector-like algorithms, including the "Grand Challenge" scientific problems. Caching is not the sole solution for these applications due to the poor temporal and spatial locality of their data accesses. Moreover, the nature of memories themselves has changed. Achieving greater bandwidth requires exploiting the characteristics of memory components "on the other side of the cache" — they should not be treated as uniform access-time RAM. This paper describes the use of hardware-assisted *access ordering*, a technique that combines compile-time detection of memory access patterns with a memory subsystem that decouples the order of requests generated by the processor from that issued to the memory system. This decoupling permits the requests to be issued in an order that optimizes use of the memory system. Our simulations show significant speedup on important scientific kernels.

1 Increasing Vector Memory Bandwidth

As processor speeds increase, memory bandwidth is becoming the limiting performance factor for many applications, particularly scientific computations. Although the addition of cache memory is often a sufficient solution to the memory latency and bandwidth problems in general-purpose scalar computing, the vectors used in scientific computations are normally too large to cache, and many are not reused soon enough to derive much benefit from caching. For computations in which vectors are reused, iteration space tiling [5, 21, 41] can partition the problem into cache-size blocks, but the technique is difficult to automate. Caching non-unit stride vectors may actually reduce a computation's effective memory bandwidth by fetching extraneous data. Thus, as noted by Lam *et al* [21], "while data caches have been demonstrated to be effective for general-purpose applications …, their effectiveness for numerical code has not been established".

---

1. Current address: Department of Mathematics and Computer Science, Emory University, Atlanta, GA 30322.
2. Current address: Fostex R&D, 2 Buck Rd., Suite 2, Hanover, NH 03755.

The traditional scalar processor concern has been to minimize memory latency in order to maximize processor performance. For scientific applications, however, the processor is not the bottleneck, and as processor speeds continue to increase relative to memory speeds, optimal system performance will leave the processor idle at times. Bridging this performance gap requires changing the way we think about the problem — to maximize bandwidth for scientific applications, we need to minimize *average* latency over a coherent set of accesses.

While many scientific computations are limited by memory bandwidth, they are by no means the only such computations. Any computation involving linear traversals of vector-like data, where each element is typically visited only once during lengthy portions of the computation, can suffer: examples include string processing, image processing and other DSP applications, some database queries, some graphics applications, and DNA sequence matching.

After defining *access ordering*, our technique for improving vector memory bandwidth, we describe a hardware Stream Memory Controller (SMC) used to perform access ordering dynamically at run time, and discuss how this technique relates to other methods for improving memory system performance. We then describe the simulation environment used to evaluate SMC systems, and present results demonstrating the effectiveness of our technique. For long vectors, an SMC achieves nearly the full bandwidth that the memory system can deliver.

2 RAM *Isn't*

The assumptions made by most memory architectures simply don't match the physical characteristics of the devices used to build them. Memory components are usually presumed to require about the same amount of time to access any random location; it was this notion of uniform access time that originally gave rise to the term RAM, for Random Access Memory. Many computer architecture textbooks ([2, 14, 15, and 26] among them) specifically cultivate this view. Others skirt the issue entirely [25, 38].

Somewhat ironically, this assumption no longer applies to modern memory devices: most components manufactured in the last ten to fifteen years provide special capabilities that make it possible to perform some access sequences faster than others. For instance, nearly all current DRAMs implement a form of page-mode operation [32]. These devices behave as if implemented with a single on-chip cache line, or *page* (this should not be confused with a virtual memory page). A memory access falling outside the address range of the current DRAM page forces a new page to be accessed. The overhead time required to set up the new page makes servicing such an access significantly slower than one that hits the current page.

Other common devices offer similar features (nibble-mode, static column mode, or a small amount of SRAM cache on chip) or exhibit novel organizations (such as Rambus [33], Ramlink, and the new synchronous DRAM designs [16]). The order of requests strongly affects the performance of all these components. For instance, Rambus devices provide high bandwidth for large transfers, but offer little performance benefit for single-word accesses.

For multiple-module memory systems, the order of requests is important on yet another level: successive accesses to the same memory bank cannot be performed as quickly as accesses to different banks. To get the best performance out of such a system, we must take advantage of the architecture's available concurrency.

Most computers already have memory systems whose peak bandwidth is matched to the peak processor bus rate. But the nature of an algorithm, its data sizes, and placement all strongly affect memory performance; an architecture that works well on one problem may perform quite poorly on another. This was put in sharp focus for the authors while attempting to optimize numerical libraries for the iPSC/860. On some applications, even with *painstakingly* handcrafted code, inadequate memory bandwidth limited us to 20% of peak processor performance [30]. Our experience is not unique; results similar to ours have been reported by Lee [24], for example.

To illustrate one aspect of the bandwidth problem — and how it might be addressed at compile time —consider the effect of executing the fifth Livermore Loop (tridiagonal elimination) using non-caching accesses to reference a single bank of page-mode DRAMs. Figure 1(a) represents the natural reference sequence for a straightforward translation of the computation:

$$\forall i \qquad x_i \leftarrow z_i \times (y_i - x_{i-1})$$

This computation occurs frequently in practice, especially in the solution of partial differential equations by finite difference or finite element methods [9]. Since it contains a first-order linear recurrence, it cannot be vectorized. Nonetheless, the compiler can employ the recurrence detection and optimization algorithm of [6] to generate streaming code: each computed value x_i is retained in a register so that it will be available for use as x_{i-1} on the following iteration. Except in the case of very short vectors, elements from x, y, and z are likely to reside in different pages, so that accessing each vector in turn incurs the page miss overhead on each access; memory references likely to generate page misses are highlighted in the figure.

```
loop:                       loop:
    load z[i]                   load z[i]
    load y[i]                   load z[i+1]
    stor x[i]                   load y[i]
    jump loop                   load y[i+1]
                                stor x[i]
                                stor x[i+1]
                                jump loop

        (a)                         (b)
```

Figure 1 *tridiag* Code

In the loop of Figure 1(a), a page miss occurs for every reference. Unrolling the loop and grouping accesses to the same vector, as in Figure 1(b), amortizes the page-miss cost over a number of accesses; in this case three misses occur for every six references. Reducing the page-miss count increases processor-memory bandwidth significantly. For example, consider a device for which the time required to service a page miss is four times that for a page hit, a miss/hit cost ratio that is representative of current technology. The natural-order loop in Figure 1(a) only delivers 25% of the attainable bandwidth, whereas the unrolled, reordered loop in Figure 1(b) delivers 40%. External effects such as bus turnaround delays are ignored for the sake of simplicity.

Figure 2 illustrates effective memory bandwidth versus depth of unrolling, given a page-miss/page-hit cost ratio of four. For the bottom curve, the loop body of Figure 1(a) is essentially replicated the appropriate number of times, as is standard practice; for the middle curve, accesses have been arranged as per Figure 1(b); the top curve depicts the bandwidth attainable if all accesses were to hit the current DRAM page. Reordering the accesses realizes a performance gain of almost 130% at an unrolling depth of four, and over 190% at a depth of eight. Although in theory we could improve performance almost 240% by unrolling to a depth of sixteen, in most cases the size of the register file won't permit unrolling that far.

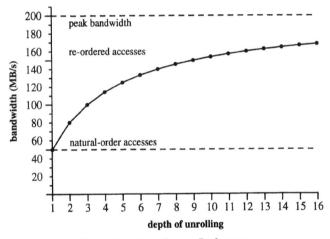

Figure 2 *tridiag* Memory Performance

A comprehensive, successful solution to the memory bandwidth problem must exploit the richness of the *full* memory hierarchy, both its architecture and its component characteristics. One way to do this is via *access ordering*, which we define as any technique for changing the order of memory requests to increase bandwidth. Here we are especially concerned with ordering a set of vector-like "stream" accesses.

As our example illustrates, the performance benefits of doing such static access ordering can be quite dramatic. Unfortunately, without the kinds of address alignment information that are usually only available at run time, the compiler can't generate the optimal access sequence. As pointed out above, the extent to which a compiler can perform this optimization is further constrained by such things as the size of the processor register file [31]. The beneficial impact of access ordering on effective memory bandwidth along with the limitations inherent in implementing the technique statically motivate us to consider an implementation that reorders accesses dynamically at run time.

There are a number of hardware and software techniques that can help manage the imbalance between processor and memory speeds. These include altering the placement of data to exploit concurrency [11], reordering the computation to increase locality (as in "blocking" [21]) address transformations for conflict-free access to interleaved memory [13, 34, 39], software prefetching data to the cache [4, 20, 37], and hardware prefetching vector data to cache [1, 8, 18, 35]. The main difference between these techniques and the complementary one we propose here is that we *reorder* stream accesses to exploit the architectural and component features that make memory systems sensitive to the sequence of requests.

3 A Taxonomy of Access Ordering Techniques

There are a number of options for when and how access ordering can be done, so first we provide a brief taxonomy of the design space. Access ordering systems can be classified by three key components:
- stream detection (*SD*), the recognition of streams accessed within a loop, along with their parameters (base address, stride, etc.),
- access ordering (*AO*), the determination of that interleaving of stream references that most efficiently utilizes the memory system, and
- access issuing (*AI*), the determination of when the load/store operations will be issued.

Each of these functions may be addressed at compile time, *CT*, or by hardware at run time, *RT*. This taxonomy classifies access ordering systems by a tuple (SD, AO, AI) indicating the time at which each function is performed.

Davidson [6] detects streams at compile time, and Moyer [31] has derived access-ordering algorithms relative to a precise analytic model of memory systems. Moyer's approach unrolls loops and orders memory operations to exploit architectural and device features of the target memory system. As our *tridiag* example illustrates, this (CT, CT, CT) system can improve bandwidth significantly, but is limited by the size of the processor register file and lack of vector alignment information available at compile time.

The purely compile-time approach can be augmented with an enhanced memory controller that provides buffer space and that automates vector prefetching, producing a (CT, CT, RT) system. Doing this relieves register pressure and decouples the sequence of accesses generated by the processor from the sequence observed by the memory components: the compiler determines a sequence of vector references to be issued and buffered, but the actual access issue is executed by the memory controller.

Both of these solutions are *static* in the sense that the order of references seen by the memory is determined at compile time; static techniques are inherently limited by the lack of alignment information. *Dynamic* access ordering systems introduce logic into the memory controller to determine the interleaving of a set of references.

For a dynamic (CT, RT, RT) system, stream descriptors are developed at compile time and sent to the memory controller at run time, where the order of memory references is determined dynamically and independently. Determining access order dynamically allows the controller to optimize behavior based on run-time interactions. Our results illustrate the dramatic impact this has on bandwidth.

Fully dynamic (RT, RT, RT) systems implement access ordering without compiler support by augmenting the previous controller with logic to induce stream parameters. Whether or not such a scheme is superior to a (CT, RT, RT) system depends on the relative quality of the compile-time and run-time algorithms for stream detection and relative hardware costs. Proposals for (RT, RT, RT) "vector prefetch units" have recently appeared [1, 35], but these do not order accesses to fully exploit the underlying memory architecture.

4 The Stream Memory Controller

Based on our analysis and simulations, we believe that the best engineering choice is to detect streams at compile time, but to defer access ordering and issue to run time — (CT, RT, RT) in our notation. Choosing this scheme over an (RT, RT, RT) system follows a philosophy that has guided the design of RISC processors: move work to compile time whenever possible. This speeds processing and helps minimize hardware. Here we describe in general terms how such a scheme might be incorporated into an overall system architecture.

The approach we suggest is generally applicable to any uniprocessor computing system, but will be described based on the simplified architecture of Figure 3. Memory is interfaced to the processor through a controller labeled "MSU" for Memory Scheduling Unit. The MSU includes logic to issue memory requests as well as logic to determine the order of requests during streaming computations. For non-stream accesses, the MSU provides the same functionality and performance as a traditional memory controller. This is crucial — the access-ordering circuitry of the MSU is *not* in the critical path to memory and doesn't affect scalar processing.

The MSU has full knowledge of all streams currently needed by the processor: given the base address, vector stride, and vector length, it can generate the addresses of all elements in a stream. The scheduling unit also knows the details of the memory architecture, including interleaving and device characteristics. The access-ordering circuitry uses this information to issue requests for individual stream elements in an order that attempts to optimize memory system performance.

A separate Stream Buffer Unit (SBU) provides high-speed buffers for stream operands and control registers that the processor uses to specify stream parameters (base address, stride, length, and data size). As with the stream-specific parts of the MSU, the SBU is not on the critical path to memory, and the speed of non-vector accesses is not adversely affected by its presence. Together, the MSU and SBU comprise a Stream Memory Controller (SMC) system.

There are a number of options for the internal architecture of the SBU: here we describe one feasible organization. A set of memory-mapped registers provides a processor-independent means of specifying stream parameters. Setting these registers allows the processor to initiate an asynchronous stream of memory access operations for a set of string operands. Data retrieval from the streams (loads) and insertion into streams (stores) may be done in any of several ways; for instance, the SBU could appear to be a traditional cache, or the model could include a set of FIFOs, as illustrated in Figure 3. Each stream is assigned to one FIFO, which is asynchronously filled from (or drained to) memory by the access/issue logic. The "head" of the FIFO is another memory-mapped register, and load instructions from or store instructions to a particular stream reference the FIFO head via this register, dequeueing or enqueueing data as is appropriate.

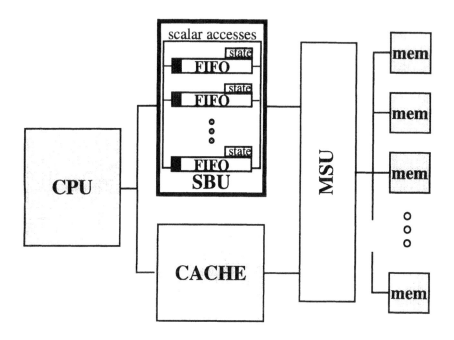

Figure 3 Stream Memory Controller

This organization is both simple and practical from an implementation standpoint: similar designs have been built. In fact, the organization is almost identical to the "stream units" of the WM architecture [42], or may be thought of as a special case of a decoupled access-execute architecture [10, 36]. Another advantage is that this combined hardware/software scheme doesn't require heroic compiler technology —the compiler need only detect the presence of streams, and Davidson's streaming algorithm [6] can be used to do this.

Continuing the *tridiag* algorithm and memory system example introduced earlier, the performance effect of such an SMC is illustrated by Figure 4. The details of this and other results are discussed later, but the gestalt is simple — performance on very short vectors is about 2.5 times that of a system without an SMC; performance on moderate length vectors is about triple that of the non-SMC system; for long vectors and deep FIFOs, bandwidth reaches 98.5% of peak.

94

vector length	w/o SMC	Percentage of Peak Bandwidth					
		SMC FIFO depth					
		8	16	32	64	128	256
10		63.83	62.5	62.5	62.5	62.5	62.5
100	25.0	78.53	85.71	87.98	80.43	73.53	73.53
10000		79.94	88.78	93.97	96.75	98.11	98.53

Figure 4 Bandwidth for the *tridiag* Illustration

5 Complementary Technologies

As mentioned above, there are a number of hardware and software techniques that can help manage the imbalance between processor and memory speeds. Most of these are complementary to access ordering.

Traditional Caching: Traditional caches retain their importance for code and non-vector data in a system equipped with an SMC. Furthermore, if algorithms can be blocked [5, 41] and data aligned to eliminate significant conflicts [21], the cache and SMC can be used in a complementary fashion for vector access. Under these conditions multiple-visit vector data can be cached, with the SMC used to reference single-visit vectors.

To illustrate this, consider implementing the matrix-vector multiply operation:

$$\bar{y} = (A + B)\bar{x}$$

where A and B are $n \times m$ matrices and $\bar{y}$ and $\bar{x}$ are vectors. Figure 5(a) depicts code for a straightforward implementation using matrices stored in column-major order; the code in Figure 5(b) strip-mines the computation to reuse elements of $\bar{y}$. Partition size depends on cache size and structure [21]. Elements of $\bar{y}$ are preloaded into cache memory at the appropriate loop level, and the SMC is then used to access elements of A and B, since each element is accessed only once. The reference to $\bar{x}$ is a constant within the inner loop, and is therefore preloaded into a processor register.

Software Prefetching: Some architectures include instructions to prefetch data from main memory into cache. Using these instructions to load data for a future iteration of a loop [4, 20, 37] can improve processor performance by overlapping memory latency with computation, but prefetching does nothing to actually *improve* memory performance. Note that prefetching can be used in conjunction with an SMC to help hide latency in FIFO references.

Software Access Ordering: Software techniques such as reordering [30] and "vectorization" via library routines [24, 29] can improve bandwidth by reordering

```
   do 20 j = 1,m
      do 10 i = 1,n
         y(i) = y(i) + (A(i,j) + B(i,j)) * x(j)
10    continue
20 continue
```

(a) straightforward implementation

```
   do 30 IT = 1,n,IS
      load y(IT) through y(min(n,IT+IS-1)) into cache
      do 20 j = 1,m
         load x(j) into processor register
         do 10 i = IT,min(n,IT+IS-1)
            y(i) = y(i) + (A(i,j) + B(i,j)) * x(j)
10       continue
20    continue
30 continue
```

(b) strip-mined implementation

Figure 5 Combining Caching and Non-Caching Accesses: $\bar{y} = (A + B)\bar{x}$

requests at compile time. Such techniques cannot exploit run-time information and are limited by processor register resources, hence they cannot outperform hardware-assisted techniques such as the SMC.

Data Placement: An SMC can provide near-optimal bandwidth for a given memory architecture, algorithm, and data placement, but cannot compensate for an unfortunate placement of operands — a vector stride that results in all elements placed in a single bank of a multi-bank memory, for example. An SMC and data placement are complementary; the SMC will perform better given a good placement.

New DRAM interfaces: Rambus [33] is a new, high-speed DRAM interface that provides both higher bandwidth for sequential accesses and true caching of two DRAM pages on the chip. Other sophisticated memory interfaces, such as RamLink and the JEDEC synchronous DRAM, provide similar benefits [16]. The more sophisticated such interfaces become, the more important it is to exploit them intelligently with controllers such as the SMC.

Alternative Storage Schemes: Skewed storage [3, 12] and dynamic address transformations [13, 34] have been proposed as methods for increasing concurrency, and hence bandwidth, in parallel memory systems. Unfortunately, these techniques do not work for interspersed multiple streams, and they do not exploit memory component features.

6 Simulation Environment

We have simulated a wide range of SMC configurations and benchmarks, varying
- FIFO depth,
- dynamic order/issue policy,
- number of memory banks,
- DRAM speed,
- benchmark algorithm, and
- vector length, stride, and alignment with respect to memory banks.

Only samples of our results are given here; complete results can be found in [28]. In particular, the results given below involve the following restrictions.

The simulations here use stride-one vectors aligned to have no DRAM pages in common, but starting in the same bank unless otherwise specified. The SMC is very robust in its ability to optimize memory bandwidth regardless of stride and alignment, so this restriction does not materially affect the results.

We model the processor as a generator of load and store requests only — arithmetic and control are assumed never to be a computational bottleneck. This places the maximum stress on the memory system by assuming a computation rate that out-paces the memory's ability to transfer data. Scalar and instruction accesses are assumed to hit in the cache for the same reason.

All memories modeled here consist of interleaved banks of page-mode DRAMs, where each page is 2K double words.

The only order/issue policy considered is exceedingly simple. The SMC looks at each FIFO in round-robin order, issuing accesses for the same FIFO stream while:

1) not all elements of the stream have been accessed, and
2) there is room in the FIFO for another read operand, or another write operand is present in the FIFO.

Results reported here are for the four kernels described in Figure 6. *Daxpy* and *swap* are from the BLAS (Basic Linear Algebra Subroutines) [22, 7], *tridiag* is the fifth Livermore Loop from our earlier example[27], and *vaxpy* is a vector *axpy*[1] computation that occurs in matrix-vector multiplication by diagonals. These benchmarks were selected because they are representative of the access patterns found in real scientific codes, including the inner loops of blocked algorithms. Nonetheless, our results show that the actual reference sequence has little effect on SMC performance.

Non-SMC results are for the "natural" reference sequence for each benchmark, using non-caching loads and stores.

SMC initialization requires two writes to memory-mapped registers for each stream; this small overhead has no significant effect on results, and is not included here.

The DRAM page-miss cycle time is four times that of a DRAM page hit, unless otherwise noted.

---

1. Here "axpy" refers to a computation involving some entity *a* times a vector *x* plus a vector *y*: for *daxpy*, *a* is a double; for *vaxpy*, *a* is a vector.

daxpy:	$\forall i$	$y_i \leftarrow ax_i + y_i$		
tridiag:	$\forall i$	$x_i \leftarrow z_i \times (y_i - x_{i-1})$		
swap:	$\forall i$	$tmp \leftarrow y_i$	$y_i \leftarrow x_i$	$x_i \leftarrow tmp$
vaxpy:	$\forall i$	$y_i \leftarrow a_i x_i + y_i$		

Figure 6   Benchmark Algorithms

7 Results

Figure 7 through Figure 10 illustrate the relative performance of the four kernels for a variety of memory systems using an SMC. The SMC's ability to optimize bandwidth is relatively insensitive to vector access patterns, hence the shape of the performance curves is similar for all benchmarks — asymptotic behavior approaches 100%.

Figure 7 shows SMC performance for long vectors (length 10,000) as a function of FIFO depth and number of memory banks (available concurrency) compared to the analogous non-SMC systems. On the *daxpy* benchmark, for example, an SMC system with two memory banks achieves 98.2% of peak bandwidth, compared to 18.8% for a non-SMC system. In general, SMC systems with deep FIFOs achieve in excess of 92% of peak bandwidth for all benchmarks and memory configurations. Even with FIFOs that are only sixteen double-words deep, the SMC systems consistently deliver over 75% of peak bandwidth.

Note that increasing the number of banks reduces *relative* performance, a somewhat counter-intuitive and deceptive effect. This is due in part to our keeping both the peak memory system bandwidth and the DRAM page-miss/hit delay ratio constant. Thus, the eight-bank system has four times the DRAM page-miss latency of the two-bank system. If, alternatively, we hold the performance of the memory banks constant and assume a faster bus, the peak bandwidth of the total system increases proportionally to the number of banks. Although the percentage of bandwidth delivered is still smaller in this case, the total bandwidth is much larger. The other reason for this effect is that increasing the number of banks decreases the number of accesses to each bank, thus page-miss cost is amortized over fewer accesses.

Figure 8 represents SMC results for medium-length (100 element) vectors compared with non-SMC performance. These SMC results depict the net effect of two competing performance factors. With deeper FIFOs, DRAM page misses are amortized over a larger number of total accesses, which can increase performance. At the same time, the processor has to wait longer to complete its first loop iteration while the SMC prefetches numerous operands to be used in the following loop iterations. This can decrease performance, as evidenced by the tail-off beyond length-32 FIFOs. Optimum FIFO depth could, and probably should, be run-time selectable in the SMC, since it is so closely related to stream length.

Note that performance of non-SMC systems depicted in Figure 7 and Figure 8 is independent of vector length. Since these systems employ no dynamic access ordering,

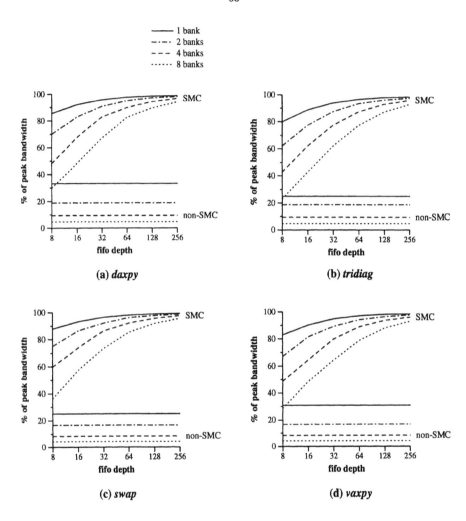

Figure 7 Long Vector Performance

the number of requests issued and the resulting percentage of total bandwidth obtained
are constant for each loop iteration. This is true of any system in which access issue is
determined at compile time, including those that use prefetching.

Lack of dynamic ordering renders the performance of non-SMC systems
particularly sensitive to vector placement. In Figure 7 and Figure 8, the vectors are
aligned so that they both compete for the same bank on each iteration; this has little
effect on SMC performance (because it reorders requests), but it prevents the non-SMC
systems from taking advantage of the potential concurrency.

Figure 9 represents the performance of non-SMC and SMC systems on medium-
length vectors with better alignment. In these experiments the vectors are staggered
such that the ith vector in the pattern begins in bank (i mod n), where n is the number of
banks. In spite of the more favorable alignment, non-SMC *daxpy* performance is limited

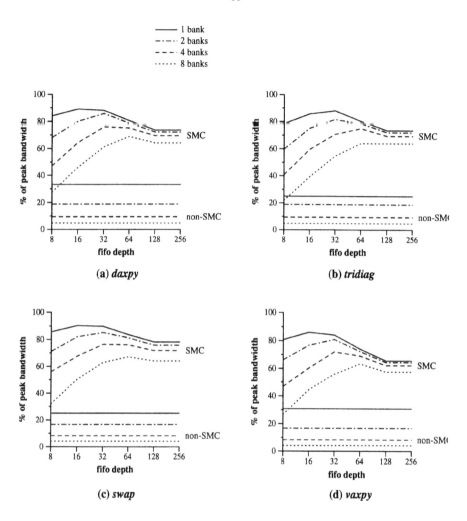

Figure 8 Medium Vector Performance

to 30.0% of total bandwidth for a two-bank memory; *tridiag*, *swap*, and *vaxpy* are limited to 18.8%, 40.0%, and 25.0%, respectively. SMC performance remains essentially the same as in Figure 8.

Figure 10 illustrates SMC performance on very short (10-element) vectors. Performance improvements are not as dramatic as for longer vectors, since there are even fewer accesses over which to amortize page-miss costs. Nonetheless, short vector computations benefit significantly from an SMC: *daxpy* run on a two-bank architecture with an SMC achieves 53.6% of the attainable bandwidth, whereas the same benchmark run on a similar non-SMC system is limited to 18.8%. The other kernels enjoy similar increases in bandwidth. Non-SMC performance is as in Figure 8 or Figure 9, depending on vector alignment; those lines are omitted here for clarity.

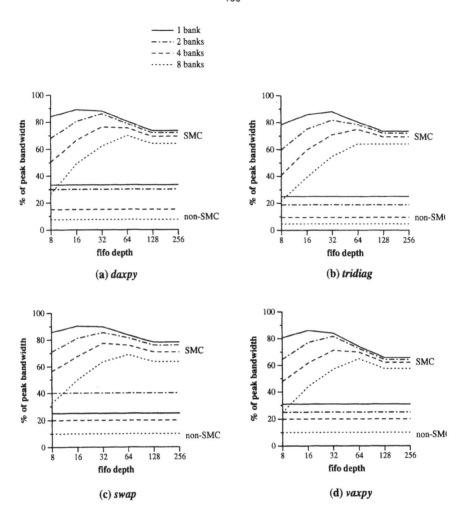

Figure 9 Medium Vector Performance for Better Vector Alignment

Figure 11(a) shows SMC performance for *daxpy* using long vectors (10,000 elements) as the page-miss to page-hit cost ratio increases. This figure may be a bit misleading: the miss/hit ratio is likely to increase primarily as the result of a reduction of the page-hit time, rather than an increase in the page-miss time. Thus, at a ratio of sixteen the SMC is delivering a somewhat smaller percentage of a *much* larger available bandwidth — resulting in a significant net increase.

If we hold the number of modules fixed and increase component performance, deeper FIFOs are required in order to amortize the page-miss costs. As evidenced by the slope of the curves in Figure 11(b), relative performance is approximately constant if we scale FIFO depth linearly with miss/hit cost. Note that the faster systems still require only modest amounts of buffer storage.

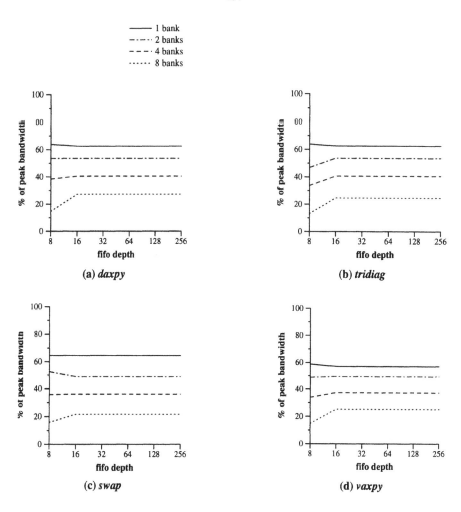

Figure 10 Very Short Vector Performance

8 Conclusions

Memory bandwidth is rapidly becoming the performance bottleneck in the application of high performance microprocessors to vector-like algorithms, including many of the "grand challenge" scientific problems. Caching is not the sole solution for these applications due to their poor temporal and spatial locality.

Achieving greater bandwidth requires exploiting the characteristics of the entire memory hierarchy; it cannot be treated as though it were uniform access-time RAM. Moreover, exploiting the memory's properties will have to be done dynamically — essential information (such as alignment) will generally not be available at compile time.

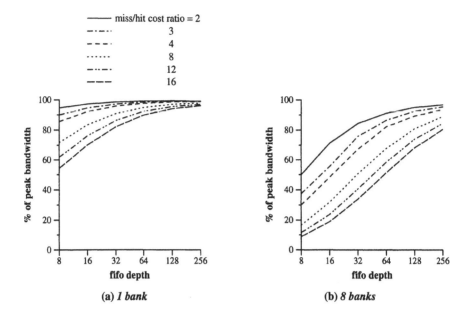

Figure 11 *daxpy* Performance for Various Page-miss/Page-hit Cost Ratios

Reordering can optimize accesses to exploit the underlying memory architecture. By combining compile-time detection of streams with execution-time selection of the access order and issue, we achieve near-optimal bandwidth for vector-like accesses relatively inexpensively. This complements more traditional cache-based schemes, so that overall effective memory performance need not be a bottleneck.

Here we have reported the basic design of a Stream Memory Controller (SMC) and have demonstrated its performance for a variety of FIFO depths, memory configurations, etc. Using current memory parts and only a few hundred words of buffer storage, an SMC system can deliver nearly the full memory system bandwidth. Moreover, it does so with naive code, and performance is independent of the alignment and stride of the operands.

9. Acknowledgments

Thanks go to the other members of Bill Wulf's research group for their valuable feedback: Scott Briercheck, Rob Craighurst, Katie Oliver, Ramesh Peri, and Alec Yasinsac. This work has been supported in part by a grant from Intel Supercomputer Division and by NSF contract MIP-9114110.

References

1. Baer, J. L., Chen, T. F., "An Effective On-Chip Preloading Scheme To Reduce Data Access Penalty", Supercomputing'91, November 1991.
2. Baron, R.L., and Higbie, L., *Computer Architecture*, Addison-Wesley, 1992.
3. Budnik, P., and Kuck, D., "The Organization and Use of Parallel Memories", IEEE Trans. Comput., 20, 12, 1971.
4. Callahan, D., et. al., "Software Prefetching", Fourth International Conference on Architectural Support for Programming Languages and Systems, April 1991.
5. Carr, S., Kennedy, K., "Blocking Linear Algebra Codes for Memory Hierarchies", Proc. Fourth SIAM Conference on Parallel Processing for Scientific Computing, 1989.
6. Davidson, Jack W., and Benitez, Manuel E., "Code Generation for Streaming: An Access/Execute Mechanism", Fourth International Conference on Architectural Support for Programming Languages and Operating Systems, April 1991.
7. Dongarra, et. al., "Linpack User's Guide", SIAM, Philadelphia, 1979.
8. Fu, J. W. C., and Patel, J. H., "Data Prefetching in Multiprocessor Vector Cache Memories", 18th International Symposium on Computer Architecture, May 1991.
9. Golub, G., and Ortega, J.M., *Scientific Computation: An Introduction with Parallel Computing*, Academic Press, Inc., 1993.
10. Goodman, J. R., et al, "PIPE: A VLSI Decoupled Architecture", Twelfth International Symposium on Computer Architecture, June 1985.
11. Gupta, R., and Soffa, M., "Compile-time Techniques for Efficient Utilization of Parallel Memories", SIGPLAN Not., 23, 9, 1988, pp. 235-246.
12. Harper, D. T., Jump., J., "Vector Access Performance in Parallel Memories Using a Skewed Storage Scheme", IEEE Trans. Comput., 36, 12, 1987.
13. Harper, D. T., "Address Transformation to Increase Memory Performance", 1989 International Conference on Supercomputing.
14. Hayes, J.P., *Computer Architecture and Organization*, McGraw-Hill, 1988.
15. Hwang, K., and Briggs, F.A., *Computer Architecture and Parallel Processing*, McGraw-Hill, Inc., 1984.
16. "High-speed DRAMs", Special Report, IEEE Spectrum, vol. 29, no. 10, October 1992.
17. *i860 XP Microprocessor Data Book*, Intel Corporation, 1991.
18. Jouppi, N., "Improving Direct-Mapped Cache Performance by the Addition of a Small Fully Associative Cache and Prefetch Buffers", 17th International Symposium on Computer Architecture, May 1990.
19. Katz, R., and Hennessy, J., "High Performance Microprocessor Architectures", University of California, Berkeley, Report No. UCB/CSD 89/529, August, 1989.
20. Klaiber, A., et. al., "An Architecture for Software-Controlled Data Prefetching", 18th International Symposium on Computer Architecture, May 1991.
21. Lam, Monica, et. al., "The Cache Performance and Optimizations of Blocked Algorithms", Fourth International Conference on Architectural Support for Programming Languages and Systems, April 1991.
22. Lawson, et. al., "Basic Linear Algebra Subprograms for Fortran Usage", ACM Trans. Math. Soft., 5, 3, 1979.

23. Lee, K., "Achieving High Performance On the i860 Microprocessor Using Naspack Subroutines", NAS Systems Division, NASA Ames Research Center, July 1990.

24. Lee, K., "On the Floating Point Performance of the i860 Microprocessor", RNR-90-019, NAS Systems Division, NASA Ames Research Center, October 1990.

25. Maccabe, A.B., *Computer Systems: Architecture, Organization, and Programming*, Richard D. Irwin, Inc., 1993.

26. Mano, M.M., *Computer System Architecture*, 2nd ed., Prentice-Hall, Inc., 1982

27. McMahon, F.H., "The Livermore Fortran Kernels: A Computer Test of the Numerical Performance Range", Lawrence Livermore National Laboratory, UCRL-53745, December 1986.

28. McKee, S.A, "Hardware Support for Access Ordering: Performance of Some Design Options", University of Virginia, Department of Computer Science, Technical Report CS-93-08, July 1993.

29. Meadows, L., Nakamoto, S., and Schuster, V., "A Vectorizing, Software Pipelining Compiler for LIW and Superscalar Architectures", RISC'92, February 1992.

30. Moyer, S.A., "Performance of the iPSC/860 Node Architecture," University of Virginia, IPC-TR-91-007, 1991.

31. Moyer, S., "Access Ordering and Effective Memory Bandwidth", Ph.D. Dissertation, Department of Computer Science, University of Virginia, Technical Report CS-93-18, April 1993.

32. Quinnell, R., "High-speed DRAMs", EDN, May 23, 1991.

33. "Architectural Overview", Rambus Inc., Mountain View, CA, 1992.

34. Rau, B. R., "Pseudo-Randomly Interleaved Memory", 18th International Symposium on Computer Architecture, May 1991.

35. Sklenar, Ivan, "Prefetch Unit for Vector Operation on Scalar Computers", Computer Architecture News, 20, 4, September 1992.

36. Smith, J. E., et al, "The ZS-1 Central Processor", The Second International Conference on Architectural Support for Programming Languages and Systems, Oct. 1987

37. Sohi, G. and Manoj, F., "High Bandwidth Memory Systems for Superscalar Processors", Fourth International Conference on Architectural Support for Programming Languages and Systems, April 1991.

38. Tomek, I., *The Foundations of Computer Architecture and Organization*, Computer Science Press, 1990.

39. Valero, M., et. al., "Increasing the Number of Strides for Conflict-Free Vector Access", 19th International Symposium on Computer Architecture, May 1992.

40. Wallach, S., "The CONVEX C-1 64-bit Supercomputer", Compcon Spring 85, February 1985.

41. Wolfe, M., "Optimizing Supercompilers for Supercomputers", MIT Press, Cambridge, MA, 1989.

42. Wulf, W. A., "Evaluation of the WM Architecture", 19th Annual International Symposium on Computer Architecture, vol 20, no. 2, May 19-21, 1992.

The Advantages of Machine-Dependent Global Optimization

Manuel E. Benitez and Jack W. Davidson

Department of Computer Science
University of Virginia
Charlottesville, VA 22903 U. S. A.

Abstract. Using an intermediate language is a well-known, effective technique for constructing interpreters and compilers. This paper describes a retargetable, optimizing compilation system centered around the use of two intermediate languages (IL): one relatively high level, the other a low level corresponding to target machine instructions. The high-level IL (HIL) models a stack-based, hypothetical RISC machine. The low-level IL (LIL) models target machines at the instruction-set architecture level. All code improvements are applied to the LIL representation of a program. This is motivated by the observation that most optimizations are machine dependent, and the few that are truly machine independent interact with the machine-dependent ones. This paper describes several 'machine-independent' code improvements and shows that they are actually machine dependent. To illustrate how code improvements can be applied to a LIL, an algorithm for induction variable elimination is presented. It is demonstrated that this algorithm yields better code than traditional implementations that are applied machine-independently to a high-level representation.

1 Introduction

A retargetable, optimizing compiler must perform a comprehensive set of code improvements in order to produce high-quality code for a wide range of machines. A partial list of code improvements that must be included in the compiler's repertoire is:

- register assignment and allocation,
- common subexpression elimination,
- loop-invariant code motion,
- induction variable elimination,
- evaluation order determination,
- constant folding,
- constant propagation,
- dead code elimination,
- loop unrolling,
- instruction scheduling, and
- inline function expansion.

This list of code improvements traditionally is divided into two groups: those that are considered to be *machine-independent* and those that are *machine-dependent*. Machine-independent code improvements are those that do not depend on any features or characteristics of the target machine. Examples of code improvements included in this group are constant folding, dead code elimination, and constant propagation. Because of their machine-independence, these code improvements are often applied to the high-level intermediate language representation of the program.

The proper application of machine-dependent code improvements, on the other hand, requires having specific information about the target machine. Obviously, code improvements such as register allocation and instruction scheduling are machine dependent. In the case of register allocation, the types and number of registers available affect the promotion of data to registers. Similarly, effective instruction scheduling requires information about the operation of the target machine's pipeline. Somewhat less obvious, but no less machine dependent are the code improvements inline function expansion and loop unrolling. Inline function expansion can be performed most effectively when details of the target machine's instruction cache is available. Similarly, the amount of loop unrolling performed depends on the number of target machine registers available, characteristics of the instruction pipeline as well as the size of the instruction cache.

The belief that some code improvements are machine-independent and some are machine-dependent and the use of a single high-level intermediate representation results in a compiler with a structure shown in Figure 1a. Unfortunately, most code improvements are not machine-independent, and the few that truly are machine independent interact with those that are machine dependent causing phase-ordering problems. For example, dead code elimination is machine independent. However, it interacts with machine-dependent code improvements such as inline function expansion. Expanding functions inline exposes new opportunities for dead code elimination by effectively propagating constants across calls. Hence, essentially *there are no machine-independent code improvements*. In Section 2, some code improvements that typically are viewed as being machine independent are examined and shown to be machine dependent. This section also provides examples of how true machine-independent code improvements interact with machine-dependent ones. Section 3 describes a compiler structure that uses two intermediate languages: a high level intermediate language (HIL) that serves to isolate the language-dependent portion of the compiler from target machine details, and a low-level intermediate language (LIL) that supports the application of global code improvements. Section 4 contains a detailed description of an induction variable elimination algorithm that operates on a low-level representation of a program. The algorithm is largely machine independent, and requires no modification when the compiler is retargeted, yet it generates superior code when compared to a traditional HIL implementation. Section 5 evaluates the effectiveness of the LIL implementation of induction variable elimination on a set of representative benchmark programs.

2   The Case for Machine-Dependent Global Optimization

To illustrate the point that all code improvements are effectively machine dependent, consider constant propagation. This deceptively simple code improvement involves propagating a constant that has been assigned to a variable (a definition) to points in the

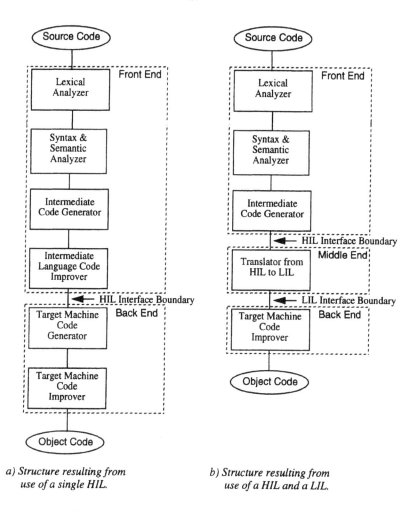

a) Structure resulting from
use of a single HIL.

b) Structure resulting from
use of a HIL and a LIL.

Figure 1. Structure of two compiler organizations.

program where the variable is used and the definition reaches. After constant propagation is performed often the assignment to the variable becomes useless and can be eliminated. Additionally, knowing the value of a constant at a particular point in the program permits other code improvements to be performed. Constant propagation typically is considered to be a machine-independent code improvement and is performed in the machine-independent front end portion of the compiler. Unfortunately, constant propagation is not machine independent. To be done most effectively, characteristics of the target machine must be known.

To explain some of the machine-dependent complications that arise when performing constant propagation, consider the C code in Figure 2a. Assuming that variable y is not modified elsewhere in the function, should the value 10.0 be

108

```
    void foo()                      main()
    {                               {
        double y;                       double y;

        y = 10.0;                       y = 10.0;
        ...                             foo(y);
        baz (y);                        ...
        ...                             foo(y + 32.56);
        bar (y);                        ...
        ...                         }
    }
```

a) Constant propagation   c) Constant propagation
         is not worthwhile.   is worthwhile.

```
_foo:                           _main:
    ...                             ...
    # load 10.0                     sethi %hi(L20),%o0
    sethi %hi(L20),%o0              # call foo with 10.0
    ldd   [%o0+%lo(L20)],%f0        call  _foo,2
    # store in y                    ldd   [%o0+%lo(L20)],%o0
    std   %f0,[%fp-8]               ...
    ...                             sethi %hi(L21),%o0
    # call baz with y              # call foo with 42.56
    call  _baz,2                    call  _foo,2
    ldd   [%fp-8],%o0              ldd   [%o0+%lo(L21)],%o0
    ...                             ...
    # call bar with y          L20: .double 0r10.0
    call  _bar,2               L21: .double 0r42.56
    ldd   [%fp-8],%o0
    ...
L20: .double 0r10.0
```

      b) SPARC assembly code for   d) SPARC assembly code for
         fragment in Figure 2a.   fragment in Figure 2c.

Figure 2. Code illustrating complications of machine-independent constant propagation.

propagated to each of the uses of the variable? If constant propagation is performed at a high-level, before details of the target machine are known, the choice is simple; propagate the constant. However, the correct action depends on the target machine and how the constant is being used. On the SPARC architecture three factors affect whether a floating-point constant should be propagated. First, there is no direct data path between the fixed-point registers and the floating-point registers. Moving a value from one register set to another requires going through memory. Second, the SPARC calling sequence requires that the first 24 bytes of arguments be passed in the fixed-point registers %o0 through %o5 regardless of their type. Third, the only way to load a floating-point constant into a register is by fetching it from global memory (i.e., there is no load immediate for floating-point values). With the current conventions this requires two instructions.

Figure 2b contains the SPARC assembly code generated for the C code fragment in Figure 2a using Sun's optimizing compiler with the highest level of optimization. Recall that call and branch instructions on the SPARC have a single-instruction delay slot. In this example, the constant was not propagated. Indeed, if it had, inferior code would have been produced. The load double instructions in the call delay slots would each be a two instruction sequence to load the constant from global memory. The astute reader might argue that the compiler, in this case, should not have propagated the constant, but rather allocated it to a floating-point register, and used the register at each point in the code where y is referenced. Unfortunately, this would result in even poorer code. Because the only path from a floating-point register to a fixed-point register is through memory, this approach would have required storing the contents of the floating-point register in memory and reloading it in the appropriate output registers. Because of these complications and because it performs constant propagation at a high-level, it appears Sun's SPARC compiler is forced to follow the simple rule: never propagate floating-point constants.

Is it always best not to propagate floating-point constants on the SPARC? To answer this question, consider the C code fragment in Figure 2c. Here it would be beneficial to propagate the constant. If the constant is propagated, constant folding can be done for the argument in the second call to foo and the resulting constant can be loaded directly into %o0 and %o1. The code is shown in Figure 2d. This code is 50% smaller than the code that would be produced without constant propagation.

As another example, consider the code improvement loop-invariant code motion (LCM). Again, many compilers perform this transformation at a high-level under the assumption that machine-specific information is not required. However, this is not the case. Consider the code fragment in Figure 3a. limit is an external global variable and update is an external function that can potentially alter the value of limit. Figure 3b shows typical HIL for this code. With this representation, there is no visible loop invariant code. The code generated for the SPARC is shown in Figure 3c. An inspection of this code reveals that the computation of limit's address was loop invariant.

It is tempting to say that this problem can be solved by changing the HIL so that computation of addresses is decoupled from the actual reference. Figure 3d shows a HIL version of the code using this approach. Now the computation of the address of limit is visible, and a code improver operating on the HIL would move it out of the loop. This code is shown in Figure 3e. Unfortunately, this still does not yield the best possible machine code. On the SPARC, the calculation of the address of a global requires two instructions. However, it is possible to fold part of the address calculation into the instruction that does the memory reference. By taking into account the machine's addressing modes and the costs of instructions, a code improver that operates on a LIL representation produces the code of Figure 3f. While the loops of Figure 3e and Figure 3f have the same number of instructions, the overall code in Figure 3f is one instruction shorter. If this is a function that is called many times, the impact on execution time will be noticeable.

110

```
extern int limit;

void find(value)
int value;
{
    extern void update();

    while (value < limit)
        update();
}
```
*a) Code with loop-invariant
address calculation.*

```
FUNC    void   find
LABEL   16
LOAD    int    value   PARAM
LOAD    int    limit   EXTERN
JMPGE   int    17
CALL    void   update  EXTERN
GOTO    16
LABEL   17
EFUNC   void   find
```
b) HIL for a.

```
_find:
    save   %sp,-96,%sp
L16:
    sethi  %hi(_limit),%o0
    ld     [%o0+%lo(_limit)],%o0
    cmp    %i0,%o0
    bge    L17
    call   _update
    ba     L16
L17:
    ret
    restore
```
c) SPARC code generated from b.

```
FUNC    void   find
LABEL   16
ADDR    int    value   PARAM
DEREF   int
ADDR    int    limit   EXTERN
DEREF   int
JMPGE   int    17
CALL    void   update  EXTERN
GOTO    16
LABEL   17
EFUNC   void   find
```
*d) HIL with address calculations
exposed.*

```
_find:
    save   %sp,-96,%sp
    sethi  %hi(_limit),%l0
    add    %l0,%lo(_limit),%l0
L16:
    ld     [%l0],%o0
    cmp    %i0,%o0
    bge    L17
    call   _update
    ba     L16
L17:
    ret
    restore
```
*e) SPARC code generated
from d with LCM.*

```
_find:
    save   %sp,-96,%sp
    sethi  %hi(_limit),%l0
L16:
    ld     [%l0+%lo(_limit)],%o0
    cmp    %i0,%o0
    bge    L17
    call   _update
    ba     L16
L17:
    ret
    restore
```
*f) SPARC code with machine-
dependent LCM.*

Figure 3. Example illustrating the machine dependence of loop-invariant code motion (LCM).

As a last example, consider dead code elimination (DCE). This transformation is truly machine independent. That is, any code that will never be executed should always be eliminated no matter what the target machine. Unfortunately, machine-dependent code improvements create opportunities for DCE, and therefore, to be most effective,

```
void daxpy(n, da, dx, incx, dy, incy)
int n, incx, incy;
double da, dx[], dy[];
{
      if (n <= 0)
         return;
      if (da = 0.0)
         return;
      if (incx != 1 || incy != 1) {
         /* Code for unequal increments or */
         /* equal increments other than one */
         }
      else
         for (i = 0; i < n; i++)
            dy[i] = dy[i] + da * dx[i];
}
```

Figure 4. *daxpy* routine from *linpack*.

DCE should also be performed at the machine level. To see this, first consider the machine-dependent code improvement inline code expansion. This transformation replaces calls to functions with the body of the called function. It eliminates call/return overhead, may improve the locality of the program, and perhaps most importantly, can enable other code improvements which includes, among others, dead code elimination. Inline code expansion is machine dependent because the decision to inline depends on the characteristics of the target machine. One important consideration is the size of the instruction cache [11]. Inlining a function into a loop and possibly causing the loop to no longer fit in the cache can result in a serious drop in performance. To illustrate how DCE interacts with inlining, consider the *daxpy* function from the well-known *linpack* benchmark. The code is shown in Figure 4. Generally, all calls to *daxpy* set incx and incy to one. Thus, by first performing inlining and constant propagation (both machine dependent code improvements), a dead code eliminator that operates on LIL after inline code expansion and constant propagation will eliminate the code for handling increments that are not both one.

3   A HIL/LIL Compiler Organization

These observations lead to the conclusion that more effective code improvement can be performed if all transformations are done on a low-level representation where target machine information is available. To accomplish this requires two intermediate representations: a HIL that isolates, as much as possible, the language-dependent portions of the compiler from the target machine specific details, and a LIL that supports applying code improvements. The use of two intermediate languages yields a compiler with the structure shown in Figure 1b. It is significantly different from that of the traditional, single intermediate language representation shown in Figure 1a. The influence of the use of two intermediate languages is pervasive—affecting the design of the HIL, as well as the code generation algorithms used in the front end. The following sections discuss these effects.

3.1 The High-Level Intermediate Language

In most modern compilers the front end is decoupled from the back end through the use of an intermediate representation. The goal is to make the front end machine independent so that it can be used for a variety of target architectures with as little modification as possible. One popular approach is to generate code for an abstract machine. Well-known abstract machines include P-code (used in a several Pascal compilers) [12], U-code (used in the compilers developed by MIPS, Inc. for the R2000/ R3000 family of microprocessors) [3], and EM (used in the Amsterdam compiler kit) [17, 18]. In the quest for efficiency, the abstract machine often models the operations and addressing modes found on the target architectures. For a retargetable compiler, with many intended targets, this can yield a large and complex abstract machine. Such abstract machines have been termed 'union' machines as they attempt to include the union of the set of operators supported on the target architectures [6]. The Berkeley Pascal interpreter, for example, has 232 operations [10].

There is an equally compelling argument for designing a small, simple abstract machine. Small, simple instruction sets are faster and less error prone to implement than a large complex instruction set. Abstract machine designers have long recognized this dilemma. In 1972, Newey, Poole, and Waite [13] observed that

> 'Most problems will suggest a number of specialized operations which could possibly be implemented quite efficiently on certain hardware. The designer must balance the convenience and utility of these operations against the increased difficulty of implementing an abstract machine with a rich and varied instruction set.'

Fortunately, applying all code improvements to the LIL removes efficiency considerations as a HIL design issue. The abstract machine need only contain a set of features roughly equivalent to the intersection of the operations included in typical target machines. The result is a small, simple abstract machine. Such abstract machines are termed 'intersection' machines. The analogy between union/intersection abstract machines and CISC/RISC architectures is obvious.

There are other reasons for preferring a small abstract machine instruction set. First, a small instruction set is more amenable to extension. Adding additional operations to support a new language feature (for example, a new opcode to support the pragma feature of ANSI C was recently added to the CVM instruction set) is generally not a problem. However, for large instruction sets, this may cause problems. For example, many abstract machines have over 200 operations. Adding more operations may require changing the instruction format (a byte code may not be sufficient). Second, intersection machines are more stable. If a machine appears with some new operation, the operation must be added to the union machine. The intersection machine, on the other hand, need only be changed if the new operation cannot be synthesized from the existing operations. Third, if the compiler is to be self-bootstrapping (a lost art), a small intermediate language can significantly reduce the effort to bootstrap [15, 13]. For additional justification for preferring a small, simple abstract machine over a large, complex one see [6].

The HIL described here is called CVM (C Virtual Machine), and it supports most imperative languages, although it was motivated mainly by the desire to support variants of the C language (K&R C, ANSI C, and C++). The CVM instruction set contains 51 executable instructions and 17 pseudo operations. Similar to the abstract machines mentioned above, CVM is a stack architecture as opposed to a register architecture. CVM is stack-oriented for a couple of reasons. First, algorithms for generating code for a stack machine are well understood and easy to implement. Second, it was important to be able to specify the semantics of operation of CVM. This is done operationally through an interpreter. Implementing an interpreter for a stack-based machine is quite simple, easy to understand, and reasonably efficient [8].

3.2 The Low-Level Intermediate Language

The LIL representation of a program is what will be manipulated by all code improvement algorithms. Thus, while it is necessary that the LIL encode machine-specific details so that the code improvement algorithms can produce better code, it must be done in such a way that the implementation of the algorithms does not become machine-dependent.

The LIL representation described here is based on register transfer lists (RTLs), which are derived from the Instruction Set Processor (ISP) notation developed by Bell and Newell. Essentially, RTLs describe the effects of machine instructions and have the form of conventional expressions and assignment's over the hardware's storage cells. Each RTL corresponds to a single target machine instruction in the same way that each traditional assembly code line corresponds to a single instruction. Unlike assembly language syntax, which varies from machine to machine, the RTL specification of an operation is identical across machines. For example, the following list shows register-to-register add instructions on various machines using their assembly language syntax:

```
a       r1=r1,r2        -- IBM RS/6000
ar      1,2             -- IBM 370
add     1,2             -- DecSystem 10
ix1     x1+x2           -- CDC 6600
add     %o2,%o1,%o1     -- Sun SPARC
add.1   d2,d1           -- Motorola 68020
addl2   r2,r1           -- VAX-11
addu    $1,$1,$2        -- MIPS R3000
```

Using the RTL notation, each of these instructions is represented by the following RTL:

```
r[1] = r[1] + r[2];
```

In contrast to the assembly language specification of the instruction, the RTL unambiguously describes the action of the instruction. Thus, the above RTL clearly indicates which register receives the result of the addition operation. Assembly language instructions, on the other hand, use conventions to designate source and destination operands.

Most machines have instructions that perform several actions. These are specified using lists of transfers (hence the name register transfer *lists*). A common occurrence in some machines are instructions that perform an arithmetic operation and set bits in a condition code register to indicate some information about the result (e.g. equal to zero, negative, overflow, etc.). Such multi-effect instructions are specified using RTLs as:

```
r[1] = r[1] + r[2]; Z = (r[1] + r[2]) == 0;
```

Each register transfer in the list is performed concurrently. Thus, the above RTL specifies the same addition operation as the previous RTL and also sets the zero bit in the condition code register (specified by Z) if the result of the operation is zero or clears it, otherwise.

RTLs have been used successfully to automate machine-specific portions of a compiler such as instruction selection, common subexpression elimination, and evaluation order determination [5, 6, 7] These are all local transformations that do not require information beyond that contained in a basic block. To better support global code improvements such as loop-invariant code motion, induction variable elimination, constant propagation, loop unrolling, and inline function expansion, we represent RTLs using a binary tree structure that allows each component of a register transfer (e.g. a register, a memory reference, an operation, a constant, etc.) to contain information that is specific to that component. For example, tree node representing operators include a type specifier and register nodes contain a pair of pointers linking them to the next and previous reference of the register. This LIL is more than a language, it is a representation that allows global code improvements algorithms to take into account the characteristics of the target machine in a machine-independent fashion.

A simplified diagram of a program fragment represented using this LIL is shown in Figure 5a. Details are shown only for basic block C and references to registers r[8] and r[9]. The representation consists of a control-flow graph of basic blocks. Associated with each basic block is a list of RTLs that represent the machine instructions that will be executed when flow of control passes through this basic block. The corresponding RTLs in string form and in SPARC assembly language for the code in this basic block are shown in Figure 5b and Figure 5c, respectively. Also associated with each basic block is information about which loops the basic block is a member of. This information includes the location of the preheader block of the loop (if one exists), dominance relations, induction variable information, and loop-invariant values.

For each reference to a register or a memory location, a def-use chain is maintained. Thus, from any reference the code improver can find either the previous reference or the next reference. Previous reference involving merging flow can be found through φ functions [4]. These functions and the def-use links form the SSA form for the code. This form is used to determine a canonical value for each component of the code. These are similar to value numbers [1] and are used to perform common-subexpression elimination as well as code motion.

For each memory reference, information about the memory partition affected by the reference is maintained [2]. This structure is used to hold information that is vital for

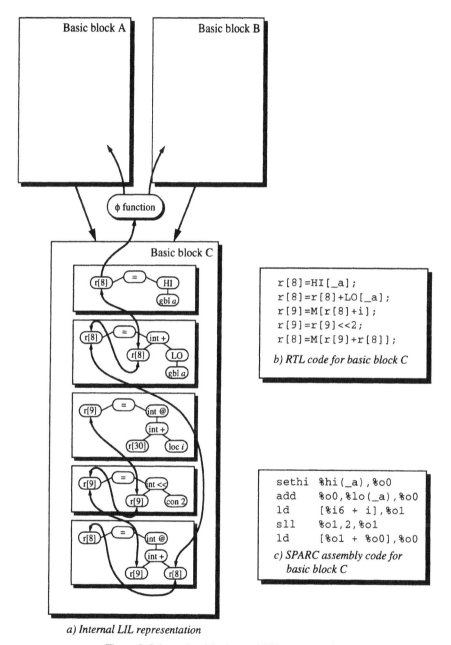

b) RTL code for basic block C

```
r[8]=HI[_a];
r[8]=r[8]+LO[_a];
r[9]=M[r[8]+i];
r[9]=r[9]<<2;
r[8]=M[r[9]+r[8]];
```

c) SPARC assembly code for basic block C

```
sethi   %hi(_a),%o0
add     %o0,%lo(_a),%o0
ld      [%i6 + i],%o1
sll     %o1,2,%o1
ld      [%o1 + %o0],%o0
```

a) Internal LIL representation

Figure 5. Schematic of the internal LIL representation.

performing induction variable elimination (IVE). For instance, if the memory reference is via an induction variable, information about the induction variable such as the scale (sometimes called the *cee* value) and displacement (sometimes called the *dee* value) is

```
int cmp(a, b)
int a[], b[];
{
        int i;

        for (i = 0; i < 100; i++)
          if (a[i] != b[i])
              return(1);
        return(0);
}
```

a) C code with two induction expressions

```
_cmp:
1.    add    %o0,400,%o2
L16:
2.    ld     [%o0],%o3
3.    ld     [%o1],%o4
4.    cmp    %o3,%o4
5.    bne    L17
6.    add    %o0,4,%o0
7.    add    %o1,4,%o1
8.    cmp    %o0,%o2
9.    bl     L16
10.   mov    0,%o0
11.   retl
L17:
12.   mov    1,%o0
13.   retl
```

```
_cmp:
1.    sub    %o1,o0,%o1
2.    add    %o0,400,%o2
L16:
3.    ld     [%o0],%o3
4.    ld     [%o0 + %o1],%o4
5.    cmp    %o3,%o4
6.    bne    L17
7.    add    %o0,4,%o0
8.    cmp    %o0,%o2
9.    bl     L16
10.   mov    0,%o0
11.   retl
L17:
12.   mov    1,%o0
13.   retl
```

b) SPARC code produced by a high-level code improver

c) SPARC code produced by a low-level code improver

Figure 6. High-level versus low-level code improvement.

maintained. This structure also contains information that allows the code improver to resolve potential aliasing problems.

The above structure is very flexible and supports the implementation of all common, and some not so common, code improvements. The following section describes one of these code improvements in detail with emphasis on how it is accomplished in a machine-independent way, yet takes into account machine-dependent information.

4 Machine-Dependent Induction Variable Elimination

An induction variable is a variable that is used to induce a sequence of values. In the code of Figure 6a, the variable i is an induction variable because it is being used to induce a series of addresses (those of the array elements). If this is the only use of the variable, it can be beneficial to eliminate the induction variable altogether and just compute the sequence of addresses. The sequence of values being computed from the induction variable is called the *induced sequence*.

Simple induction variables are used to compute induced sequences of the form $scale \times i + displacement$, where i is the basic induction variable. In the example in Figure 6a, the sequences being computed are $4 \times i + a$ and $4 \times i + b$, where a and b are the starting addresses of the arrays. Using well known algorithms [1], the induction variable i can be eliminated and be replaced by the computation of the induced sequence of the addresses. The SPARC code produced by a code improver operating on a HIL is shown in Figure 6b. Notice that the sequences of addresses are being computed using two registers. The sequence for referencing a is being computed in %o0, and the sequence for b is being computed in register %o1. As argued in Section 2, no code improvement is really machine independent. Better IVE can be performed if it is done on a LIL where target machine information is available. The loop in Figure 6c is one instruction shorter than the loop in Figure 6b. On the SPARC, machine-dependent IVE results in one instruction being saved for every induction expression that can be computed via a difference from the basic induction variable. A systematic inspection of source code shows that approximately 22 percent of loops with induction variables contain multiple references using the same basic induction variable. Figure 7 contains a high-level description of the algorithm that is used to perform IVE on RTLs.

As the algorithm is explained, one key point should be kept in mind: the algorithm is machine-independent! That is, no changes are necessary to it when a new machine is accommodated. The algorithm obtains needed machine-dependent information via calls to a small set of machine-dependent routines that are constructed automatically from a description of the target architecture. These calls are underlined. This is a subtle point, but very important. It is possible to implement code improvements in a machine-independent way, yet take into account machine-dependent information.

Basic information needed to perform IVE is collected (lines 2-4). This includes the loop invariant values, the basic induction variables, and the induction expressions. The induction expressions are expressions that involve the use of the same basic induction variable. The list of induction expressions include the basic induction variables. This information is stored with the loop information that is accessible from each block in the loop.

The *while* loop starting at line 5 processes each of the induction expressions. For a particular induction expression, all induction expressions that depend on that one are collected into a list at lines 8 and 9. This list is sorted by the machine-dependent routine, *OrderInductionExprs* according to the capabilities of the target machine. For example, if the target machine allows only positive offsets in the displacement addressing mode, it is best to have the expression in order of increasing value. On the other hand, if the machine has a limited range of offset, yet supports both negative and positive offsets, the list should be ordered so that expressions in the middle of the range are first so smaller offsets can be employed.

```
 1   proc ImproveInductionExprs(LOOP) is
 2     LOOP.InvariantVals = FindLoopInvariantVals(LOOP)
 3     LOOP.InductionVars = FindInductionVars(LOOP, LOOP.InvariantVals)
 4     LOOP.InductionExprs = FindInductionExprs(LOOP, LOOP.InductionVars, LOOP.InvariantVals)
 5     while LOOP.InductionExprs ≠ ∅ do
 6       IND = FirstItem(LOOP.InductionExprs)
 7       EXPR = ∅
 8       for each E where E ∈ LOOP.InductionExprs ∧ E.Family = IND.Family ∧ E.Scale = IND.Scale do
 9         EXPR = EXPR ∪ E
10       endfor
11       OrderInductionExprs(EXPR)
12       IND = FirstItem(EXPR)
13       R = NewRegister(ADDRESS_TYPE)
14       if LOOP.Preheader = ∅ then
15         BuildPreheader(LOOP)
16       endif
17       InsertCalculation(LOOP.Preheader, "R = IND.Family × IND.Scale + IND.Displacement")
18       for each E where E ∈ EXPR do
19         DIFF = CalculateDifferenceExpression(E.Displacement, IND.Displacement)
20         UPDATED = FALSE
21         if DIFF = 0 then
22           NEW = ReplaceExpression(E.Inst, E, "R")
23           if IsValidInstruction(NEW) then
24             UPDATED = TRUE
25           endif
26         endif
27         if ¬UPDATED ∧ IsLiteralConstant(DIFF) then
28           NEW = ReplaceExpression(E.Inst, E, "R + DIFF")
29           if IsValidInstruction(NEW) then
30             UPDATED = TRUE
31           endif
32         endif
33         if ¬UPDATED ∧ IsLoopInvariant(DIFF, LOOP.InvariantVals) then
34           DR = NewRegister(ADDRESS_TYPE)
35           NEW = ReplaceExpression(E.Inst, E, "R + DR")
36           if IsValidInstruction(NEW) then
37             UPDATED = TRUE
38             InsertCalculation(LOOP.Preheader, "DR = E.Displacement - IND.Displacement")
39           endif
40         endif
41         if UPDATED then
42           ReplaceInstruction(E.Instruction, NEW)
43         endif
44         if UPDATED ∨ (DIFF = 0) then
45           LOOP.InductionExprs = LOOP.InductionExprs - E
46         endif
47       endfor
48     endwhile
49   endproc
```

Figure 7. High-level description of machine-dependent induction variable elimination algorithm.

To accommodate computing the induced sequences, a preheader is added to the loop at line 14 if one does not exist, and the machine instructions needed to generate the first value of the sequence are inserted in the preheader. This routine is machine-dependent because it must generate the LIL code that represents the target machine instructions needed to compute the value. The *for* loop at line 18 processes the induction expressions (including the first one selected outside the loop). Lines 21 through 40 of this loop determine, for each induction expression, the best way to compute the expression for the target machine. The difference between the displacement value of first induction expression selected and the current one is determined at line 19. If the difference is zero, then the register holding the induction value can be used. The routine, *ReplaceExpression*, replaces the reference to the expression with the reference to the register containing the induction value. This new instruction is checked to see whether it is valid on the target machine by the call to *IsValidInstruction* at line 23.

If the difference was not zero, the difference is checked at line 27 to see if it is a literal constant. If it is, then this expression can potentially be computed using a displacement address mode. An RTL expression is constructed at line 28 and is substituted for the expression. This new instruction is checked to see whether it is a valid machine operation. Whether it is a target machine instruction depends on the addressing modes supported by the target machine and the size of the displacement.

If the previous two alternatives do not succeed, the difference is checked to determine if it is loop invariant. If it is, then the induction expression potentially can be computed by adding the difference to the basic induction variable. If the target machine supports this addressing mode, a calculation is placed in the loop preheader to compute the difference. In the example in Figure 6c, the difference between the starting addresses of a and b is calculated by the instruction at line 1 of Figure 6c. The induction expression is replaced with this register plus register computation.

If one of the alternatives succeeds, then *UPDATED* will be true, and the instruction that used the induction expression will be replaced by the new instruction at line 41. If this induction expression can be calculated from the induction expression selected at line 12, it is removed from the list of induction expressions. If not, it will be handled by a subsequent iteration of the *while* loop. That is, a single register will be allocated to be used to induce the sequence. After the algorithm completes, a pass is made over the loop and instruction selection is repeated on all changed instructions. This ensures that the most efficient target machine instructions are used. Again, this is an advantage of applying code improvements to a LIL. This pass also updates use-def chain information.

The algorithm is guaranteed to terminate because the initial set of induced expressions in *LOOP.InductionExprs* is finite and because line 45 removes at least one item from this set each time through the *while* loop. Note that *IND* represents one of the elements in the set *EXPR* upon entry to the *for* loop at line 18. Consequently, there will be at least one case where the difference value calculated at line 19 will be zero. Thus, even if *UPDATED* is not set for any of the items in the *EXPR* set, the condition in line 44 will be true for at least the item with the zero difference.

Name	Description	Source	Type	Lines
cache	Cache simulation	User code	I/O, Integer	820
compact	Huffman file compression	UNIX utility	I/O, Integer	490
diff	Text file comparison	UNIX utility	I/O, Integer	1,800
eqntott	PLA optimizer	SPEC benchmark	CPU, Integer	2,830
espresso	Boolean expression translator	SPEC benchmark	CPU Integer	14,830
gcc	Optimizing compiler	SPEC benchmark	CPU, Integer	92,630
li	LISP interpreter	SPEC benchmark	CPU, Integer	7,750
linpack	Floating-point benchmark	Synthetic benchmark	CPU, FP	930
mincost	VLSI circuit partitioning	User code	CPU, FP	500
nroff	Text formatting	UNIX utility	I/O, Integer	6,900
sort	File sorting and merging	UNIX utility	I/O, Integer	930
tsp	Traveling salesperson problem	User code	CPU, Integer	450

Table I. Benchmark programs.

An astute reader will note that this algorithm does not necessarily produce the best code sequences for all possible architectures. For example, on a machine with an indexed displacement mode [*base_reg* + *index_reg* + *displacement*], the algorithm would not realize that only one index register needs to be incremented and that the difference between the base address of two arrays is not needed. This deficiency, however, can be easily overcome by adding a test for this addressing mode similar to those at lines 27 and 33. This additional test does not prevent the algorithm from working on architectures that do not have an indexed displacement mode, but produces better code for the machines that do. This extensibility greatly simplifies the task of retargeting the compiler and often allows the effort invested in improving the code for one architecture to be amortized across many machines.

5 Results

A retargetable optimizing C compiler has been constructed with the structure shown in Figure 1b that operates on the LIL described in Section 3.2. The compiler is fully operational for six architectures.† These are:

- VAX-11
- Intel 80386
- MIPS R2000/R3000
- Motorola 68020
- Sun SPARC
- Motorola 88100

To determine the effectiveness of machine-dependent IVE, the SPARC architecture was chosen as the platform to run experiments. A set of experiments were performed using the benchmark programs described Table I. This set of programs includes the four C programs in the SPEC suite [16] along with some common Unix utilities and user code. Together the programs comprise approximately 130,000 lines of source code.

---

†Actually over ten different architectures have been accommodated, but only these six are maintained.

Program Name	Percent Speedup with Machine-Independent IVE (Column A)	Percent Speedup with Machine-Dependent IVE (Column B)	Column b - Column a (Column C)
cache	-2.39	-0.08	2.31
compact	0.58	3.74	3.16
diff	-3.26	0.56	3.82
eqntott	-4.05	3.68	7.53
espresso	-0.78	-8.51	-7.73
gcc	-1.39	-0.88	0.51
iir	25.30	40.00	14.70
li	-5.20	-0.80	4.40
linpack	-7.43	1.34	8.86
mincost	-1.97	3.49	5.46
nroff	-3.46	0.80	4.26
sort	0.37	4.02	3.65
tsp	3.98	4.27	0.29

Table II. Comparison of the effectiveness of machine-independent and machine-dependent induction variable elimination on the SPARC2.

The first experiment determined the overall effectiveness of IVE. The programs in Table I were compiled with and without IVE enabled. For the runs with IVE enabled the machine-dependent aspects of the algorithm were disabled effectively making it mimic a high-level machine-independent implementation of IVE. The resulting executables were run five times on a lightly loaded SPARC2 and an average execution time was computed. From this average, the speedup due to machine-independent IVE was computed (see the Column A of Table II). Surprisingly, most programs slowed down with machine-independent IVE enabled. Because the effect was most pronounced for *linpack*, the code for this program was examined to determine what was happening. Most of *linpack*'s execution time is spent in *daxpy*. Comparison of the two versions of this loop revealed why machine-independent IVE ran slower. Without IVE, the loop was 9 instructions long. With machine-independent IVE, the loop was also 9 instructions, but the preheader contained instructions that copied the addresses of the arrays to temporaries, and computed the value needed to test against for loop termination ($dx + 400 \times n$). Because the routine is called tens of thousands of times during the course of a run, the extra overhead lowered performance. For one program, *iir*, machine-independent IVE showed a large benefit. Inspection of this code revealed that this was because IVE produced an opportunity for recurrence detection and optimization [2] to take effect, and a large percent of the benefit was from this improvement. These results confirm that it is difficult to apply code improving transformations to a HIL because the cost/benefit analysis is so dependent on the target machine.

To determine the effectiveness of machine-dependent IVE, the same programs were compiled and run, but this time the machine-dependent aspects of the IVE algorithm were enabled. Column B of Table II shows the speedup when machine-dependent IVE was performed compared to when no IVE was performed. The improvement due to machine-dependent IVE is similar to that reported elsewhere in the literature averaging two or three percent [14].

The one anomaly in Column B is the serious loss of performance for *espresso*. Using a measurement tool called *ease* [9], the execution behavior of the three versions of this program was examined. First, it was observed that several of the routines that were called frequently had loops with very low iteration counts (50% of the loops in these routines had iteration counts of less than two). This explained why IVE was producing poor results. The preheader overhead was not being offset by savings in the loops. However, this did not explain why machine-dependent IVE, with smaller preheader loop overhead, ran slower than machine-independent IVE. The measurement tool revealed that the version of the program produced by compiling the program with machine-dependent IVE performed fewer instructions (less preheader overhead), but more memory references than the version produced by compiling it with machine-independent IVE. Inspection of the optimized loops showed that because the loop was tighter (i.e. fewer instructions), the scheduler had, in order to fill the delay slot of the branch at the end of the loop, resorted to using an annulled branch and had placed a load in the delay slot and replicated it in the preheader. Apparently, these extra (useless) loads caused performance to suffer.

Column C shows the performance difference in machine-independent IVE and machine-dependent IVE. For all but the anomalous *espresso*, performing IVE at a low level where machine-specific information is available appears to be worthwhile, and performs better than machine-independent IVE. Experience with the compiler indicates that other code improvements yield similar benefits when applied at a low-level.

These experiments show, in general, that any single code improvement will affect only a subset of the programs to which it is applied. For some programs the effect will be small and for others it will be large. Thus, a good optimizing compiler uses a collection of code improvements where each transformation produces a small benefit most of the time and a large benefit occasionally. The results also show how difficult it is to measure the effects of a code improvement. Each code improvement can affect what another does and it sometimes difficult to isolate the effect of a single transformation.

To determine the compilation times between a production compiler structured as shown in Figure 1a and one using the structure in Figure 1b, the amount of time spent in the middle end was measured and compared against the amount of time required to perform the entire compilation process. Obtaining these measurements for the benchmark suite shown in Table I revealed that the extra translation step from HIL to LIL increases the compilation time of the compiler by an average of 3.1%. This value ranged from a low of 1.9% for the *linpack* program to a high of 5.5% for *li*. This slight increase in compilation time is the primary disadvantage of using a structure that utilizes both an HIL and a LIL.

6 Summary

To be applied most effectively, most global optimizations require information about the target machine. For those few transformations where this is not true, it is likely that they interact with those that do and thus, effectively, they are also machine dependent. This paper has described the structure of a compiler that is designed so that code improvomonto oan bo appliod whon machine specific information is available. The compiler has two intermediate representations: one that is a target for intermediate code generation, and a second one that is designed to support the machine-specific application of global code improvements such as code motion, induction variable elimination, and constant propagation.

Using one transformation as an example, this paper showed that it is possible to implement global code improvements that operate on a LIL representation of the program and that it is beneficial to do so. The implementation of the algorithm is itself kept machine-independent by carefully isolating the access to target-specific information via a few routines that can be generated automatically from a specification of the target architecture. The results presented show that the benefits of such a structure are worth the effort in spite the modest compilation time penalty that it incurs.

Acknowledgments

This work was supported in part by National Science Foundation grant CCR-9214904.

References

1. Aho, A. V., Sethi, R., and Ullman, J. D., *Compilers Principles, Techniques and Tools*, Addison-Wesley, Reading, MA, 1986.
2. Benitez, M. E., and Davidson, J. W., "Code Generation for Streaming: an Access/ Execute Mechanism", *Proceedings of the Fourth International Symposium on Architectural Support for Programming Languages and Operating Systems*, Santa Clara, CA, April 1991, pp. 132—141.
3. Chow, F. C., *A Portable Machine-Independent Global Optimizer—Design and Measurements*, Ph.D. Dissertation, Stanford University, 1983.
4. Cytron, R., Ferrante, J., Rosen, B. K., Wegman, M. N., and Zadeck, F. K., "Efficiently Computing Static Single Assignment Form and the Control Dependence Graph", *ACM Transactions on Programming Languages and Systems*, 13(4), October 1991, pp. 451—490.
5. Davidson, J. W. and Fraser, C. W., "Register Allocation and Exhaustive Peephole Optimization", *Software—Practice and Experience*, 14(9), September 1984, pp. 857—866.
6. Davidson, J. W. and Fraser, C. W., "Code Selection Through Peephole Optimization", *Transactions on Programming Languages and Systems*, 6(4), October 1984, pp. 7—32.

7. Davidson, J. W., "A Retargetable Instruction Reorganizer", *Proceedings of the '86 Symposium on Compiler Construction*, Palo Alto, CA, June 1986, pp. 234—241.

8. Davidson, J. W. and Gresh, J. V., "Cint: A RISC Interpreter for the C Programming Language", *Proceedings of the ACM SIGPLAN '87 Symposium on Interpreters and Interpretive Techniques*, St. Paul, MN, June 1987, pp. 189—198.

9. Davidson, J. W. and Whalley, D. B., "A Design Environment for Addressing Architecture and Compiler Interactions", *Microprocessors and Microsystems*, **15**(9), November 1991, pp. 459—472.

10. Joy, William N. and McKusick, M. Kirk, "Berkeley Pascal PX Implementation Notes Version 2.0—January, 1979", Department of Engineering and Computer Science, University of California, Berkeley, January 1979.

11. McFarling, S., "Procedure Merging with Instruction Caches", *Proceedings of the ACM SIGPLAN '91 Symposium on Programming Language Design and Implementation*, Toronto, Ontario, June 1991, pp. 71—79.

12. Nelson, P. A., "A Comparison of PASCAL Intermediate Languages", *Proceedings of the SIGPLAN Symposium on Compiler Construction*, Denver, CO, August 1979, pp. 208—213.

13. Newey, M. C., Poole, P. C., and Waite, W. M., "Abstract Machine Modelling to Produce Portable Software—A Review and Evaluation", *Software—Practice and Experience*, **2**, 1972, pp. 107—136.

14. Powell, M. L., "A Portable Optimizing Compiler for Modula-2", *Proceedings of the SIGPLAN '84 Symposium on Compiler Construction*, Montreal, Canada, June 1984, pp. 310—318.

15. Richards, M., "The Portability of the BCPL Compiler", *Software—Practice and Experience*, **1**(2), April 1971, pp. 135—146.

16. Systems Performance Evaluation Cooperative, Waterside Associates, Fremont, CA, 1989.

17. Tanenbaum, A. S., Staveren, H. V., and Stevenson, J. W., "Using Peephole Optimization on Intermediate Code", *Transactions on Programming Languages and Systems*, **4**(1), January 1982, pp. 21—36.

18. Tanenbaum, A. S., Staveren, H. V., Keizer, E. G., and Stevenson, J. W., "A Practical Tool Kit for Making Portable Compilers", *Communications of the ACM*, **26**(9), September 1983, pp. 654—660.

Dependence-Conscious Global Register Allocation

Wolfgang Ambrosch   M. Anton Ertl   Felix Beer   Andreas Krall

Institut für Computersprachen
Technische Universität Wien
Argentinierstraße 8, A-1040 Wien
{anton,fbeer,andi}@mips.complang.tuwien.ac.at
Tel.: (+43-1) 58801 {4459,3036,4462}

Abstract. Register allocation and instruction scheduling are antagonistic optimizations: Whichever is applied first, it will impede the other. To solve this problem, we propose dependence-conscious colouring, a register allocation method that takes the dependence graph used by the instruction scheduler into consideration. Dependence-conscious colouring consists of two parts: First, the interference graph is built by analysing the dependence graphs, resulting in fewer interference edges and less spilling than the conventional preordering approach. Second, during colouring the register selection keeps dependence paths short, ensuring good scheduling. Dependence-conscious colouring reduces the number of interference edges by 7%–24% and antidependences by 46%–100%.

1   Introduction

Global register allocation and instruction scheduling are two standard compiler techniques. Register allocation reduces the traffic between the processor and memory by keeping frequently-used variables in registers. Instruction scheduling reduces the number of pipeline stalls (wait cycles) by reordering instructions.

However, these techniques are antagonistic: Instruction scheduling tends to move dependent instructions apart. This lengthens the lifetimes of the values and increases register pressure, which in turn may cause more memory traffic. On the other hand, the register allocator can assign the same register to two different temporary values. This can reduce the opportunities for reordering instructions and it can increase the number of pipeline stalls. So, the technique that is applied first will reduce the effectivity of the other technique. This problem is especially important for pipelined and superscalar implementations of register-starved architectures, e.g., the Pentium and the 68060.

As an example, consider Figure 1: Conventional register allocation before scheduling can introduce dependences that cause bad scheduling. The last instruction stalls for two cycles waiting for the result of the multiply. Scheduling before allocation uses more than the four available registers (the other registers hold global values) and causes spilling to memory.

To solve this dilemma, Goodman and Hsu developed DAG-driven register allocation, a local register allocation algorithm that avoids introducing additional dependences if possible [GH88]. Inspired by DAG-driven register allocation, we created dependence-conscious colouring, a global register allocator that takes its effect on

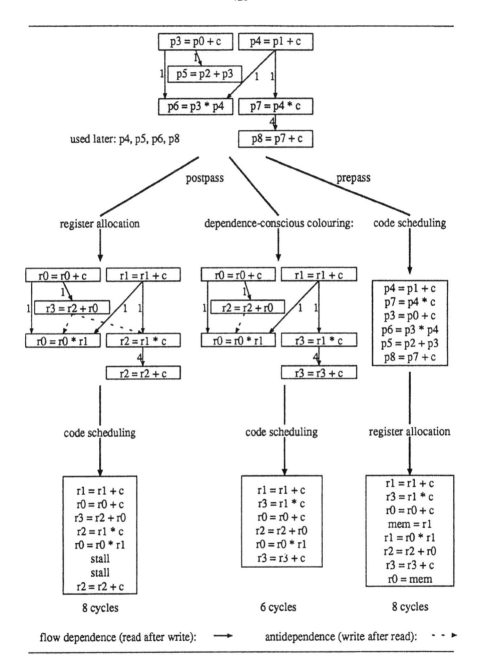

Fig. 1. Postpass scheduling vs. dependence-conscious colouring vs. prepass scheduling

scheduling into account. Figure 1 (middle) shows how this method can avoid ineffi-
ciencies: The register allocator is aware of the dependence graph and avoids introducing expensive dependences by selecting the right registers. Note that, while this
example considers only one basic block, dependence-conscious colouring is a global
register allocator.

The rest of the paper explains the data dependence graphs used in scheduling
(Section 2), graph colouring register allocation (Section 3) and the two innovations
of dependence-conscious colouring: the minimal interference graph (Section 4) and
dependence-conscious register selection (Section 5). Finally, preliminary results are
presented and dependence-conscious colouring is compared with other work.

2 The Data Dependence Graph

Current RISC processors achieve their high performance by exploiting parallelism
through pipelining. As a consequence, the results of previous instructions are sometimes not available when the next instruction can be executed. If the next instruction
uses the result, it has to wait and the pipeline stalls. The problem of arranging the
instructions in a way that reduces the number of wait cycles is known as instruction
scheduling. In this paper we consider only instruction scheduling within basic blocks.

The basic data structure for instruction scheduling is the dependence graph
[GM86]. Figure 1 shows (several variations of) a dependence graph. An edge from
instruction a to instruction b indicates that a must be executed before b to preserve
the correctness of the overall program. Dependence edges must be drawn from writes
to reads of the same register or memory location (flow dependences), from reads to
writes (anti dependences), and from writes to writes (output dependences). The dependence graph is essentially the expression evaluation graph (drawn up-side-down),
with some edges added due to dependencies between memory accesses. Register allocation can add antidependences (write-after-read dependences) by allocating the
same register to several live ranges.

Path lengths of the dependence graph play an important role in instruction scheduling: There can be no schedule that is shorter than the critical path length. The
edge length of a flow (read-after-write) dependence is the latency of the parent
instruction. The length of an antidependence is zero or one cycle. However, if it
connects two long paths, an antidependence can increase the critical path length
and the execution time. Therefore, antidependences should be avoided or placed
well.

3 Graph Colouring Register Allocation

The compiler front end and the optimizer can use an infinite number of live ranges (also known as pseudoregisters) for the variables and temporary values of the
program. The task of register allocation is to map these live ranges onto a finite register set. The standard approach to register allocation is graph colouring
[CAC+81, Cha82, BCKT89, CH90, Bri92b]. As the basis for dependence-conscious
colouring, we used the algorithm presented in [BCKT89]. Figure 2 presents the phases of a graph colouring register allocator.

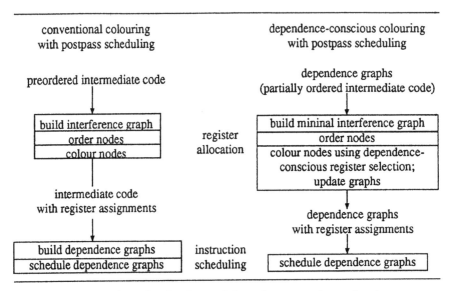

Fig. 2. Conventional graph colouring and dependence-conscious colouring

3.1 The Interference Graph

The basic data structure used by graph colouring register allocators is the interference graph. Every live range is represented by a node. There is an edge between two nodes if the live ranges interfere, i.e. if they must not stay in the same register, because they overlap. So, the problem of assigning registers to live ranges is mapped into the problem of "colouring" the nodes of the interference graph with registers such that directly connected (i.e., interfering) nodes do not get the same "colour" (register).

3.2 Colouring

Once the interference graph is complete, registers can be allocated in the following way: select an uncoloured node and give it a register that differs from the registers of the neighbours. There are several open points in this algorithm:

- In what order are the nodes coloured? Since dependence-conscious colouring does not differ from other colouring register allocators in this respect, the ordering will not be discussed here.
- What register is given to the live range? For register allocation purposes, it does not make much of a difference, which of the legal registers is selected for colouring a node [Bri92a]. However, for scheduling it makes a big difference (see Section 5).
- What happens if there is no register available for a live range? The live range has to be spilled into memory. Dependence-conscious colouring does not differ from other colouring register allocators in this respect.

4 The Minimal Interference Graph

The right side of Figure 2 shows the overall structure of dependence-conscious colouring.

In conventional graph colouring, the interference graph is computed from totally ordered code. In a postpass approach, this ordering usually is the coincidental result of some earlier phase. It could also be produced by a register friendly scheduler. In any case, the ordering will cause some interference edges that need not be valid for the final schedule. These edges needlessly restrict register allocation.

To avoid this problem, dependence-conscious colouring computes the interference edges of a basic block from its dependence graph. I.e., the notions of "before" and "after" in a totally ordered basic block are replaced by the data dependence relation, which is a partial ordering.

4.1 Building the Minimal Interference Graph

The conventional method of computing live information and computing the interferences from that cannot be generalized straightforwardly to dependence graphs. Therefore we go back to the roots: Two live ranges interfere, if one is defined before the other is used and the other is defined before the first one is used. Formally:

$$interfere\text{-}with(l) = used\text{-}later\text{-}out(definition(l)) \cap \bigcup_{i \in uses(l)} defined\text{-}earlier\text{-}in(i)$$

$interfere\text{-}with(l)$ is the set of live ranges that interferes with the live range l in the basic block; $definition(l)$ is the node of the dependence graph where l is defined; $uses(l)$ is the set of nodes where l is used. The $used\text{-}later$ and $defined\text{-}earlier$ information can be computed by applying data flow analysis techniques to the data dependence graph (Figure 3 shows an example):

$$used\text{-}later\text{-}out(i) = \bigcup_{j \in successors(i)} used\text{-}later\text{-}in(j)$$
$$used\text{-}later\text{-}in(i) = used\text{-}later\text{-}out(i) \cup use(i)$$
$$defined\text{-}earlier\text{-}in(i) = \bigcup_{j \in predecessors(i)} defined\text{-}earlier\text{-}out(j)$$
$$defined\text{-}earlier\text{-}out(i) = defined\text{-}earlier\text{-}in(j) \cup def(i)$$

This method is only used for computing the interferences within basic blocks. Conventional data flow techniques are used for computing the global interferences, i.e., interferences between variables that are live at the same control flow graph edges. In order to get correct interference edges for non-local live ranges that become live or dead in the basic block, we insert into every dependence graph a top node $\top$ that is the definition point of all live ranges accessed in the basic block that are defined before the start of the basic block, and a bottom node $\bot$ that is a use point of live ranges accessed in the basic block that are used after end of the basic block.

The interference computation given above is only correct if the live range is contiguous within the basic block (otherwise it computes too many interferences).

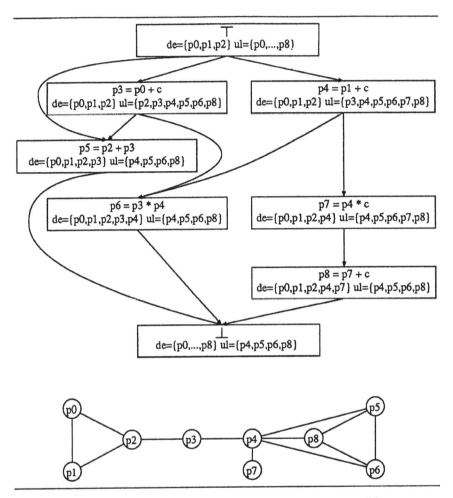

Fig. 3. Dependence graph of Figure 1 with *defined-earlier-in* and *used-later-out* sets and the resulting interference graph

Therefore, two definitions should be treated as belonging to two separate live ranges, even if they belong to the same global live range. There is one exception: If the second definition is in the same instruction as a use of the live range, the live range is contiguous in the basic block and should be treated as one live range.

Using this method, we get all interference edges that will be valid for every schedule and only those. The complexity for computing all interferences of a basic block is $O(e + u)$, where e is the number of dependence edges and u the number of register uses in the basic block.

131

4.2 Maintaining the Minimal Interference Graph

During colouring, the same register may be assigned to several live ranges in the basic block. Similarly, during coalescing[1] separate live ranges can be united. These actions may cause additional interferences:

Assigning the same register to different live ranges introduces antidependence edges in the dependence graph. If the antidependences are not redundant[2], new *defined-earlier* and *used-later* information may be propagated through them, possibly resulting in new interference edges. In other words, due to the antidependences the scheduler has less freedom to adapt to the register allocation, therefore the register allocator also has less freedom. In the example of Figure 3, allocating p8 to the same register as p2 causes an antidependence from p5=p2+p3 to p8=p7+c, which causes an interference between p5 and p7. Our prototype implementation handles these cases by recomputing the interference edges for the whole basic block, but we are investigating incremental recomputation. Fortunately, our register selection algorithm (see Section 5) avoids introducing non-redundant antidependences if possible.

Coalescing two live ranges introduces interferences between the united live range and live ranges that before coalescing could be assigned the same register as either of the coalesced live ranges. This can also occur if the same register is assigned to two live ranges and the definition of the second live range is in the same instruction as a use of the first. The interferences for the combined live range can be computed by

$$interfere\text{-}with(l_{12}) = used\text{-}later\text{-}out(definition(l_1)) \cap \bigcup_{i \in uses(l_2)} defined\text{-}earlier\text{-}in(i)$$

where l_{12} is the combined live range, l_1 is the first one and l_2 is the second.

Note that Chaitins original algorithm cannot handle adding interference edges during colouring, because it performs spill decisions before colouring. However, Briggs' modification [BCKT89] can handle it.

5 Dependence-Conscious Register Selection

For register allocation, it makes little difference, which of the available registers is selected [Bri92a]. But for the instruction scheduler the difference is important: Antidependences introduced by colouring can produce long dependence paths, which result in bad scheduling.

Therefore, dependence-conscious colouring carefully selects the registers. Colouring a live range with a register should introduce no antidependences or only redundant ones. If everything else fails, the introduced antidependences should connect only short paths. The problem is complicated by the fact that several basic blocks have to be considered at the same time: all basic blocks where the live range is born or dies. For all allowed registers the cost of the dependences introduced by selecting

---

[1] Coalescing is an optimization that is performed immediately after interference graph construction. It eliminates copies.

[2] A dependence between two instructions is redundant, if the ordering between the instructions is already enforced by other dependences.

the register is computed over all basic blocks; the register with the lowest weighted sum of costs is selected. The weights are the expected execution frequencies of the basic blocks.

The cost of a dependence d is the expected increase in the execution time of the basic block due to adding the dependence. It is 0 for redundant dependences. For non-redundant dependences we use the following cost function:

$$cost(d) = \max(\frac{path\text{-}length(d)}{expected\text{-}time}, path\text{-}length(d) - expected\text{-}time)$$

$path\text{-}length(d)$ is the length of the longest path containing the dependence d. It can be computed in constant time, if the earliest issue and finish times [GH88] are precomputed. $expected\text{-}time$ is the expected execution time of the basic block before adding the dependence:

$$expected\text{-}time = \max(cycles, critical\text{-}path\text{-}length)$$

where $cycles$ is the number of cycles the basic block would need if its instructions were independent (i.e., the naive expected execution time), and $critical\text{-}path\text{-}length$ is the critical path length of the dependence graph of the basic block before adding the antidependence.

The cost of of an edge d is quite low as long as $path\text{-}length(d)$ is smaller than $expected\text{-}time$, but is high otherwise.

E.g., consider the basic block in Figure 3 after assigning registers to p4, p5, p6 and p8. Figure 4 shows the costs of antidependences introduced by selecting a register for p3. Since p3s lifetime is restricted to that basic block, these are the total costs and r1 or r2 is selected for p3.

register	antidependences	path lengths	costs
r0 (p4)	–	–	interferes
r1 (p5)	p6=p3*p4 → p5=p2+p3	2	0.33
r2 (p6)	p5=p2+p3 → p6=p3*p4	2	0.33
r3 (p8)	p5=p2+p3 → p8=p7+c, p6=p3*p4 → p8=p7+c	2, 2	0.67

Fig. 4. The costs of selecting a register for p3

This selection process may seem to be expensive. But graph colouring register allocation is dominated by the time for building the interference graph (e.g., 90% of the register allocation time in [BCKT89]), so making the colouring slower does not make much of a difference for the whole algorithm.

Register selection is independent of the interference graph building method described in Section 4. Either method can be used separately.

6 Preliminary Results

We implemented a prototype dependence-conscious colouring allocator by modifying the register allocator of a C compiler for the Mips R3000. Since the C compiler was under development, the results are preliminary. They are shown in Table 1. We are currently reimplementing dependence-conscious colouring in the finished compiler.

We compiled two programs, a fast Fourier transform (FFT) and the Dhrystone integer benchmark (Dhry). We compared a conventional colouring register allocator ([BCKT89]) to dependence-conscious colouring (DCC). In both cases the compiler schedules after the register allocation (postpass scheduling). The programs consist of several procedures that are compiled one at a time. The presented data is cumulated.

| program | FFT | | Dhry | |
register allocator	[BCKT89]	DCC	[BCKT89]	DCC
initial interference edges	1024	948	1383	968
additional interference edges		0		81
all antidependences	407	0	713	382
redundant antidependences		0		23

Table 1. Results

As expected, dependence-conscious colouring produces fewer interference edges while building the interference graph (initial interference edges). During colouring, dependence-conscious colouring inserts additional edges in the interference graph (additional interference edges), but there are still fewer interferences than from preordered code. Both allocators do not produce spill code. The conventional register allocator introduces a considerable number of antidependences. Dependence-conscious register allocation produces no antidependences for FFT and halves Dhry's antidependences. This means that for FFT dependence-conscious colouring achieves the best behaviour possible: no spilling and full scheduling freedom. Unfortunately we do not have speedup numbers, since the compiler back end was still under development and did not produce fully functional code.

7 Related Work

The standard approach to the problem is to more or less ignore it. Scheduling is performed either before register allocation (prepass, [AH82]) or afterwards (postpass, [HG83]). With the prepass approach, scheduling has to be repeated after register allocation to schedule spill code. Prepass schedulers usually employ a register-saving heuristic, but only as low-priority secondary heuristic. Our approach is postpass scheduling, but our register allocator takes scheduling into consideration.

The Harris C compiler uses postpass scheduling, but reallocates registers during scheduling to remove harmful dependences [Beu92]. In contrast, our register allocator tries to do it right the first time.

In [GH88] two techniques for integrating local register allocation and instruction scheduling are introduced: Integrated prepass scheduling switches between scheduling for pipelining and scheduling for register allocation based on the number of available registers. DAG-driven register allocation tries to avoid introducing long paths into the dependence graph. The performance of both methods is about equal, with DAG-driven register allocation being simpler. Dependence-conscious colouring can be seen as the global version of DAG-driven register allocation.

[BEH91] presents integrated prepass scheduling in a global register allocation setting and introduces register allocation with schedule estimates (RASE). In RASE, the global register allocator leaves a number of registers to the local allocator, which also performs instruction scheduling. The number of registers left for local allocation is determined by the estimated costs of scheduling with x registers. The estimates are computed from practice runs of the scheduler. RASE and integrated prepass scheduling are about equal in code quality; integrated prepass scheduling is simpler. Like RASE, dependence-conscious colouring makes the register allocator aware of scheduling. But dependence-conscious colouring does all of the register allocation, including local allocation. It directly sees the data dependence graphs and the effects of allocation decisions on it instead of just heeding a register limit.

[Pin93] proposes using postpass scheduling with a modified register allocator. The register allocator uses an interference graph that contains all interference edges that could be introduced by scheduling. I.e., even more interference edges and more spilling than in a prepass scheduling approach. Pinter proposes heuristics for removing edges from that interference graph to avoid excessive spilling, but does not give results. In contrast, dependence-conscious colouring uses a minimal interference graph to minimize spilling and preserves scheduling freedom through its register selection heuristics.

[PF91] gives an optimal algorithm. Unfortunately it solves a very limited and unrealistic problem: scheduling and stupid register allocation for binary expression trees with single-delay-slot loads at the leaves. I.e., no unary operators, no constants or register variables, no common subexpression elimination and the algorithm is restricted to one expression. In contrast, dependence-conscious colouring does not have these restrictions and performs global register allocation.

In [FR91] instruction scheduling and register allocation are performed tracewise[3], starting with the most frequently executed trace. This approach is similar to coagulation [Mor91]. In contrast to dependence-conscious colouring, the first phase does not consider the needs of the second; instead, the phases are interleaved, so possible problems are pushed into low-frequency code.

[RLTS92] discusses register allocation for globally scheduled loops. In contrast, dependence-conscious register allocation can handle general control structures, but is restricted to basic block instruction scheduling.

8 Conclusion

Dependence-conscious colouring is a global register allocation method based on graph colouring, that takes the needs of instruction scheduling into account. It consists of two independent techniques:

---

[3] Traces are parts of possible execution paths. You can think of them as multiple-entry, multiple-exit basic blocks.

- The *minimal interference graph* is built by analysing dependence graphs of basic blocks instead of preordered code. This eliminates unnecessary interference edges, making colouring easier.
- *Dependence-conscious register selection* is aware of the antidependences that it can introduce. It keeps dependence graph path lengths short by selecting the right registers for assignment. This provides for good instruction scheduling.

Dependence-conscious colouring reduced the number of interference edges by 7%–24% and antidependences by 46%–100% in the two programs we measured.

Acknowledgements

We want to thank the referees, Bjarne Steensgaard, and Franz Puntigam for their helpful comments on earlier versions of this paper.

References

[AH82]   Marc Auslander and Martin Hopkins. An overview of the PL.8 compiler. In SIGPLAN '82 [SIG82], pages 22–31.

[BCKT89] Preston Briggs, Keith D. Cooper, Ken Kennedy, and Linda Torczon. Coloring heuristics for register allocation. In *SIGPLAN '89 Conference on Programming Language Design and Implementation*, pages 275–284, 1989.

[BEH91]   David G. Bradlee, Susan J. Eggers, and Robert R. Henry. Integrating register allocation and instruction scheduling for RISCs. In *Architectural Support for Programming Languages and Operating Systems (ASPLOS-IV)*, pages 122–131, 1991.

[Beu92]   Paul Beusterien. Personal communication, 1992.

[Bri92a]   Preston Briggs. Personal communication, 1992.

[Bri92b]   Preston Briggs. *Register Allocation via Graph Coloring.* PhD thesis, Rice University, Houston, 1992.

[CAC+81] Gregory J. Chaitin, Marc A. Auslander, Ashok K. Chandra, John Cocke, Martin E. Hopkins, and Peter W. Markstein. Register allocation via coloring. *Computer Languages*, 6(1):45–57, 1981. Reprinted in [Sta90].

[CH90]   Fred C. Chow and John L. Hennessy. The priority-based coloring approach to register allocation. *ACM Transactions on Programming Languages and Systems*, 12(4):501–536, October 1990.

[Cha82]   G. J. Chaitin. Register allocation & spilling via graph coloring. In SIGPLAN '82 [SIG82], pages 98–105.

[FR91]   Stefan M. Freudenberger and John C. Ruttenberg. Phase ordering of register allocation and instruction scheduling. In Robert Giegerich and Susan L. Graham, editors, *Code Generation — Concepts, Tools, Techniques*, Workshops in Computing, pages 146–170. Springer, 1991.

[GH88]   James R. Goodman and Wei-Chung Hsu. Code scheduling and register allocation in large basic blocks. In *International Conference on Supercomputing*, pages 442–452, 1988.

[GM86]   Phillip B. Gibbons and Steve S. Muchnick. Efficient instruction scheduling for a pipelined architecture. In *SIGPLAN '86 Symposium on Compiler Construction*, pages 11–16, 1986.

[HG83]   John Hennessy and Thomas Gross. Postpass code optimization of pipeline cons-
           traints. *ACM Transactions on Programming Languages and Systems*, 5(3):422–
           448, July 1983.

[Mor91]   W. G. Morris. CCG: A prototype coagulating code generator. In SIGPLAN '91
           [SIG91], pages 45–58.

[PF91]   Todd A. Proebsting and Charles N. Fischer. Linear-time, optimal code schedu-
           ling for delayed-load architectures. In SIGPLAN '91 [SIG91], pages 256–267.

[Pin93]   Shlomit S. Pinter. Register allocation with instruction scheduling: A new ap-
           proach. In *SIGPLAN '93 Conference on Programming Language Design and
           Implementation*, pages 248–257, 1993. SIGPLAN Notices 28(6).

[RLTS92]   B. R. Rau, M. Lee, P. P. Tirumalai, and M. S. Schlansker. Register allocation
           for software pipelined loops. In *SIGPLAN '92 Conference on Programming Lan-
           guage Design and Implementation*, pages 283–299, 1992.

[SIG82]   *SIGPLAN '82 Symposium on Compiler Construction*, 1982.

[SIG91]   *SIGPLAN '91 Conference on Programming Language Design and Implementa-
           tion*, 1991.

[Sta90]   William Stallings, editor. *Reduced Instruction Set Computers*. IEEE Computer
           Society Press, second edition, 1990.

Type Test Elimination using Typeflow Analysis

Diane Corney and John Gough*

Queensland University of Technology, Brisbane, Australia

Abstract. Programs written in languages of the *Oberon* family usually contain runtime tests on the dynamic type of variables. In some cases it may be desirable to reduce the number of such tests. Typeflow analysis is a static method of determining bounds on the types that objects may possess at runtime. We show that this analysis is able to reduce the number of tests in certain plausible circumstances. Furthermore, the same analysis is able to detect certain program errors at compile time, which would normally only be detected at program execution. This paper introduces the concepts of typeflow analysis and details its use in the reduction of runtime overhead in *Oberon-2*.

1 Introduction

Implementations of object-oriented languages based on type extension[1] such as *Oberon*[2] and *Oberon-2*[3] must perform tests on the dynamic type of objects. These type tests are comparable in runtime cost to the index bounds checks of Pascal-like languages[4]. Just as it is possible to eliminate much of the overhead of index bounds checking by careful static analysis of programs (see [5], for example, for a recent review), so it is possible to eliminate some of the runtime cost of type testing. In this paper we introduce the concept of typeflow analysis, which has some similarities to dataflow analysis, and explain how these concepts have been implemented to achieve this objective in an *Oberon-2* compiler.

Section 2 discusses type extension as a form of inheritance and introduces a notation for type extension and explains its use. Section 3 shows how conventional common subexpression elimination may be used to eliminate some type tests. However, we show that this method falls short of detecting some obvious redundancies, demonstrating the need for a more careful analysis.

A discussion is given in Section 4 of the concepts of typeflow analysis and its relation to control flow. The notion of a *manifest type* is introduced and its use in tracking variable types during this form of static analysis is explained. For different control flow constructs, the manner in which the manifest type may be updated is discussed, and examples given.

Section 5 discusses the testing of an implementation of typeflow analysis limited to some simple cases. Consideration is given as to how typeflow analysis might be fully implemented and utilised.

---

* This work is supported by the Cooperative Research Centre for Distributed Systems Technology

2 Inheritance and Type Extension

Type extension is a data structuring mechanism introduced by Niklaus Wirth in the language *Oberon*. It allows record types to be declared which extend previously declared record types by the addition of extra fields. This mechanism is a form of single inheritance and allows a type hierarchy to be constructed.

2.1 Notation

If there is a type T_0, which we shall call a base type, and T_1 is an extension of T_0, then we denote this $T_1 \triangleright T_0$. We say that T_1 is a *direct extension* of T_0. However, T_1 could then be further extended, so that $T_2 \triangleright T_1$, and so on. The reflexive transitive closure of the extends relation is denoted $\triangleright^*$. For example, $T_n \triangleright^* T_m$ states that T_n is a (possibly indirect) extension of T_m, or else $T_n = T_m$.

The language semantics provide that references to objects may be assigned bound objects of any type which is an extension of the statically declared bound type of the reference. When a program makes an access to a selected component belonging only to certain extensions of a type, the designator of the selector component contains a type guard. This type guard is an assertion that the dynamic type of the object will be in the 'extends' relation of the nominated type, and causes a runtime test to be generated. Syntactically the type guard is denoted by a type name, appearing in parenthesis, following the denoter of the referenced object. For example, the designator *exp(Binop).op* asserts that the object referenced by *exp* is dynamically of a type which is in the $\triangleright^*$ relation with the type *Binop*.

It should be noted that the word "type" in this connection has its conventional meaning in strongly typed languages. This is in contrast to the use of the word to refer to a *finite set of classes*, as discussed in [8], for example. In our notation, all types are distinct, and an object has a unique type at every epoch of its lifetime.

3 The Limitations of Common Subexpression Elimination

Some type tests may be eliminated by conventional common subexpression elimination. Consider a situation where a type *Expr* has several extensions including a *Binop* type, and the *Binop* type has extensions *Plus* and *Star*. The relevant part of the type hierarchy is shown in Figure 1.

A typical fragment of code might be -

```
... exp(Binop).leftExp ...;
... exp(Binop).rightExp ...;
```

It is clear that if the type guard is treated as a unary operator on the object designator, the second of the two runtime tests may be eliminated. In effect, the tested value becomes a common subexpression.

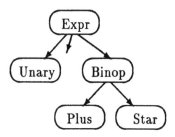

Fig. 1. Fragment of type hierarchy

Unfortunately, this mechanism is insufficient to eliminate all unnecessary tests. Consider the fragment -

```
... exp(Plus).field1 ...;
... exp(Binop).field2 ...;
```

In this case the second test is redundant (since if an object is an extension of type *Plus* it is necessarily an extension of type *Binop*), although there is no common subexpression.

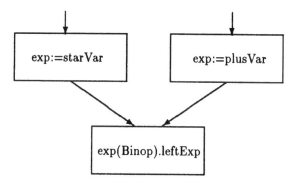

Fig. 2. Merging control flow

Global flow of type information must also be taken into account. Consider the control flow graph fragment in Figure 2, where the variables have static types as indicated by their names. In this case, understanding the semantics of the type system, we can deduce that the type guard in the successor block is redundant, since all paths leading to that block set the type to an extension of the *Binop* type.

4 Typeflow Analysis

If a particular reference object is declared as being of some static type S, then the static semantic checks of *Oberon-2* ensure that the dynamic type D is always such that $D \vartriangleright^* S$. However, we wish to place some tighter bound on the type by computing a *manifest type* M for the object so that, at every point in the program

$$D \; \triangleright^* \; M \; \triangleright^* \; S$$

For any given variable S is constant, while D and M may vary from place to place in the program. If we use the manifest type as a bound on the dynamic type of a variable, then the manifest type can be used to decide if a type guard is redundant. At any point in the program a type assertion $D \; \triangleright^* \; t$ is redundant if the statically computed manifest type $M \; \triangleright^* \; t$, since we know that $D \; \triangleright^* \; M$.

For technical reasons which become clear later we lift the domain of types to include two additional, extremal values. The special type value $\perp$ is an error type, and is an extension of every type. All types in a program are considered to be an extension of the other special value $\top$. Formally, for every type T, $\perp \; \triangleright^* \; T \; \triangleright^* \; \top$.

It is not difficult to see that every "object oriented language" which admits (a possibly limited form of) polymorphic assignment could be subjected to static analysis to compute the set of possible types which an object might take at runtime. In general, such a computation requires the finding of a least fixed point of a set of set-valued equations[9]. In this general case the computations are over vectors of elements which are members of the powerset of program types. It is the major contribution of this paper to show that for *Oberon-2* the computation may be carried out in terms of vectors of elements of a denumerable type, of cardinality equal to the number of types in the program. Furthermore, the equations have a formal correspondence to the familiar equations of dataflow analysis, although the meaning of the symbols is somewhat different.

4.1   Control Flow

Typeflow analysis involves the tracking of the type of variables across the control flow graph. At entry to the control flow graph the manifest type of each reference object is set equal to its static type, that is, $M = S$. The static type does not change, but the manifest type is updated as the effect of each statement in the block is computed. The manifest type of a variable into a particular block is denoted M^{in}. At the end of the block the new manifest type, that is, the manifest type out of the block, is denoted M^{out}. In straight line control flow the M^{in} of one block is equal to the M^{out} for the predecessor block. However, in more complex control flow the M^{in} of a block requires more computation. For a merge block, that is, a block with several predecessor blocks, M^{in} is the *lowest common ancestor* of the set of M^{out} values of all the predecessor blocks.

The *lowest common ancestor* (LCA) of a set of types is the most extended type in the type hierarchy which includes all of the types in the set. For example, using the type hierarchy fragment in Figure 1, the LCA of the set $\{Plus, Star\}$ is *Binop*. The LCA of the set $\{Unary, Plus, Star\}$ is *Expr*, and the LCA of the set $\{Binop, Plus, Star\}$ is *Binop*.

In a companion paper[6], we develop the formal theory of typeflow based on an algebra of *clades*, a term borrowed from evolutionary biology to denote a species and all the species evolved from that species. For any type T we define the corresponding clade as the set

$$\text{Clade}(T) = \{t \mid t \rhd^* T\}$$

In the companion paper we show that the manifest type into each single input, single output region is given by the following equations

$$
\begin{aligned}
M_{start}^{in} &= S \\
M_i^{in} &= \bigvee_{j \in \text{pred}(i)} A_j \vee (B_j \wedge M_j^{in})
\end{aligned}
\tag{1}
$$

In these equations, $pred(i)$ is the set of blocks which are control flow predecessors of block i, and $\vee$ denotes the LCA operation. $\wedge$ denotes the more extended of its two operands if one is an extension of the other, and the special value $\bot$ if the two are incommensurate. The value $\bot$ denotes an error value, the root of the empty clade. The local information from any block j appears in the coefficients A_j, B_j. These encapsulate the effect of downwardly visible Assignments and type Bounding tests respectively for the block. For the basic block j, if k is an index over program objects, and a_{jk}, b_{jk} are the k-th elements of the vectors A_j, B_j, then one of a_{jk}, b_{jk} will be an extremal value for every index k. Blocks with assignments to a particular variable with index k will have the corresponding element, b_{jk} equal to $\bot$, while blocks with no assignments to variable k will have the corresponding element, a_{jk} equal to $\top$.

The equations 1 must be solved in the forward control flow direction, so as to obtain a least (most constrained) value for M. In dataflow terms this is a *forward flow, all paths* problem. However, it should be noted that the meaning of the $\wedge, \vee$ are quite distinct to the bitwise, Boolean *and* and *or* of conventional dataflow analysis.

Strict analysis would show that at every point in a program the strongest possible assertion regarding the dynamic type of an object would constrain the type to be a member of some set of types. However, there are no operations in *Oberon-2* which test membership in such an arbitrary set of types. Instead, all explicit and implicit tests are based on the $\rhd^*$ relation. Suppose that D is the type of an object at some point in a program, and it is known that $D \in S$, where S is some set of types. In this case, the condition that a test $D \rhd^* T$ is redundant is $S \subseteq \text{Clade}(T)$. What we have called the manifest type M is simply the smallest clade covering S, and the redundancy condition is $M \rhd^* T$.

The approximation which we make is to compute the smallest covering clades (that is, the manifest types) directly and test the $\rhd^*$ predicate, rather than compute the exact sets and test the subset predicate. It is this mechanism of representing arbitrary sets by their smallest covering clade which results in the reduction of cardinality in the entities which we compute. In a program with N types there are 2^N sets of types, but there are only $N + 2$ clades.

In common with the formally similar dataflow equations, the system of equations (1) possess a "loop breaking rule". The practical consequence of this rule is that loops need only be traversed twice to determine the manifest type at each point in the body.

4.2 Manifest Types and the Control Flow Graph

Any reducible control flow graph can be derived from a small number of constructs. They are straight line control flow, merging control flow, and control flow with back edges. The following describes how the equations are solved for each case.

Straight Line Control Flow. This is a trivial case for typeflow. This structure is produced by simple sequential code. The M^{out} of a block becomes the M^{in} to the following block, see Figure 3.

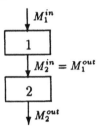

Fig. 3. A sequence construct

Merging Control Flow. Merging control flow is produced by $WITH$, IF and $CASE$ statements in $Oberon$-2. Figure 4 shows a control flow graph fragment which contains merging control flow[2]. In this situation the M^{out} of block 1 is the M^{in} for blocks 2–5. The M^{in} to block 6 is the LCA of the manifest types out of blocks 2–5. The equations are also shown in Figure 4.

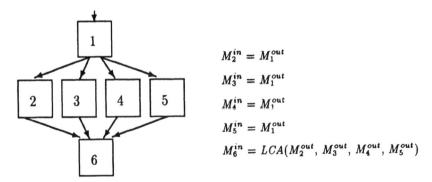

$$M_2^{in} = M_1^{out}$$

$$M_3^{in} = M_1^{out}$$

$$M_4^{in} = M_1^{out}$$

$$M_5^{in} = M_1^{out}$$

$$M_6^{in} = LCA(M_2^{out}, M_3^{out}, M_4^{out}, M_5^{out})$$

Fig. 4. Example of merging control flow

[2] This diagram strictly applies only to those cases where the flow of control is not predicated on the result of type tests. The modifications required to handle such predicates are dealt with in a later section.

Control Flow With Back Edges. The control flow graph fragment in Figure 5 contains a back edge. This may require the traversal of the block more than once. As can be seen, the M^{in} to block 2 is the LCA of the manifest types out of block 1 and block 2. To calculate the M^{out} of block 2 it is first traversed with an

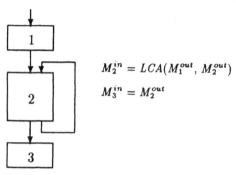

$$M_2^{in} = LCA(M_1^{out}, M_2^{out})$$
$$M_3^{in} = M_2^{out}$$

Fig. 5. Fragment with a back edge

initial M^{in} equal to the M^{out} of block 1. Then the block is traversed a second time with M^{in} equal to the LCA of the M^{out} of block 1 and the M^{out} of block 2 which was calculated on the first traversal. The M^{out} of block 2 is the M^{in} for block 3 in the usual way. The *Oberon-2* statements which produce this type of control flow are the *WHILE, FOR* and *REPEAT* loops.

The *LOOP* statement produces the slightly different control flow, shown in Figure 6. The fragment shown is for a loop with two exits. The M^{in} into block 2 is calculated by the same method as for other loops, that is, blocks 2, 3 and 4 are traversed with the M^{in} to block 2 being the M^{out} of block 1. Once the M^{out} of block 4 is calculated, blocks 2, 3 and 4 are traversed again using the LCA of the manifest types out of blocks 1 and 4 as the M^{in} for block 2. The M^{in} to block 5 is then calculated as the LCA of the manifest types out of blocks 2 and 3.

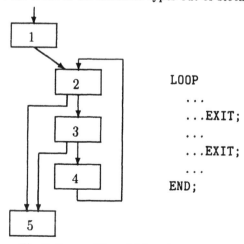

```
LOOP
  ...
  ...EXIT;
  ...
  ...EXIT;
  ...
END;
```

Fig. 6. Loop with multiple exits

4.3   Local Typeflow

In order to compute the relationship between M^{in} and M^{out} in any particular block, we must take account of the effect of the statements of the block. Three constructs affect the manifest type of variables within a block. These are assignment statements, procedure calls and type guards.

Assignment Statement. The *Oberon-2* language semantics allow for variables of a particular type to be assigned expressions which are extensions of that type. For reference objects the dynamic type then becomes equal to the dynamic type of the expression. With this is mind, consider the assignment statement $var := exp$; where the static type of exp is T_e. The static semantic rules of *Oberon* demand that T_e must be in the extends relations with T_v, the static (declared) type of the variable. After this assignment statement has been executed it can be asserted that the type of the variable is equal to the type of the expression. During typeflow analysis this means that the manifest type of the variable becomes the manifest type of the expression.

Procedure Call. When a reference object is sent as a variable parameter of a procedure call, the called procedure may change the type of that parameter. Without inter-procedural type tracking, the only assertion that can be made about the type of the variable after the procedure call is that it is an extension of the static type of the formal parameter. The manifest type of the actual is changed to reflect this.

Type Guards. As mentioned previously, a type guard is an assertion that the dynamic type of a variable is in the extends relation of the guard type. A type guard produces a runtime test for this. Since the goal of typeflow analysis is to remove redundant type tests, each type guard must be compared to the manifest type of the variable it is guarding. The type guard $var(T).field$ has different effects depending on the manifest type of the variable var. First, suppose that the manifest type of var is M, and this is an extension of T. In this case the type guard may be removed. We know that the dynamic type of var will be at least as extended as M, so if M is an extension of T, the test is bound to evaluate to true. In this case, the manifest type of var is unchanged by the test, as M is already more extended than T.

On the other hand, if the guard type is an extension of the manifest type of var, $T \triangleright^* M$, then the test cannot be removed. However, following the test it is known that the type of var is at least as extended as T, and so the manifest type of var is thus updated to equal T.

4.4   Type Predicates

There are two constructs in *Oberon-2* in which the flow of control is explicitly mediated by the dynamic type of objects. These are *IS* expressions and *WITH* statements, also called regional type guards.

IS Expression. In *Oberon* the *IS* expression is a boolean expression of the form *var IS T*, which gives the result of the runtime type test $D \, \triangleright^* \, T$ where D is the dynamic type of *var*. When this expression is used as the condition for a loop or an if statement, then the manifest type of *var* can be updated to be the more extended of M and T, where M is the existing manifest type of *var*. Note that the manifest type of *var* should not be simply set to T, as the existing manifest type may already be more extended than T. For example, again referring to the type hierarchy in Figure 1, for the if statement -

```
IF exp IS Binop THEN
   StatementSequence
ELSE
   StatementSequence
END;
```

If *exp* is statically of the type *Expr*, and its existing manifest type is also *Expr* then the manifest type of *exp* should be changed to *Binop* before the *then-part* statement sequence is traversed. However, if the existing manifest type of *exp* is *Plus*, say, which is an extension of *Binop*, then the manifest type is unchanged for the traversal. In either case, the manifest type entering the *else-part* statement sequence is the same as the manifest type on entry to the if statement.

WITH Statement. Syntactically, a *WITH* statement in *Oberon-2* has the form—

```
WITH var : T1 DO
    StatementSequence;
| var : T2 DO
    StatementSequence;
| var : T3 DO
    StatementSequence;
ELSE
    StatementSequence;
END;
```

The *WITH* statement is equivalent to an *IF* statement containing multiple *ELSIF*s with a number of *IS* expressions as the condition on each. It sequentially tests each option for the *var* $\triangleright^* \, T_n$ relation, where T_n is the type in the option, until it finds a test which succeeds. If none of the tests succeed, then the statements in the *ELSE* clause are executed. The *ELSE* clause is optional and, if it is not present and no other test succeeds, then the program aborts. Each option in the *WITH* statement is treated as an *IS* expression. Before traversal of each statement sequence the manifest type of the variable is changed to be the lower of the existing manifest type of *var* and T_n.

This *WITH* statement produces the control flow found in Figure 7. The manifest type of *var* after the *WITH* statement, that is, M^{in} for the merge block (block 8), is changed to the LCA of the manifest type of *var* after traversal of each of the options of the *WITH*.

146

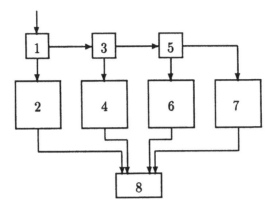

Fig. 7. A *WITH* statement fragment

4.5 A Bonus — Detection of Incorrect Programs

We have described how typeflow analysis can be used to detect type tests which will always evaluate true at runtime. At the same time, and at no extra cost, type tests which are bound to fail at runtime may be detected.

Incorrect programs can be found when carrying out the tests for type guards. As suggested previously in Section 4.3, the manifest type following a type test for *Typeof(var)* $\rhd^*$ *T* is the lower of *M* and *T*. If *M* $\rhd^*$ *T* the test is redundant and may be removed, otherwise the test must remain. If *T* and *M* are incommensurate, then the new manifest type is $\bot$, the *error type* defined in section 4.1. In this case the program is in error, and will certainly fail at runtime if control ever reaches the test. As an example, consider the following fragment of code, referring to the type hierarchy in Figure 1, where *exp* is of type *Expr* and *unary* is of type *Unary–*

```
...   exp := unary;
      exp(Binop).leftExp; ...
```

This code is statically correct. In the first statement *exp* is assigned a variable which has a type which is an extension of the static type of *exp*. The second statement contains a type guard which tests that *exp* is of some type which is an extension of its static type. However, it is obvious that the type guard in the second statement will fail at runtime, as *exp* is dynamically of type *Unary*, and *Binop* and *Unary* are roots of non-overlapping clades[3].

Runtime errors also occur in *Oberon-2* if the dynamic type of a record on the left of an assignment is not equal to its static type. This is automatically the case for record variables, since their type cannot change. For *references to records* an implicit runtime test must be generated. If it is found during typeflow

---

[3] If we set the initial manifest type of unallocated (*NIL*) objects to $\top$ rather than to the static type, then the same computation will detect programs where the object is not allocated along every path leading to the type guard.

analysis that the manifest type of the record being referenced is not equal to its static type, then a warning can be issued that the program will fail at runtime.

Dead code can also be detected in *WITH* statement options. If the same two comparisons of the manifest type and the option type are carried out as for type guards, $M \vartriangleright^* T_n$ and $T_n \vartriangleright^* M$, and both fail, then code for that option of the *WITH* statement will never be executed. Of course, if all the options of a *WITH* statement are found to be dead code and there is no *ELSE* clause, then the program is incorrect and must fail at runtime. For example, consider the *WITH* statement:

```
WITH var : Plus DO
    ...
| var : Binop DO
    ...
END;
```

If the manifest type of *var* is *Star* then the code in the *Plus* option is dead, whereas if the manifest type of *var* is *Plus* then the code in the *Binop* option is dead code. If *var* has the manifest type *Unary* then the code for both options is dead, and the program will give a runtime error. An *ELSE* clause added to this *WITH* statement would prevent the runtime error, if the manifest type were *Unary*, but the code in the *Plus* and *Binop* options would still be dead.

WHILE loops and *IF* statement branches which are never executed can be found by the same method. This only applies for those statements where the condition is an *IS* expression. The same tests used for the *WITH* options are applied to the *IS* expression to determine if the conditional code will ever be executed.

5 An Implementation of Typeflow Analysis

We have described the computation of manifest types in terms of manipulation of the control flow graph. However, typeflow analysis can be carried out without the need for constructing the control flow graph. Since *Oberon-2* is a structured language and only produces a limited number of control flow constructs, it is not necessary to solve the manifest type equations for arbitrary control flow. In fact, the equations may be solved "on the fly" during a traversal of the abstract syntax tree of the program. This method is similar to that described by Mössenböck for computing static single assignment form[7].

It is probably possible to carry out typeflow analysis during the same traversal which checks static semantic rules. However, our initial implementation of typeflow analysis is done during a separate traversal, after the static semantic checks have been completed.

To achieve this, each variable descriptor contains a manifest type attribute as well as the expected static type attribute. This attribute is updated during the traversal for assignments, type guards and so on. When a statement is encountered which has diverging control flow (see Figure 4) then as variables are encountered in each of these blocks, a record is kept of the manifest type emerging from the predecessor block. As each divergent path is traversed the M^{in} of modified variables are reset before the traversal of the next path. Referring to Figure 4, this means that before traversing each of blocks 2, 3, 4 and 5 the manifest type is reset to the M^{out} of block 1. At the end of each block the M^{out} of the variable is 'collected' so as to be able to compute M^{in} for the successor block. After traversing the first block with an in edge to the merge block (block 2) the M^{in} is initialised. For subsequent blocks M^{in} is replaced by the lowest common ancestor of the previous M^{in} and the M^{out} on the new in edge. After all predecessor blocks have been traversed, the final value for M^{in} is known.

5.1 Implementation Details

In the analysis so far, it has been assumed that the manifest type out of each block is the same along each edge. However, this is not strictly correct for either *WITH* statements, or other conditional forms which include an explicit type test. In our formal analysis[6] we introduce type assertions in synthetic blocks placed on edges to regularise the analysis. Here however we may take this information directly into account. The *WITH* statement serves as a good example of this[4]. Figure 7 shows the control flow graph representing the *WITH* statement in Section 4.4 which has three options and an *ELSE* clause. Blocks 1, 3 and 5 are the tests for executions of the different options and blocks 2,4 and 6 respectively, contain the corresponding statements for each option. Block 7 is the *ELSE* statements. Because of the semantics of the *WITH* statement the manifest types along the two out edges from block 1 are not the same. The statements in block 2 are only executed if the test in block 1 evaluates true, that is, *WITH var : T_1 DO* .... Thus on this path we assert that M is T_1. On the path from block 1 to block 3, the manifest type of *var* is unchanged from M^{in} to block 1.

In detail, the steps of the typeflow analysis for the *WITH* statement structure in Figure 7 for a particular variable are as follows–

- block 1 is traversed. This sets the manifest type of M_{var} to the type of the first *WITH* option
- block 2 is traversed. M^{in} for block 8 is set to the current manifest type of the variable M_{var} (i.e. M^{out} for block 2)
- block 3 is traversed, which sets M_{var} to the type of the second *WITH* option
- block 4 is traversed. M^{in} for block 8 is adjusted to the *LCA* of the interim M^{in} and the M^{out} of block 4
- block 5 is traversed, which sets M_{var} to the type of the third *WITH* option
- block 6 is traversed. M^{in} for block 8 is adjusted to the *LCA* of the interim M^{in} and the M^{out} of block 6

---

[4] For simplicity of exposition, we shall assume that neither dead code nor errors are discovered during typeflow analysis in our example.

- M_{var} is reset to M^{in} for block 1. Block 7, which is the block for the *ELSE* clause is then traversed. Again, M^{in} for block 8 is updated to the *LCA* of the interim M^{in} and the M^{out} of block 7
- all of the blocks of the *WITH* statement have now been traversed. The M^{in} of block 8 is now the *LCA* of its predecessor blocks and analysis may proceed with block 8

Because the control flow produced by each statement is known, the required information about types can be stored as state variables of the recursive traversal procedures. All structures produced by *Oberon-2* statements are traversed in a similar manner.

An extra step must be included when traversing loop structures. As mentioned earlier, the block(s) in a loop must be traversed twice so that the correct M^{in} to the loop can be computed using the M^{out} of the loop block(s). It is a consequence of the "loop breaking rule" that the types *out* of a loop can be correctly computed without knowing the manifest types on the back edge. Once the M^{out} values are known, the correct M^{in} may be computed and the necessity of any type tests evaluated.

A first traversal is performed, without removing any apparently redundant type tests. The M^{out} value obtained is then used to calculate the final M^{in} value for a second traversal of the loop. Since the correct M^{in} value is known during the second traversal, redundant type tests may be removed on this pass. In the case of nested loops, the same reasoning shows that the statements of the innermost loop may be traversed a number of times equal to the loop nesting-level plus one.

5.2 Future Work

Our initial implementation of typeflow analysis computes manifest types on a procedure basis. There is no inter-procedural analysis. It also deals only with entire variables, and does not track *fields* of objects. For example, for the designator–

```
exp(Binop).leftExp(Unary).op
```

the variable *exp* is tracked, and the *Binop* type guard will be checked for redundancy and thus might be removed. However, the field *leftExp*, which is also a reference object, is not tracked and the *Unary* type guard is not redundancy checked. This represents a limitation of our current descriptor structure. The implementation is currently being extended to track fields of records.

Inter-procedural typeflow analysis should be able to be carried out, at least for procedures local to the module being compiled. This can be done by type traversing a procedure before traversing any code which calls that procedure. The manifest type of formal parameters at procedure exit can then be used to refine manifest type information in the caller. For each called procedure an attribute indicates if the formal is modified within the procedure. If a formal parameter is flagged as type modified, the manifest type of the corresponding actual parameter can be updated to the manifest type already calculated for the formal at exit. If the formal parameter is not so flagged, then the manifest type of the actual will be the lower of the previous value and the manifest type of the formal at exit.

6 Summary

The typeflow analysis outlined in this paper has been implemented and incorporated into an *Oberon-2* compiler. Some preliminary testing has been carried out, and removal of redundant type tests as well as the incorrect program detection have been achieved.

References

1. N. Wirth: Type Extension, *ACM Trans. on Prog. Lang. and Systems* Vol 10, No 2 pp 204–214, April 1988.
2. N. Wirth: The Programming Language Oberon, *Software Practice and Experience*, Vol 18, No 7, pp 671–690, July 1988.
3. H. Mössenböck and N. Wirth: The Programming Language Oberon-2, Computer Science Report 160, ETH Zurich, May 1991
4. N. H. Cohen: Type-Extension Type Tests Can Be Performed in Constant Time, (Technical Correspondence), *ACM Trans. on Prog. Lang. and Systems*, Vol 13, No 4, pp 626–629, October 1991.
5. J. Asuru: Optimisation of Array Subscript Range Checks, *ACM Letters on Prog. Lang. and Systems*, Vol 1, No 2, June 1992.
6. J. Gough and D. Corney: Typeflow Analysis by Clade Algebra, Information Technology Report 9/93, QUT Brisbane, July 1993, (preprint)
7. H. Mössenböck: private communication March 1993.
8. J. Palsberg and M. Schwartzbach: Static Typing for Object-Oriented Programming, To appear in *Science of Computer Programming*.
9. N. Oxhøj, J. Palsberg and M. Schwartzbach: Making Type Inference Practical, ECOOP '92, *Lecture Notes in Computer Science*, Vol 615, Springer Verlag, 1992.

Where Concurrent Processes Originate[*]

Stanislaw Chrobot,
Kuwait University, Department of Mathematics, P.O. Box 5969 Safat, 13060 Kuwait
chrobot@mcc.sci.kuniv.edu.kw

Abstract: Direct control over short- and medium- term scheduling algorithms is of primary importance in systems programming, real-time programming and multiprocessor parallel programming. Such algorithms can be implemented in high level language programs if low level synchronisation primitives are incorporated into the language. The objective of the paper is to discuss an extension of the Modula-2 low level synchronisation primitives (meant for a uni-processor system) to multi-processor systems. Its aim is to investigate how elementary the low level primitives can be without losing their integrity. The solution introduces idling processes and two rudimentary operations for the delay and the immediate resumption of a process inside the monitor.

1. Introduction

The art of operating system design emerged in the late sixties when Dijkstra [1] introduced semaphores as a tool for concurrent process synchronisation. Very soon some other synchronisation concepts like Brinch Hansen's conditional regions [2] and Hoare's monitors [3] were introduced. On this basis, concurrent programming languages like CONCURRENT PASCAL [5] were proposed. As a result of this, the operating system kernel, which used to be designed as a monolithic module programmed in assembly language, could now be designed as a well-structured collection of concurrent processes co-operating with each other in a harmonious way using synchronisation variables [4]. These systems were implemented on uni- or multi-processors where the concurrent processes could access one common memory. Such systems are called *shared-memory systems* or *centralised systems.*

A new era was initialised in 1978 by Hoare's proposal [6] to base the co-operation of concurrent processes on information exchange. Interprocess message passing could now be applied to a set of processes which did not share any common memory. These processes could be spread over several computers interconnected with communication lines, creating in this way a *distributed system*. A large variety of interprocess communication concepts has been suggested, for example, the rendezvous in occam [14] and ADA [10], remote procedure call [11], and distributed shared data [15].

---

[*] This work was supported by Kuwait University under the grant number SM 087

Massive parallelism on multi-processors introduced two-level concurrency problem. Operating system supported processes, too heavy to be used for fine-grained massive parallelism were split into lightweight threads in the user address space [7], [8], [9]. Operating system synchronisation concepts - monitor and its condition - were adapted for the thread synchronisation. Several improvements were introduced to them, e.g.:
- various forms of spinning and blocking to reduce the monitor's entry overhead or
- various forms of resumption and broadcasting to increase the long-term synchronisation flexibility.

The problem of integration of the two level concurrency emerged. If the user threads are implemented simply by splitting an operating system process, all the threads are delayed when the process supporting them is blocked in the operating system kernel. An elegant solution for the problem - scheduler activation - was implemented on the Firefly multi-processor [19].

With the increasing number of processors in the system the importance of dynamic allocation policy becomes increasingly important for the system performance [20]. Both concurrency levels have to co-operate for dynamic processor allocation: operating system allocates processors for applications which schedule their threads on these processors informing the operating system about their current needs.

The synchronisation concepts described above are implemented at the bottom parts of the operating system kernel and the application. They have similar structure and functionality. We will call them *microkernels*. User level microkernel takes a form of user library or run-time support. Both microkernels are usually written in assembly language. Thus some effort was taken to structure the microkernel itself.

One solution is to define so called *low level concepts* as a part of a high level language. These low level concepts still have to be implemented in an assembly language. Once it is done they can be used to implement *high level concepts* using high level language.

Such a low level solution was proposed by Wirth [13] for shared memory systems. In his MODULA-2 language the concepts of coroutine and two low level primitives (TRANSFER and IOTRANSFER) to operate in a *uni-processor environment* were suggested. The advantage of this approach is that the programmer may design his own scheduling algorithm according to the problem needs [12], as a part of his program.

A closer look at Wirth's primitives reveals that the expansion of these *uni-processor* environment oriented primitives to multi-processor and multi-computer systems is not a straightforward problem. In the case of *multi-processors* it is not enough to disable interrupts to ensure mutual exclusion of concurrent processes. Processes running concurrently on different processors can attempt to access shared variables at the same time even though the interrupts are disabled. With *multi-computers* a more powerful primitive for interrupt handling has to be introduced to implement selective waiting of processes running on one processor for messages incoming from other processors. Selective waiting is a common feature of modern distributed programming languages. In turn, for *massive parallelism,* new primitives for two level concurrency integration as well as for dynamic processor allocation are needed.

The objective of the paper is to propose an extension of Wirth's low level primitives to cover multiprocessor systems. Actually a part of the area outlined above is touched in the paper - delay and resumption of processes in the monitor. Some other author's attempts raise the hopes that other parts of the problem can be approached in the same way. The paper also discusses the conditions necessary for low level primitive design, to investigate how elementary the low-level primitives can be.

Usually the synchronisation primitives allow the processors to work in a time multiplexed mode. As a result, a number of logical programmed processes can run quasi-concurrently on a much smaller number of physical processors. Thus it is in the microkernel where the degree of concurrency is increased and where the logical processes originate. The Modula-2 microkernel seems to be the simplest possible microkernel, but it proved that it cannot be used to implement high level primitives for multiprocessors and for multicomputers.

The paper consists of 7 sections. Section 2 presents the Modula-2 TRANSFER primitive and formulates the problem of extension of this primitive in multi-computer systems. In sections 3 to 5 three steps are presented in formulating the solution: (1) an attempt to use the TRANSFER primitive in a multi-processor system; (2) discussion why this attempt fails and what the conditions are for the correct solution, and (3) presentation of the final solution. Section 6 reviews the obtained results in the context of the evolution of the shared memory synchronisation tools. The last section drafts the directions for future investigations in the area of parallel processing and multi-computer systems.

2. The Modula-2 TRANSFER Primitive and its Usage

Wirth introduces in Modula-2 [13] a kind of logical process called *coroutine.* A set of coroutines can run on a single processor in an interleaving mode, one at a time. The coroutines pass control explicitly from one to the other using the TRANSFER(a, b) operation. It suspends the coroutine *a* in such a way that *a* can be later on resumed by another TRANSFER operation, and resumes the coroutine *b* previously suspended by a TRANSFER operation. The TRANSFER procedure is imported from the standard module SYSTEM. Standard module means in this case a pseudo-module which cannot be written in Modula-2 itself and is in fact a part of the language definition.

To discuss the implementation aspects of the TRANSFER operation, we assume the following model: each coroutine has a so-called *coroutine control block* allocated in the memory where the coroutine status can be saved. The coroutine can stay in one of two states: *running* - when the coroutine status is loaded to the processor registers or *suspended* - when the coroutine status has been saved in its control block.

Inside the TRANSFER(a, b) operation, the control is passed from coroutine *a* to coroutine *b* as a result of
- storing the current contents of the processor registers in the control block of the current coroutine (the current coroutine identifier is maintained inside the SYSTEM module),
- assigning its value to *a*,
- making coroutine b to be the current coroutine, and
- restoring the contents of the new current coroutine in the processor registers.

Let us first implement the TRANSFER operation using two pseudo-operations: SAVE(a), which saves the process status registers in the coroutine *a* control block and RESTORE(b), which restores the control block of coroutine *b* in the process status registers.

```
VAR current: ADDRESS;
PROCEDURE TRANSFER(VAR a, b: ADDRESS);
VAR new: ADDRESS;
BEGIN
   SAVE(current); new:= b; a:= current; current:= new; RESTORE(current)
END TRANSFER;
```
Note[13]: assignment to *a* occurs after identification of the process *b*; hence the actual parameters may be identical.

We have called both the SAVE and RESTORE pseudo-operations since they can not be implemented as procedures invoked from inside the TRANSFER primitive. The SAVE is actually a continuation of the procedure call action, and the RESTORE is a prelude to the procedure return action. They store and restore the coroutine status as valid just at the point of the TRANSFER call.

To eliminate these pseudo-operations we introduce a new language construct called EXPROCEDURE (for EXtracode PROCEDURE). Both the call and return actions of exprocedure involve the SAVE and RESTORE automatically.

```
VAR current: ADDRESS;
EXPROCEDURE TRANSFER(VAR a, b: ADDRESS);
VAR new: ADDRESS;
BEGIN new:= b; a:= current; current:= new END TRANSFER;
```

It is worth noticing that such an exprocedure is seen at the coroutine level as one instruction, in the sense that the internal status of the exprocedure is not interpretable at the coroutine level.

Using the TRANSFER operation, Wirth builds a more powerful synchronisation tool called Signals. Signals are defined in the module Processes [13]. There are two procedures available for coroutines to deal with Signals: Wait and Send. The Wait(s) procedure delays the calling coroutine in the queue associated with the

Signal s and transfers control to one of the ready coroutines. The Send(s) procedure immediately resumes one of the coroutines delayed in the queue associated with s. The calling coroutine becomes ready and the resumed coroutine - running. While using Signals, the coroutines do not transfer control to each other explicitly any more. This decision is taken by the scheduling algorithm hidden in the Wait and Send operations.

In APPENDIX 1 another implementation of the module Processes is presented. It is a restructured version of the original module. The queue implementation details (which are of no importance to this paper) have been moved to another module.

Signals, when used inside monitors, correspond to Hoare's condition synchronisation variables [3]. The conditions can be used easily to implement semaphores as well as many other synchronisation variables [3]. Such variables are defined for logical processes with no restriction regarding the number of processors the logical processes are implemented on. Unfortunately, the same assumption does not apply to the Signals which are based on the TRANSFER primitive. Wirth clearly states his solution is valid for uni-processor systems only.

There are two main problems with expanding this solution to multi-processors. The first lies in the Wait operation when the *Ready queue is empty*. Wirth considers such a situation to be faulty since it leads to a deadlock. In such situations the Wait simply halts. This solution is correct as long as no interrupt in a uni-processor environment is assumed. When interrupts are taken into account, (Wirth never uses the module Processes in the interrupt context) all the processes can be delayed, waiting for device transfer completion. The Ready queue stays empty, but there is no deadlock. A similar situation appears in a multi-processor system when all the processors share one Ready queue. If the number of undelayed coroutines is less than the number of processors, some of the processors will find the Ready queue empty.

The second problem concerns the way the *critical region* of the monitor is implemented. In a uni-processor system without interrupts the problem does not exist: one coroutine is running at a time on the processor until it transfers control to another coroutine. On a uni-processor with interrupts it is enough to disable interrupts at the entry to the critical region and enable them at the exit. In a multi-processor, however, the coroutines on different processors try to access the monitor regardless of the states of their interrupt systems.

In this paper the solution is confined to the no-interrupt case. We consider this case a good basis for further investigation in the area of distributed systems where the interrupts are used as a basic hardware support for inter-processor message passing.

To solve the first problem we introduce a so-called Idling process in each processor. When a processor finds the Ready queue empty, it resumes its Idling process. The Idling process, by definition, is never delayed. It keeps looping until a regular process is found in the Ready queue. To solve the second problem we introduce an instance of a busy semaphore which is shared by all the processors.

Our solution is discussed in three steps. In step 1 we create a hypothetical expanded module HProcesses where the TRANSFER primitive is used in a multi-processor system. In step 2 we analyse why this solution fails and what the conditions are for the correct solution. In step 3 we formulate the final solution.

3. The HProcesses Module. Using the TRANSFER Primitive in a Multi-Processor Environment.

In the module HProcesses presented in ALGORITHM 1 we assume a set of identical processors shares a common memory. Each processor has its own unique identifier. Its value is returned by the function THIS of the PROCESSOR type. Both the THIS and the PROCESSOR identifiers are imported from the module SYSTEM:

```
TYPE PROCESSOR;
PROCEDURE THIS: PROCESSOR;
(* It returns the identifier of the calling processor *)
```

Since each processor can run a coroutine, we assume the identifiers of these running coroutines for all processors are collected in the array run:

```
VAR run: ARRAY PROCESSOR OF ADDRESS;
```

The processors use a busy waiting semaphore based on the Test-and-Set [16] indivisible operation. We assume the type BusySemaphore and the three associated operations: initbs, wait, and signal are implemented in the module BusySemaphoreHandler;

```
DEFINITION MODULE BusySemaphoreHandler;
   TYPE BusySemaphore;
   PROCEDURE initbs(VAR bs: BusySemaphore; i: CARDINAL);
   PROCEDURE wait(VAR bs: BusySemaphore)
   PROCEDURE signal(VAR bs: BusySemaphore)
END BusySemaphoreHandler.
```

The module HProcesses uses an instance of BusySemaphore called mutex to assure a critical region for operations accessing the Ready queue and all the Signal queues:

```
VAR mutex: BusySemaphore;
```

The Wait and Send operations can be invoked from inside the monitor only. The monitor is entered and left by the Enter and Exit operations. We have added these two operations to the module HProcesses.

The procedure Idle defines the body of the Idling coroutine. A private instance of Idling coroutine for each processor is started when the HProcesses module is initiated. The identifiers of these coroutines are collected in the array Idling:

VAR Idling: ARRAY PROCESSOR OF ADDRESS;

The Ready queue and the Signal queues are of the ProcessQueue type. This type is imported from the QueueHandler module. Its definition is given in APPENDIX 1.

To focus on the monitor with immediate resumption of delayed processes, we simplified the procedure StartProcess by removing the TRANSFER call from it.

As we mentioned earlier, the module HProcesses from ALGORITHM 1 contains a hypothetical solution for a multi-processor system using the TRANSFER primitive. The areas which need analysis are the ones where the TRANSFER primitive is used. In ALGORITHM 1 they are marked as (a), (b), and (c).

The pair (a), (b):

signal(mutex); TRANSFER(old, run[THIS]);   (* (a) *)
TRANSFER(old, run[THIS]); wait(mutex);   (* (b) *)

reflects the classical scheme of delay and immediate resumption of a process in the monitor. The delayed process releases the critical region (in(a)) and the resumed process takes over the critical region from the resuming one (in (b)). In the sequence (c) an idling process resumes a ready process:

signal(mutex); TRANSFER(Idling[THIS], run[THIS]).

ALGORITHM 1

```
IMPLEMENTATION MODULE HProcesses;
  FROM QueueHandler IMPORT ProcessQueue, Initq, Insert, Remove, Empty;
  FROM SYSTEM IMPORT ADDRESS, TRANSFER, NEWPROCESS,
    INITIAL, THIS, PROCESSOR;
  FROM Storage IMPORT Allocate;
  FROM BusySemaphoreHandler IMPORT BusySemaphore, initbs, wait, signal;

TYPE Signal = ProcessQueue;
VAR run, Idling: ARRAY PROCESSOR OF ADDRESS;
      Ready: ProcessQueue;
      mutex: BusySemaphore;
  p: PROCESSOR;
```

```
PROCEDURE Enter;
BEGIN wait(mutex) END Enter;

PROCEDURE Exit;
BEGIN signal(mutex) END Exit;

PROCEDURE StartProcess(P: PROC, n: CARDINAL);
VAR new, wsp: ADDRESS;
BEGIN Allocate(wsp, n);  NEWPROCESS(P, wsp, n, new);  Insert(Ready, new);
END StartProcess;

PROCEDURE Wait(VAR s: Signal);
VAR old: ADDRESS;
BEGIN Insert(s, run[THIS]); old:= run[THIS];
   IF NOT Empty(Ready) THEN Remove(Ready,  run[THIS]);
   ELSE run[THIS]:= Idling[THIS] END;
   signal(mutex); TRANSFER(old, run[THIS])                    (*(a)*)
END Wait;

PROCEDURE Send(VAR s: Signal);
VAR old: ADDRESS;
BEGIN IF NOT Empty(s) THEN old:= run[THIS]);
   Remove(s, run[THIS]); Insert(Ready, old);
   TRANSFER(old, run[THIS]); wait(mutex)                      (*(b)*)
END END Send;

PROCEDURE Init(VAR s: Signal);
BEGIN Initq(s) END Init;

PROCEDURE Idle;
BEGIN LOOP Enter;
   IF Empty(Ready) THEN Exit
   ELSE Remove(Ready, run[THIS]);
     signal(mutex); TRANSFER(Idling[THIS], run[THIS])         (*(c)*)
   END
END END Idle;

BEGIN Initq(Ready); initbs(mutex, 1);  run[THIS]:= INITIAL;
   FOR p:= MIN(PROCESSOR) TO MAX(PROCESSOR) DO
     Idling[p]:= StartProcess(Idle, 100) END
END HProcesses.
```

4. Singularities in the HProcesses Module. Conditions for the Correct Solution

The synchronisation pattern shown in (a) and (b), so natural in designing "regular" monitors, fails in this case. Let us analyse the scenario (see Fig 1.):

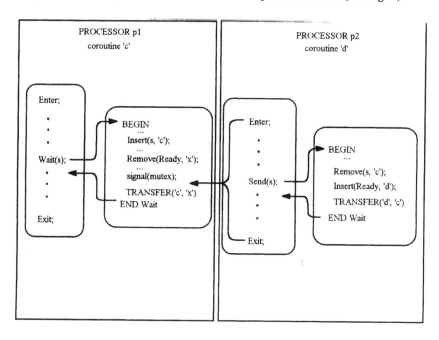

Fig. 1. Coroutine 'd' running in PROCESSOR p2 is executing its critical region after the coroutine 'c' in Processor p1 has completed signal(mutex) and before it starts TRANSFER('c', 'x') inside its Wait(s) operation.

- processor p1 executes coroutine 'c'; it enters the critical region of the monitor and calls Wait(s); inside Wait(s) it inserts 'c' to the queue s, removes a new coroutine 'x' from the Ready queue and calls signal(mutex);
- at the same time processor p2 executes coroutine 'd' which attempts to enter the critical region and to call Send(s); p2 is waiting inside the Enter operation until the processor p1 executes signal(mutex);
- when p1 completes signal(mutex) the critical region is free and the processor p2 can continue; inside Send(s) it removes 'c' from the Signal s queue, inserts 'd' to the Ready queue and calls TRANSFER with the parameters 'd' and 'c';
- p2 continues executing 'c' with the instructions following the TRANSFER call which resumed 'c' previously;

- p1 continues executing 'c' with the call TRANSFER('c', 'x') following the signal(mutex).

As a result the two processors are executing the same coroutine 'c' at the same time. The reason for this misbehaviour is that the coroutine 'c' is not being delayed in a consistent way. Its name is inserted in the Signal queue by p1, and it is resumed by p2 and continued before it has been suspended inside TRANSFER by p1.

From the above it can be inferred that

it is necessary that the coroutine currently being delayed in (a) is   (*)
suspended before the critical region of the monitor is released.

To prove that (*) is not a sufficient condition, it is enough to reverse both the signal(mutex) and the TRANSFER operations in the sequence (a) and repeat the same scenario. The coroutine 'c' which is being delayed is now suspended before the critical region is released. The newly resumed coroutine 'x' continues now in the critical region. If it has happened that 'x' had previously resumed another process inside a Send operation, (this means that 'x' is now at the semicolon in the sequence (b)) then 'x' is going to regain the critical region after the resumption (calling the wait(mutex)). The region is still occupied by 'x'. Thus it leads to a deadlock.

This is why it can be inferred that

it is necessary that the process to which the control is currently   (**)
transferred in (a) is resumed after the critical region of the monitor is released.

The conditions (*) and (**) lead directly to the conclusion that *the critical region is to be released after the coroutine 'c' has been suspended and before the coroutine 'x' has been resumed.* What it actually means is that it is necessary to introduce a new exprocedure, to be called RELEASETRANSFER.

The exprocedure RELEASETRANSFER(VAR a, b: ADDRESS) suspends the calling coroutine *a*, releases the critical region of the monitor, and resumes the coroutine *b*.

```
VAR current: ARRAY PROCESSOR OF ADDRESSES;
EXPROCEDURE RELEASETRANSFER(VAR a, b: ADDRESS);
VAR new: ADDRESS;
BEGIN
  new:= b; a:= current[THIS]; signal(mutex); current[THIS]:= new
END RELEASETRANSFER ;
```

Since the structure of the sequence (c) is identical to the structure of the sequence (a), the exprocedure RELEASETRANSFER will replace both the sequences (a) and (c) in our solution.

The above primitive suggests that a symmetrical exprocedure TRANSFERREGAIN for the sequence (b) could be designed. Its role would be to 'hide' the wait(mutex) operation.

```
EXPROCEDURE TRANSFERREGAIN(VAR a, b: ADDRESS);
VAR new: ADDRESS;
BEGIN
  new:= b;  a:= current[THIS];  wait(mutex);  current[THIS]:= new
END TRANSFERREGAIN;
```

The next scenario, however, shatters these hopes. Let us assume that
- processor p1 executes coroutine 'd' which enters the critical region of the monitor and calls the Send(s) operation; inside Send(s) the processor p1 removes 'c' from the queue s, inserts 'd' into the Ready queue, and calls our new TRANSFERREGAIN exprocedure.
- inside this primitive, the processor p1 which already occupies the critical region tries to enter it again executing wait(mutex).
This scenario leads to another deadlock. The critical region of the monitor will never be released. Thus for the sequence (b) we formulate the following necessary condition:

the coroutine transferring control in the sequence (b) has
to start after its resumption with the wait(mutex) operation   (***)
(to regain the critical region of the monitor).

The condition (***) implies that the wait(mutex) cannot be invoked from inside an exprocedure. TRANSFERREGAIN must not be an exprocedure, but it can be structured as a procedure like this:

```
PROCEDURE TRANSFERREGAIN(VAR a, b: ADDRESS);
BEGIN TRANSFER(a, b);  wait(mutex) END TRANSFERREGAIN;
```

Thus in the model, two new synchronisation primitives are needed. They are called RELEASETRANSFER and TRANSFERREGAIN. While the last primitive can be defined on the basis of the TRANSFER exprocedure, the former one has to be defined as a new exprocedure.

5. Microkernel Module for a Multi-Processor System

In Modula-2 the TRANSFER operation is a part of the module SYSTEM. This module also imports some other attributes which are outside of the focus here. This is why a new standard module called MKERNEL is introduced to solve only the coroutine switching problem. Like the module SYSTEM, our module MKERNEL is a pseudo-module in the sense that some of its attributes will be expressed   as exprocedures. The BusySemaphore mutex defined previously in the HProcesses module is moved to the MKERNEL module since the new operations use it.

A revised version of the module HProcesses from   ALGORITHM 1 is presented in APPENDIX 2 as the MProcesses module. This module uses the new procedures: RELEASETRANSFER and TRANSFERREGAIN instead of sequences (a), (b), and (c) from ALGORITHM 1.

ALGORITHM 2

```
DEFINITION MODULE MKERNEL;
   CONST INITIAL;
   TYPE PROCESSOR;
   PROCEDURE THIS: PROCESSOR;
   PROCEDURE ENTER;
   PROCEDURE EXIT;
   PROCEDURE NEWPROCESS(P: PROC; A: ADDRESS; n: CARDINAL;
                        VAR new: ADDRESS);
   EXPROCEDURE RELEASETRANSFER(VAR a, b: ADDRESS);
   PROCEDURE TRANSFERREGAIN(VAR a, b: ADDRESS);
END MKERNEL.

IMPLEMENTATION MODULE MKERNEL;
   FROM SYSTEM IMPORT ADDRESS,
   FROM BusySemaphoreHandler IMPORT BusySemaphore, initbs, wait, signal;

CONST INITIAL = ...;
TYPE PROCESSOR = ...;
PROCEDURE THIS: PROCESSOR;  BEGIN ... END THIS;
VAR current: ARRAY PROCESSOR OF ADDRESS;
      mutex: BusySemaphore;

PROCEDURE ENTER;
BEGIN wait(mutex) END ENTER;

PROCEDURE EXIT;
BEGIN signal(mutex) END EXIT;

PROCEDURE NEWPROCESS(P: PROC; A: ADDRESS; n: CARDINAL;
                     VAR new: ADDRESS);
BEGIN ... END NEWPROCESS;

EXPROCEDURE TRANSFER(VAR a, b: ADDRESS);
VAR new: ADDRESS;
BEGIN new:= b; a:= current[THIS]; current[THIS]:= new END TRANSFER;
```

```
EXPROCEDURE RELEASETRANSFER(VAR a, b: ADDRESS);
VAR new: ADDRESS;
BEGIN new:= b; a:= current[THIS]; signal(mutex); current[THIS]:= new
END RELEASETRANSFER;

PROCEDURE TRANSFERREGAIN(VAR a, b: ADDRESS);
BEGIN TRANSFER(a, b); wait(mutex) END TRANSFERREGAIN;

BEGIN initbs(mutex, 1);
END MKERNEL.
```

The primitives defined above solve the problem of synchronisation in the monitor with immediate resumption of delayed processes [3]. We can say that the RELEASETRANSFER primitive is a rudimentary form of a delaying operation inside the monitor and the TRANSFERREGAIN is a rudimentary form of the resuming operation. With the additional assumption that the TRANSFERREGAIN always appears as a last operation inside the monitor, this primitive can be simplified by stripping off its ending wait(mutex). In this case it becomes equivalent to the TRANSFER primitive.

A somewhat different solution is needed for conditional critical regions where there is no immediate resumption and the resumed process regains the critical region for itself. The primitive procedure needed in this case consists of the RELEASETRANSFER exprocedure followed by wait(mutex).

It is worth mentioning that using the Idling processes to solve the problem of the *Ready queue empty* is a commonly used trick. It gives a simple but not clear solution. It is hard to say what the nature of the Idling processes is. If the Idling processes are not used, two additional primitives are needed. One of them delays the physical or virtual processors in the system and the other one resumes these processors. Introducing these two primitives seems to be a good starting point to deal with both the dynamic processor allocation and the two level concurrency integration problems.

6. Evolution of Shared-Memory based Synchronisation Tools

The need for using exprocedure to define synchronisation primitives was clear quite early. In these early stages whole powerful synchronisation operations were structured as exprocedures. For example Brinch Hansen used them [2] to show the implementation of semaphores.

The evolution of shared-memory synchronisation tools can be seen as a process of moving consecutive parts of the medium- and short- time processor scheduling mechanism from inside the exprocedure level to the coroutine level. As a result of this process, high level language programmers were gaining more and more freedom in designing their own scheduling algorithms. But was this freedom gained free of charge? Our thesis is that the price for this increasing freedom was lower safety in using these new tools.

Let us again start with semaphores. As instances of a well-defined class they are accessed through their two operations: wait and signal only. They are safe objects in the sense that they preserve the class invariant regardless of the way their operations are used by processes. The *semaphore invariant* says that the synchronisation signals cannot be consumed in wait operation until they are sent into semaphore by the signal operation. If a semaphore implements a non-busy form of waiting, it holds a more powerful invariant reflecting the properties of the medium-term (delayed) and short-term (ready) process scheduling. It says that processes cannot be lost or multiplied as a result of a synchronisation operation (in other words: the sum of running, ready and delayed processes has to be constant). But with this safe tool the synchronisation method (sending signals via semaphore) and both the medium- and short- term scheduling algorithms are fixed for the programmer. Both the wait and signal operations are implemented as exprocedures.

Hoare's monitor with the condition variable can be seen as a result of splitting the semaphore capsule into two parts: a shell which provides mutual exclusion (monitor construct) and a mechanism for medium- and short term scheduling (condition variable). The programmer can now build objects with any synchronisation conditions in his high level program. Unfortunately these two tools are not as self-contained as the semaphore is. An object which is built using both the monitor and the condition can destroy the process scheduling mechanism if the rule: "condition is used inside the monitor only" is violated by the programmer. To implement Hoare's monitors there must be four operations:
- two procedures to enter and leave the monitor's critical region;
- two exprocedures to implement the condition's wait and signal operations.

Brinch Hansen's queues in Concurrent Pascal were the next step in this evolution. Their main aim is to make medium-term scheduling available to high level language programs (short-term scheduling is still hidden). The queues of delayed processes can be implemented as a part of Concurrent Pascal program providing the programmer takes care not to resume a process which has not been delayed yet. The two queue's operations: delay and continue are implemented as exprocedures.

Modula-2 offers the freedom of also implementing the short-term scheduling at program level leaving the responsibility for moving the processes between running, ready, and delayed states in the programmer's hand. The TRANSFER primitive seems to be a rudimentary exprocedure to be used inside the monitor to implement both the delay and resumption of concurrent processes. This operation is so simple and conceptually elegant that a question: Is it possible to go any further? stands as an obvious one.

This paper suggests, however, that the simplicity and elegance are only a result of restricting the use of these tools to uni-processor environment. In the multi-processor environment more complicated primitives for process switching are needed.

Nevertheless, these more complicated primitives from the module MKERNEL seem to reach a limit in the sense that the RELEASETRANSFER primitive cannot be expressed using the TRANSFER exprocedure and has to be defined as an exprocedure itself.

7. Conclusion

The paper discusses an extension of the Modula-2 low-level mechanism for concurrent process synchronisation in a multi-processor environment. It has proven that the Modula-2 TRANSFER primitive is not enough in such an environment. Two new primitives have been added (RELEASETRANSFER and TRANSFERREGAIN) and the Idling processes have been incorporated as the elements of the solution. To implement the primitives a special kind of procedure must be used which saves the calling coroutine's status at its entry and restores another coroutine status on its exit; it is called exprocedure.

Three correctness conditions for the solution have been formulated. They imply how the exprocedure has to be used to structure the primitives. Some of the synchronisation actions (releasing the critical region of the monitor in the RELEASETRANSFER) have to be implemented inside an exprocedure while others (regaining the critical region in the TRANSFERREGAIN) must not.

The paper proves that it is possible to introduce to MODULA-2 low-level synchronisation primitives for a multi-processor environment. The result is important
- in systems programming and in real-time programming where the programmer can design his own short- and medium- term scheduling algorithms for heavyweight processes at kernel level, as well as;
- in multiprocessor parallel programming where the programmer can design his own parallel processing model for lightweight processes at the library or application level [17],[18].

The solution also has theoretical value as a method of searching for minimal synchronisation tools for concurrent processes. The results of this paper will be used in the author's further investigation of both interrupt handling and processor allocation in multi-computer and multiprocessor environments.

The main idea for interrupt handling is to expand Wirth's IOTRANSFER primitive to cover selective waiting for interrupts. Such new primitive(s) will be used (along with TRANSFER and TRANSFERRELEASE exprocedures) to implement different types of high level message passing tools. In this way, many implementation details of these tools (among others, the message scheduling algorithms) will become available in high level language programs. The software and hardware interrupts are also closely related to pre-emptive scheduling. Interrupt handler invocation pre-empts the interrupted process. The ability to express the pre-emptive scheduling in terms of well-defined and simple primitives can shed more light on the nature of priority attribute in distributed programming languages and the two level concurrency integration problem.

The main idea for processor allocation investigation is to eliminate the idling processes and to define new primitives that the whole hardware and virtual processor allocation policy can be implemented in a high level programming language.

Acknowledgement

The final shape of the article emerged through many discussions with Zbigniew Banaszak and Krzysztof Jedrzejek at Kuwait University. The remarks of the reviewers were also very helpful.

References

1. Dijkstra, E. W. The Structure of the THE Multiprogramming System. Comm. ACM, Vol. 11, No. 5. pp. 341 - 346. (1968)
2. Brinch Hansen, P. Operating Systems Principles, Englewood Cliffs, NJ, Prentice-Hall (1973).
3. Hoare, C. A. R. Monitors: An Operating System Structuring Concept. Comm. ACM, Vol. 17, No. 10. pp. 549 - 557.(1974)
4. Brinch Hansen, P., Deamy: a Structured Operating System, Information Science Technical Report No. 11. (1974)
5. Brinch Hansen, P. The Programming Language Concurrent Pascal. IEEE Trans. on Software Engineering, SE-1, (2), pp. 190 - 207.(1975)
6. Hoare, C. A. R. Communicating Sequential Processes, Comm. ACM, Vol. 21, No. 8, pp. 666-678.(1978)
7. Rovner, P., Extending Modula-2 to Build Large, Integrated Systems, IEEE Software, Vol. 3. No 6, pp. 46 - 57, (1986).
8. Cardelli, L., Donahue, J., Glassman, L., Jordan, M., Kalsow, B., Nelson, G., Modula-3 Language Definition, ACM SIGPLAN Notices Vol. 27, No. 8, pp. 15.42, (1992).
9. Powell, M. L., Kleiman, S. R., Barton, D., Shah, D., Stein, D., Weeks, M., SunOS Multi-thread Architecture. Usenix, Winter '91, Dallas, TX. (1991)
10. United States Department of Defence Reference Manual for the ADA Programming Language, ANSI/MIL-STD 1815A. (1983)
11. Birel, A., D., Nelson, B., J. Implementing Remote Procedure Calls, ACM Trans. on Computer Systems, Vol. 2, No. 1, pp. 39-59.(1984)
12. Sewry, D., A. Modula-2 Process Facilities. ACM SIGPLAN Notices, Vol. 19, No. 11, pp. 23-41.(1984)
13. Wirth, N. Programming in MODULA-2. Springer-Verlag. (1985)
14. INMOS Limited occam2 Reference Manual, Prentice-Hall. (1988)
15. Bal, H. E., Tanenbaum, A., S. Distributed Programming with Shared Data. Comput. Lang. Vol. 16, No. 2, pp. 129-146. (1991)
16. Silbershatz, A., Peterson, J., Gelvin, P., Operating System Concepts. Addison-Wesley. (1991)

17. Scott, M. L., LeBlanc, T. J., Marsh, B. D., Multi- Model Programming in Psyche. Proceedings of the 2nd ACM SIGPLAN Symposium on Principles and Practice of Parallel Programming, pp. 70-78, (1990).
18. Morriset, J. G., Tolmach, A., Procs and Locks: A Portable multiprocessing Platform for Standard ML of New Jersey, 4th ACM PPOPP, 5/93, CA, USA, SIGPLAN Notices, Vol. 28. No. 7. (1993).
19. Anderson, E., Bershad, B. N., Lazowska, E. D., Levy, H. M., Scheduler Activation: Effective Kernel Support for the User-Level Management of Parallelism. ACM Transactions on Computer systems, Vol. 10, No. 1, .pp 53-79, (1993).
20. Mccann, C., Vaswani, R., Zahorjan, J. A Dynamic Processor Allocation Policy for Multiprogrammed Shared-Memory Multiprocessors. ACM Transaction on Computer Systems, Vol. 11, No. 2, (1993).

APPENDIX 1

Presented below is the version of the module Processes which moves the scheduling algorithm implementation details to the module QueueHandler. This module exports the type ProcessQueue and some (self-explanatory) operations associated with this type:

```
DEFINITION MODULE QueueHandler;
    TYPE ProcessQueue;
    PROCEDURE Initq(VAR q: ProcessQueue);
    PROCEDURE Insert(VAR q: ProcessQueue;  p: ADDRESS);
    PROCEDURE Remove(VAR q: ProcessQueue; VAR p: ADDRESS);
    PROCEDURE Empty(q: ProcessQueue): BOOLEAN;
END QueueHandler.
```

An instance of the ProcessQueue called Ready is used for scheduling all ready coroutines and one instance of the ProcessQueue is used for all processes waiting for a given Signal.

Compared to Wirth's original solution we have introduced an additional attribute imported from the module SYSTEM. It is called INITIAL. It denotes the name of the process initiated by the environment when the Modula program is started. It becomes the first running process when the module Processes is initiated.

```
DEFINITION MODULE Processes; ... END Processes;
```

```
IMPLEMENTATION MODULE Processes[1];
    FROM QueueHandler IMPORT ProcessQueue, Initq, Insert, Remove, Empty;
```

```
FROM SYSTEM IMPORT ADDRESS, TRANSFER, NEWPROCESS,
   INITIAL;
FROM Storage IMPORT Allocate;

TYPE Signal = ProcessQueue;
VAR run: ADDRESS;
      Ready: ProcessQueue;

PROCEDURE StartProcess(P: PROC, n: CARDINAL);
VAR old, wsp: ADDRESS;
BEGIN Allocate(wsp, n); old:= run; NEWPROCESS(P, wsp, n, run);
   Insert(Ready, old); TRANSFER(old, run);
END StartProcess;

PROCEDURE Wait(VAR s: Signal);
VAR old: ADDRESS;
BEGIN Insert(s, run); old:= run;
   IF Empty(Ready) THEN HALT END;
   Remove(Ready, run); TRANSFER(old, run);
END Wait;

PROCEDURE Send(VAR s: Signal);
VAR old: ADDRESS;
BEGIN IF NOT Empty(s) THEN old:= run;
   Remove(s, run); Insert(Ready, old);  TRANSFER(old, run) END
END Send;

PROCEDURE Init(VAR s: Signal);
BEGIN Initq(s) END Init;

   BEGIN Initq(Ready); run:= INITIAL
END Processes.
```

APPENDIX 2

A revised version of the module HProcesses called MProcesses is presented below.

```
DEFINITION  MODULE MProcesses;
  TYPE Signal;
  PROCEDURE Enter;
  PROCEDURE Exit;
  PROCEDURE StartProcess(P: PROC, n: CARDINAL);
```

```
   PROCEDURE Wait(VAR s: Signal);
   PROCEDURE Send(VAR s: Signal);
   PROCEDURE Init(VAR s: Signal);
END MProcesses.

IMPLEMENTATION MODULE MProcesses;
   FROM QueueHandler IMPORT ProcessQueue, Initq, Insert, Remove, Empty;
   FROM SYSTEM IMPORT ADDRESS,
   FROM MKERNEL IMPORT  ENTER, EXIT, RELEASETRANSFER,
      TRANSFERREGAIN,  INITIAL, THIS, PROCESSOR, NEWPROCESS;
   FROM Storage IMPORT Allocate;

   TYPE Signal = ProcessQueue;
   VAR run, Idling: ARRAY PROCESSOR OF ADDRESS;
       Ready: ProcessQueue;

   PROCEDURE Enter;
   BEGIN ENTER END Enter;

   PROCEDURE Exit;
   BEGIN EXIT END Exit;

   PROCEDURE StartProcess(P: PROC, n: CARDINAL);
   VAR new, wsp: ADDRESS;
   BEGIN
      Allocate(wsp, n); NEWPROCESS(P, wsp, n, new);  Insert(Ready, new);
   END StartProcess;

   PROCEDURE Wait(VAR s: Signal);
   VAR old: ADDRESS;
   BEGIN Insert(s, run[THIS]); old:= run[THIS];
      IF NOT Empty(Ready) THEN Remove(Ready,  run[THIS]);
      ELSE run[THIS]:= Idling[THIS] END;
      RELEASETRANSFER(old, run[THIS])
   END Wait;

   PROCEDURE Send(VAR s: Signal);
   VAR old: ADDRESS;
   BEGIN IF NOT Empty(s)
      THEN old:= run[THIS]); Remove(s, run[THIS]);
         Insert(Ready, old); TRANSFERREGAIN(old, run[THIS])
      END
   END Send;
```

```
PROCEDURE Init(VAR s: Signal);
BEGIN Initq(s) END Init;

PROCEDURE Idle;
BEGIN LOOP ENTER;
  IF  Empty(Ready) THEN  EXIT
  ELSE Remove( Ready, run[THIS]);
      RELEASETRANSFER(Idling[THIS], run[THIS])
  END
END END Idle;

  BEGIN Initq(Ready); run[THIS]:= INITIAL
    FOR p:= MIN(PROCESSOR) TO MAX(PROCESSOR) DO
      Idling[p]:= StartProcess(Idle, 100) END
END MProcesses.
```

High-Level Abstractions for Efficient Concurrent Systems

Suresh Jagannathan and James Philbin

NEC Research Institute, 4 Independence Way, Princeton, 08540, NJ, USA.
{*suresh,philbin*}*@research.nj.nec.com*

1 Introduction

Parallel symbolic algorithms exhibit characteristics that make their efficient implementation on current multiprocessor platforms difficult: data is generated dynamically and often have irregular shape and density, data sets typically consist of objects of many different types and structure, the natural unit of concurrency is often much smaller than can be efficiently supported on stock hardware, efficient scheduling, migration and load-balancing strategies vary widely among different algorithms, and sensible decomposition of the program into parallel threads of control often cannot be achieved by mere examination of the source text.

Many of these challenges are also faced by implementors of multi-threaded operating systems and kernels[2, 23, 29]. In both cases, the utility of an implementation is determined by how well it supports a diverse range of applications in terms of performance and programmability.

Implementations that hard-wire decisions about process granularity, scheduling, and data allocation often perform poorly when executing parallel symbolic algorithms. An alternative approach to building concurrent systems would permit a system's internals to be customized by users *without* sacrificing performance or security. The generality and expressivity of the abstractions used by an implementation dictate the extent to which it may be customized.

Because different applications will exercise diferent operations in different ways, systems which permit customization of these operations on a per-application basis offer the promise of greater flexibility and programmability. First-class procedures[1, 21] and continuations[13, 28] are two abstractions we have found to be extremely useful in implementing a variety of environment and control operations. Languages such as Scheme[10] or ML[25] have demonstrated that these abstractions are effective building blocks for expressing a number of interesting data, program and control structures.

It is often claimed, however, that such high-level program constructs are too inefficient to serve as a foundation for building efficient high-performance concurrent systems. In this paper, we present evidence to the contrary. Sting is a highly efficient and flexible operating system implemented in Scheme. It provides mechanisms to (a) create lightweight asynchronous threads of control, (b) build customized scheduling, migration, and load-balancing protocols, (c) support a range of execution strategies from *fully eager* to *completely lazy* evaluation, (d) experiment with diverse storage allocation policies, and (e) handle multiple persistent address spaces.

The design of the Sting implementation relies heavily on both first-class procedures and continuations. Higher-order procedures are used to implement thread creation and synchronization operations, message passing, and customizable thread schedulers. Continuations[1] are used to implement state transition operations, exception handling, and important storage optimizations.

Earlier reports on Sting [17, 18] focussed primarily on program methodology and paradigms, discussing how one might express and use lightweight threads of control in a high-level programming language such as Scheme. In this paper, we discuss systems-level concerns within the Sting context, concentrating on the role of continuations and first-class procedures in the implementation of an efficient and general-purpose multi-threaded operating system and programming environment. In particular, we discuss three components of the Sting architecture not explicated elsewhere; these features are influenced heavily by the system's pervasive use of first-class procedures and continuations:

1. *First-class Ports:* Sting allows message-passing abstractions to be integrated within a shared-memory environment. A port is a first-class data object that serves as a receptacle for messages that maybe sent by other threads. Because Sting uses a shared virtual memory model [22], any complex data structure (including closures) can be sent along a port. This flexibility permits Sting applications to implement user-level message-passing protocols transparently and to combine the best features of shared memory and message passing within a unified environment.
2. *Memory Management:* A Sting virtual address space consists of a collection of *areas*; areas are used for organizing data that exhibit strong temporal or spatial locality. Sting supports a variety of areas: thread control blocks (or continuations), stacks, thread private heaps, thread shared heaps, etc.. Data is allocated to areas based on its intended use and lifetime; different areas can thus have different garbage collectors associated with them.
3. *Exception Handling:* As is the case with a thread-level context-switch, exceptions and interrupts are always handled in the execution context of some thread. Exception handlers are implemented as ordinary Scheme procedures, and dispatching an exception primarily involves manipulating continuations.

The remainder of this paper is structured as follows. Section 2 gives a brief overview of the system. Section 3 describes Sting's first-class user-level threads. Section 4 outlines the interaction of message-passing primitives and first-class procedures. Section 5 provides details of Sting's memory model. Section 6 describes *virtual processors*, an abstraction that permit significant programmer-level control over mapping, scheduling and load-balancing of lightweight threads. Section 7 describes abstract physical machines, and Sting's exception handling mechanism. Section 8 provides benchmark results. Conclusions and comparison to related work is given in Section 9.

---

[1] A continuation is an abstraction of a program point. It is typically represented as a procedure of one argument that defines the remaining computation needed to be performed from the program point it denotes [13].

2 Sting Overview

Sting is a parallel dialect of Scheme[10] designed to serve as a high-level operating system for modern symbolic parallel programming languages.

The abstract architecture of Sting is organized hierarchically as layers of abstractions (see Fig. 1). The lowest level in this hierarchy is an *Abstract Physical Machine* (APM). An APM consists of a collection of *Abstract Physical Processors* (APP). An APP is an abstraction of an actual computing engine. An abstract physical machine contains as many abstract physical processors as there are real processors in a multiprocessor environment.

A *Virtual Machine* (VM) is an abstraction that is mapped onto a APM or a portion thereof. Virtual machines manage a single address space. They are also responsible for mapping global objects into local address spaces. In addition, each virtual machine contains the root of a graph of objects (*i.e.,* root environment) that is used to trace the set of live objects in its associated address space. A virtual machine is closed over a set of *Virtual Processors* (VPs). Virtual processors execute on abstract physical processors.

Abstract physical machines are responsible for managing virtual address spaces, and shared objects, handling hardware device interrupts and coordinating physical processors. Associated with each physical processor P is a virtual processor policy manager that implements policy decisions for the virtual processors that execute on P.

Virtual processors are responsible for managing user-created *threads*. A thread defines a separate locus of control. In addition, virtual processors also handle non-blocking I/O, synchronous exceptions (*e.g.,* invalid instructions, memory access violations), and interrupts (page refill, thread quantum expiration, etc.). Each virtual processor V is closed over a Thread Policy Manager (TPM) that (a) schedules threads executing within V, (b) migrates threads to and from other VPs, and (c) performs initial thread placement on the VPs defined within a given virtual machine.

Virtual processors are multiplexed on a physical processor in the same way that threads are multiplexed on a virtual processor. Separating the virtual machine abstraction from a specific hardware configuration allows us to map virtual topologies (in terms of virtual processors) onto any concrete architectural surface[12, 16].

Physical processors context-switch virtual processors because of preemption, or because a VP specifically requests a context switch (*e.g.,* because of an I/O call initiated by its current thread). We discuss virtual processors in greater detail in the following sections. State transitions on threads are implemented by a thread controller that implements the thread interface and thread state transitions.

Abstract physical machines, abstract physical processors, virtual machines, virtual processors and threads are all first-class objects in Sting. Since all these abstractions are implemented in terms of Scheme objects, the policy decisions of a processor (both physical and virtual) can be fully customized; for example, different virtual processors (even in the same virtual machine) may be closed over different policy managers. A processor is closed over a given policy manager by simply assigning a procedure to the

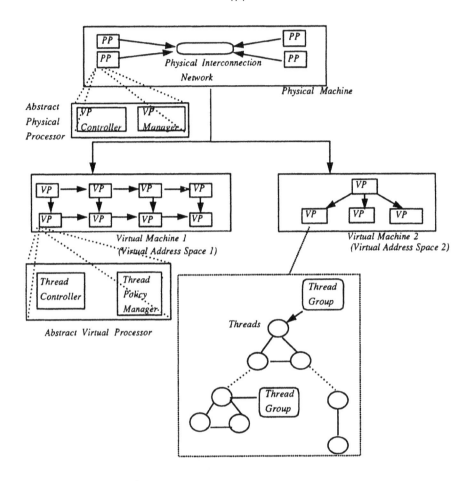

Fig. 1. The Sting Abstract Architecture.

appropriate slot of processor's data structure; this procedure defines an implementation of a scheduling or migration policy. Thus, processor abstractions serve to separate issues of policy (*i.e.*, scheduling, migration, etc.) from control (*i.e.*, blocking, suspension, termination etc.). This separation of concerns leads to important expressivity gains, and the use of first-class procedures ensures no performance penalty is incurred as a result of this flexibility.

3 Threads

Threads are first-class objects in Sting. Thus, they may be passed as arguments to procedures, returned as results, and stored in data structures. Threads can outlive the objects that create them. A thread's state contains a thunk, *i.e.*,a nullary procedure, that is invoked when the thread executes. The value of the application is stored in the thread on completion. For example, evaluating the expression:

```
(*fork-thread (lambda () (+ y (* x z))) (current-vp))
```

creates a lightweight thread of control which is enqueued for execution on its evaluating virtual processor. When run, the thread applies the procedure,

```
(lambda () (+ y (* x z)))
```

The evaluation environment of this thunk is its lexical environment. We can avoid writing explicit thunks by using the syntactic form, `fork-thread`; the expression,

```
(fork-thread E V)
```

is equivalent to,

```
(*fork-thread (lambda () E) V)
```

A thread can be either *delayed, scheduled, evaluating, absorbed* or *determined*. A *delayed* thread will never be run unless the value of the thread is explicitly demanded. A *scheduled* thread is a thread that is scheduled to be evaluated on some VP, but which has not yet been allocated storage resources. An *evaluating* thread is a thread that has started running. A thread remains in this state until the invocation of its thunk yields a result, at which point its state becomes *determined*. Absorbed threads are an important specialization of evaluating threads used to optimize fine-grained programs in which threads exhibit significant data dependencies among one another[17].

Users manipulate threads via a set of procedures and syntactic forms defined by a thread controller (TC) which implements synchronous state transitions on a thread's state. The TC is written entirely in Scheme with the exception of two primitive operations to save and restore registers.

Because Scheme supports first-class procedures seamlessly, it is easy to build variants on the functionality provided by the TC *without* altering the implementation. For example, consider the following expression:

```
(define (P op)
  (let ((v initial value))
    (op (lambda () (f v)))))
```

If P is applied thus,

```
(P (lambda (proc) (fork-thread (proc) (current-vp))))
```

the application of f to v occurs eagerly in a separate thread; in a slightly different call to P ,

```
(P (lambda (proc) (delay-thread proc)))
```

the application is evaluated lazily. It is straightforward to construct several other alternatives, none of which involve altering P 's definition. Concurrency abstraction follows naturally from procedural abstraction in this framework.

3.1 Control State

Control state in Sting is managed via *thread control blocks* (TCBs). A TCB is a generalized representation of a *continuation*[13]; it records its associated thread's current state (registers, program counter, etc.). A TCB also includes its own stack and heaps.

The Sting implementation delays the allocation of a TCB to a thread until necessary. In most thread systems, the act of creating a thread involves not merely setting up the environment for the thread to be forked, but also allocating and initializing storage. This approach lowers efficiency in three important respects: first, in the presence of fine-grained parallelism, the thread controller may spend more time creating and initializing threads than actually running them. Second, since stacks and process control blocks are immediately allocated upon thread creation, context switches among threads often cannot take advantage of cache and page locality; in fact, locality is often significantly reduced in such an implementation. Third, if threads T_1 and T_2 are both scheduled, but T_1 evaluates fully before T_2, T_2 will not reuse the storage allocated to T_1; thus, the aggregate storage requirements for such a program will be greater than necessary. Sting permits storage allocated to a one thread to be recycled and used by other threads that execute subsequently when possible [17, 18].

Thread control blocks are allocated from a pool local to each VP. This pool is organized as a LIFO queue. When a thread terminates, its TCB is recycled on the TCB pool of the VP on which it was executing; the LIFO organization of this pool guarantees that this TCB will be allocated to the next new thread chosen for execution on this VP. Since it is likely that the most recently used pieces of the TCB will be available in the physical processor's working set, cache and page locality is not compromised as a consequence of initiating evaluation of a new thread. Sting incorporates one further optimization on this basic theme: if a thread T terminates on VP V, and V is next scheduled to begin evaluation of a new thread (*i.e.,* a thread that is not yet associated with a dynamic context), T's TCB is immediately allocated to the new thread; no pool management costs are incurred in this case.

Besides local VP pools, VPs share access to a global TCB pool. Every local VP pool maintains an overflow and underflow threshold. When a pool overflows, its VP moves half the TCBs in the pool to the global pool; when the pool underflows, a certain number of TCBs are moved from the global pool to the VP-local one. Global pools serve two purposes: (1) they minimize the impact of program behavior on TCB allocation and reuse, and (2) they ensure a fair distribution of TCBs to all virtual processors. Since new TCBs are created only if both the global and the VP local pool are empty, the number of TCBs actually created during the evaluation of a Sting program is determined collectively by all VPs.

4 Message-Passing Communication

Sting uses a shared virtual memory model. Implementations of Sting on distributed memory platforms must be built on top of a distributed shared virtual memory substrate[5, 22]. Thus, the meaning of a reference does not depend on where the reference is generated, or where the object is physically located.

Message-passing has been argued to be an efficient communication mechanism on disjoint memory architectures [20], especially for parallel applications that are coarse-grained, or have known communication patterns. A *port* is a data abstraction provided in Sting to minimize the overheads of implementing shared memory on disjoint memory architectures. First-class procedures and ports exhibit an interesting and elegant synergy in this context.

Ports are first-class data structures. There are two basic operations provided over ports:

1. (put *obj port*) copies *obj* to *port*. The operation is asynchronous with respect to the sender.
2. (get *port*) removes the first message in *port*, and blocks if *port* is empty.

Objects read from a port P are copies of objects written to P. This copy is a shallow copy, *i.e.*, only the top-level structure of the object is copied, the substructure is shared. We have designed ports with copying semantics because they are designed to be used when shared memory would be inefficient. While the standard version of put does a shallow copy, there is also a version available that does a deep copy; this version not copies the top-level object, but also all its substructures.

For example, sending a closure in a message using shallow copying involves constructing a copy of the closure representation, but preserving references to objects bound within the environment defined by the closure. The choice of copy mechanisms used clearly is influenced by the underlying physical architecture, and the application domain. There are a range of message transmission implementations that can be tailored to the particular physical substrate on which the Sting implementation resides.

Thus, evaluating the expression,

```
(put (lambda () E) port)
```

transmits the closure of the procedure (lambda () E) to *port*. If we a define a receiver on *port* thus,

```
(define (receiver port)
  (let ((msg (get port)))
    (fork-thread (msg) (current-vp))
    (receiver)))
```

the procedural object sent is evaluated on the virtual processor of the receiver. By creating a new thread to evaluate messages, the receiver can accept new requests concurrently with the processing of old ones.

This style of communication has been referred to as *active messages* [31] since the action that should be taken upon message receipt is not encoded as part of the underlying implementation but determined by the message itself. There is a great deal of flexibility and simplicity afforded by such a model since the virtual processor/thread interface does not require any alteration in order to support message communication. Two aspects of the Sting design are crucial for realizing this functionality: (1) the fact that objects reside in a shared virtual memory² allow all objects (including those containing references to other objects, *e.g.*, closures) to be transmitted among virtual processors freely, and (2) first-class procedures permit complex user-defined message handlers to be constructed; these handlers can execute in a separate thread on any virtual processor. To illustrate, in the above example, E may be a complex query of a database. If a receiver is instantiated on the processor on which the database resides, such queries do not involve expensive migration of the database itself. Communication costs are reduced because queries are directly copied to the processor on which the database resides; the database itself does need not to migrate to processors executing queries. The ability to send procedures to data rather than more traditional RPC-style communication leads to a number of potentially significant performance and expressivity gains [14].

First-class procedures and lightweight threads make active message passing an attractive high-level communication abstraction. In systems that support active messages without the benefit of these abstraction, this functionality is typically realized in terms of low-level support protocols. First-class procedures make it possible to implement active message trivially; an active message is simply a procedure sent to a port. We also note that first-class ports have obvious and important utility in distributed computing environments as well and lead to a simpler and cleaner programming model than traditional RPC.

5 Memory Management

In Sting, there are three storage areas associated with every TCB. The first, a stack, is used to allocate objects created by the thread whose lifetime does not exceed the dynamic extent of its creator. More precisely, objects allocated on a stack may only refer to other objects that are allocated in a current (or earlier) stack frame, or which are allocated on some heap. Stack allocated objects can refer to objects in heaps because the thread associated with the stack is suspended while the heap is garbage collected; references contained in stacks are part of the root set traced by the garbage collector.

---

² On distributed memory machines, objects would reside in a distributed shared virtual memory.

5.1 Private Heaps

Thread private heaps are used to allocate non-shared objects whose lifetimes might exceed the lifetime of the procedure that created them. We say might exceed because it is not always possible for the compiler to determine the lifetime of an object in higher order programming languages such as Scheme or ML. Furthermore, it may not be possible to determine the lifetimes of objects in languages which allow calls to higher-order unknown procedures. References contained in a private heap can refer to other objects in the same private heap, or objects in shared heaps, but they cannot refer to objects in the stack. References in the stack may refer to objects in the private heap, but references in the shared heap may not. Private heaps lead to greater locality since data allocated on them are used exclusively by a single thread of control; the absence of interleaving allocation among multiple threads means that objects close to together in the heap are likely to be logically related to one another.

No other thread can access objects that are contained in a thread's stack or private heap. Thus, both thread stacks and private heaps can be implemented in local memory on the processor without any concern for synchronization or memory coherency. Thread private heaps are actually a series of heaps organized in a generational manner. Storage allocation is always done in the youngest generation in a manner similar to other generational collectors [30, 4]. As objects age they are moved to older generations. All garbage collection of the private heap is done by the thread itself. In most thread systems that support garbage collection all threads in the system must be suspended during a garbage collection [19]. In contrast, Sting's threads garbage collect their private heaps independently and asynchronously with respect to other threads. Thus, other threads can continue their computation while any particular thread collects its private heap; this leads to better load balancing and higher throughput. A second advantage of this garbage collection strategy is that the cost of garbage collecting a private heap is charged only to the thread that allocates the storage, rather than to all threads in the system.

5.2 Thread Groups and Shared Heaps

Sting provides *thread groups* as a means of gaining control over a related collection of threads. Every thread is associated with some thread group. A child thread is in the same group as its parent unless it is created as part of a new group. Thread groups provide operations analogous to ordinary thread operations (*e.g.,* termination, suspension, etc.) as well as operations for debugging and monitoring (*e.g.,* listing all threads in a given group, listing all groups, profiling, genealogy information, etc..) In addition, a thread group also includes a *shared heap* accessible to all its members.

A thread group's shared heap is allocated when the thread group is created. The shared heap like the private heap is actually a series of heaps organized in a generational manner. References in shared heaps may only refer to other objects in shared heaps. This is because any object that is referenced from a shared object is also a shared object and, therefore must reside in a shared heap. This constraint on shared heaps is enforced by ensuring that references stored in shared heaps refer to objects that are (a) either in a shared heap, or (b) allocated in a private heap and garbage collected into a shared one. That is, the graph of objects reachable from the referenced object must be copied into or located in the shared heap. The overheads of this memory model depend on how frequently references to objects allocated on private heaps escape; in our experience in implementing fine-grained parallel programs, we have found that most objects allocated on a private heap remain local to the associated thread, and are not shared. Those objects that are shared among threads often are easily detected either via language abstractions or by compile-time analysis.

To summarize, the reference discipline observed between the three areas associated with a thread are as follows:

1. References in the stack refer to objects in its current or previous stack frame, its private heap, or its shared heap.
2. References in the private heap refer to objects on that heap or to objects allocated on some shared heap.
3. References in the shared heap refer to objects allocated on its shared heap (or some other shared heap).

Like private heaps shared heaps are organized in a generational manner, but garbage collection of shared heaps is more complicated than that for private heaps because many different threads can simultaneously access objects contained in the shared heap. Note that as a result, shared heap allocation requires locking the heap.

In order to garbage collect a shared heap, all threads in the associated thread group (and its inferiors) are suspended. This is because any of these threads can access data in the shared heap. However, other threads in the system, *i.e.*, those not inferior to the group associated with the heap being collected, continue execution independent of the garbage collection.

Each shared heap has a set of incoming references associated with it. These sets are maintained by checking for stores of references that cross area boundaries. After the threads associated with the shared heap have been suspended, the garbage collector uses the set of incoming references as the roots for the garbage collection. Any objects reachable from the incoming reference set are copied to the new heap. When the garbage collection is complete the threads associated with the shared heap are resumed.

6 Virtual Processors and Thread Policy Managers

Virtual processors define scheduling, migration, and load-balancing decisions for the threads they execute. A virtual processor is closed over a thread policy manager (TPM) which defines a set of procedures that collectively determine a thread scheduling regime for this VP. Virtual machines or VPs can thus be tailored to handle different scheduling protocols or policies. The implementation relies on first-class procedures to attain this flexibility; VPs can execute different scheduling protocols simply by closing themselves over different TPMs.

The Sting design seeks to provide a flexible framework that is able to incorporate different scheduling regimes transparently without requiring modification to the thread controller itself. To this end, all TPMs must conform to the same interface although no constraints are imposed on the implementations themselves. The interface provides operations for choosing a new thread to run, enqueuing a scheduled or evaluating thread, setting thread priorities, and migrating threads. These procedures are expected to be used exclusively by the thread controller (TC); in general, user applications need not be aware of the thread policy manager/thread controller interface. A detailed description of policy managers and virtual processors is given in [17, 26].

7 Physical Processors

Sting is intended to serve as an operating system for modern programming languages. Like other contemporary operating systems (*e.g.*, Mach[7, 29], Chorus[27], or Psyche[23]), Sting's lowest-level abstraction is a micro-kernel called the *Abstract Physical Machine (APM)*.

The APM plays three important roles in the Sting software architecture:

1. It provides a secure and efficient foundation for supporting multiple virtual machines.
2. It isolates all other components in the system from hardware dependent features and idiosyncrasies.
3. It controls access to the physical hardware of the system.

Physical processors multiplex virtual processors just as virtual processors multiplex threads. Each physical processor in an APM includes a virtual processor controller (VPC), and a virtual processor policy manager (VPM). In this sense, physical processors are structurally identical to virtual processors. The VPC effects state changes on virtual processors in the same way that the TC effects state changes on threads. Like threads, virtual processors may be running, ready, blocked, terminating or dead. A running VP is currently executing on a physical processor; a ready VP is capable of running, but is currently not. A blocked VP is executing one or more threads waiting on some external event (*e.g.*, I/O). The VPM is responsible for scheduling VPs on a physical processor; its structure is similar to a TPM, although the scheduling policies it defines will obviously be quite different. The VPM presents a well-defined interface to the VP controller; different Sting systems can contain different VP policy managers (*e.g.*, time-sharing systems have different scheduling requirements from real-time ones).

In current micro-kernels, program code in the micro-kernel is significantly different from that found in user programs. This occurs because many of the facilities available to user programs are not available at the kernel level. Sting addresses this problem by implementing the abstract physical machine in the *root virtual machine*. The root virtual machine has all the facilities that are available to any other program (or subsystem) running in a virtual machine including, a virtual address space, virtual processors, and threads. In addition, it has abstract physical processors, device drivers, and a virtual memory manager. A Sting abstract physical machine therefore has several important characteristics that distinguish it from other operating systems:

1. There are no heavyweight threads in the system. All threads are lightweight.
2. There are no kernel threads or stacks for implementing system calls. All system calls use the execution context of the thread making the system call. This is possible because portions of the abstract physical machine are mapped into every virtual machine in the system.
3. Asynchronous programming constructs in the abstract physical machine are implemented using threads as in any other virtual machine. Threads in the abstract physical machine can be created, terminated, and controlled in the same manner that threads in any virtual machine can.
4. When a thread blocks in the kernel it can inform its virtual processor that it has blocked. The virtual processor can then choose to execute some other thread. This is true for inter-thread communication as well as for I/O (*e.g.*, page faults).

A primary responsibility for APMs is to create, destroy and manage virtual machines. A virtual machine loosely corresponds to a Unix kernel process, a Mach task, or a Topaz address space [6]. Each of these entities define a virtual address space and a thread, but in these other systems the thread is a kernel thread, and therefore, heavyweight. Furthermore, these systems have no concept of customizable virtual processors or machines; thus, constructing scheduling, migration and load-balancing protocols in user-space is not possible. Schedular activations [3] and Psyche's processor abstraction [23] address one of the problems that Sting's virtual machines are intended to solve, *i.e.*, user space threads blocking in the kernel and informing the kernel when no thread is available to run in a kernel process or task; however, scheduler activations do not address customizable virtual processors, and do not leverage off high-level language mechanisms such as continuations or first-class procedures.

Psyche kernel processes are used to implement the virtual processors that execute user level threads. When a user thread blocks in the kernel, the kernel calls the software interrupt handler for this condition. The handler may decide to block the virtual processor, run another thread, or do whatever else is appropriate for the particular thread system being implemented. Software interrupts allow the kernel to notify the virtual processor whenever an event which might be of interest to it occurs in the kernel. Given Psyche's virtual processors it is possible to implement many different thread semantics and many different scheduling policies, and thus, Psyche is fully customizable. However, Psyche does not separate control and policy mechanisms in the virtual processor and thus each thread package must implement both of these mechanisms. In Psyche, user threads are distinct from kernel threads. In Sting, lightweight threads are integrated into the kernel design so that no kernel threads are necessary. We believe the Sting approach provides a significant increase in programmer efficiency, while at the same time providing an increase in program efficiency compared to that of more traditional operating systems.

To create a new virtual machine, an abstract physical machine must:

-- create a new virtual address space,
-- map itself into the virtual address space,
-- create a root virtual processor in the virtual address space,
-- and (possibly) create other virtual processors.

In addition, it must also allocate necessary abstract physical processors and map the virtual processors of the new machine onto them. It must also schedule the root virtual processor to run on its corresponding abstract physical processor. The root virtual processor executes the root thread for the virtual machine.

Destroying a virtual machine terminates all non-root threads executing on any virtual processors in the virtual machine. All devices are closed, all persistent areas in the address space are unmapped, and the root thread of the root virtual processor notifies its abstract physical processor which de-allocates the virtual address space associated with the virtual machine.

7.1 Exceptions

Of particular interest is Sting's treatment of exceptions (both synchronous and asynchronous) within the APM. Exception handling in Sting relies fundamentally on first-class procedures and continuations, and offers a good domain to highlight the utility of these abstractions in an important systems programming application.

Associated with every exception is a handler responsible for performing a set of actions to deal with the exception. Handlers are procedures that execute within a thread. An exception raised on processor P executes using the context of P's current thread. There are no special exception stacks in the Sting micro-kernel.

When an exception (*e.g.*, invalid instruction, memory protection violation, etc) is raised on processor P, P's current continuation (*i.e.*, program counter, heap frontier, top-of-stack, etc.) is first saved. The exception dispatcher then proceeds to find the target of the exception, interrupting it if the thread is *running*, and pushing the continuation of the handler and its arguments onto the target thread's stack. Having done so, the dispatcher may choose to (a) resume the current thread by simply returning into it, (b) resume the target thread, or (c) call the thread controller to resume some other thread on this processor. When the target thread is resumed, it will execute the continuation found on the top of its stack; this is the continuation of the exception handler.

The implementation of exceptions in Sting is novel in several respects:

1. Handling an exception simply involves calling it since it is a procedure.
2. Exceptions are handled in the execution context of the thread receiving it.
3. Exceptions are dispatched in the context of the current thread.
4. Exceptions once dispatched become the current continuation of the target thread and are executed automatically when the thread is resumed.
5. An exception is handled only when the target thread is resumed.

6. Exception handling code is written in Scheme and manipulates continuations and procedures to achieve the desired effect.

The target thread of a synchronous exception is always the current thread. Asynchronous exceptions or interrupts are treated slightly differently. Since interrupts can be directed at any thread (not just the currently executing one), handling such exceptions requires the handler to either process the exception immediately, interrupt the currently running thread to handle the exception, or create a new handler thread. Since interrupt handlers are also Scheme procedures, establishing a thread to execute the handler or using a current thread for that purpose merely involves setting the current continuation of the appropriate thread to call the handler.

```
 1: (define (exception-dispatcher type . args)
 2:   (save-current-continuation)
 3:   (let ((target handler (get-target&handler type args)))
 4:     (cond ((eq? target (current-thread))
 5:            (apply handler args))
 6:           (else
 7:            (signal target handler args)
 8:            (case ((exception-priority type))
 9:              ((continue) (return))
10:              ((immediate) (switch-to-thread target))
11:              ((reschedule) (yield-processor)))))))
```

Fig. 2. Pseudo Code for the Sting Exception Dispatcher.

To illustrate, Figure 2 shows pseudo-code for the Sting exception dispatcher. In line 2, the current continuation is saved on the stack of the current thread. The continuation can be saved on the stack because it cannot escape and it will only be called once. On line 3 the dispatcher finds the thread for which the exception is intended and the handler for the exception type. Line 4 checks to see if the target of the exception is the current thread and if so does not push the exception continuation (line 5). Rather, the dispatcher simply applies the handler to its arguments. This is valid since the dispatcher is already running in the context of the exception target, i.e. the current thread. If the target of the exception is not the current thread, the dispatcher sends the exception to the target thread (line 7). Sending a thread a signal is equivalent to interrupting the thread and pushing a continuation containing the signal handler and its arguments onto the thread's stack, and resuming the thread which causes the signal handler to be executed. After signaling the target thread, the handler decides which thread to run next on the processor (line 8). It may be itself (line 9), the target thread (line 10), or the thread with the highest priority (line 11).

There is one other important distinction between Sting's exception handling facilities and those found in other operating systems. Since threads that handle exceptions are no different from other user-level threads in the system (*e.g.*, they have their own stack and heap), and since exception handlers are ordinary first-class procedures, handlers are free to allocate storage dynamically. Data generated by a handler will be reclaimed by a garbage collector in the same way that any other datum is recovered. The uniformity between the exception handling mechanism and higher-level Sting abstractions allows device driver implementors expressivity and efficiency not otherwise available in parallel languages or operating systems.

Sting is able to provide this model of exceptions because first-class procedures and threads, manifest continuations, dynamic storage allocation, and a uniform addressing mechanism are all central features of its design.

8 Benchmarks

Sting is currently implemented on Silicon Graphics MIPS R3000 shared-memory multiprocessor (cache-coherent) machines. The physical machine configuration maps physical processors to lightweight Unix threads; each node in the machine runs one such thread. We ran the benchmarks shown below using a virtual machine in which each physical processor runs exactly one virtual processor. The timings shown below were taken on an 8 processor configuration.

Fig. 3 gives baseline figures for various thread operations; these timings were derived using a single global FIFO queue. Our processor platform was an eight node 40MHz SGI cache-coherent shared-memory machine.

Case	Timings(in μseconds)
Thread Creation	40
Thread Fork and Value	86.9
Thread Enqueue/Dequeue	14.5
Synchronous Context Switch	12.8
Thread Block and Resume	27.9

Fig. 3. Baseline timings.

The "Thread Creation" timing is the cost to create a thread. "Thread Fork and Value" measures the cost to create a thread that evaluates the null procedure and returns. "Thread Enqueue/Dequeue" is the cost of inserting a thread into the ready queue of the current VP, and immediately removing it; it is thus one measure of scheduling overhead. A "Synchronous Context Switch" is the cost of context-switching a thread that resumes immediately. "Thread Block and Resume" is the cost of blocking and resuming a null thread.

Fig. 4 shows benchmark times for four applications (solid lines indicate actual wallclock times; dashed lines indicate ideal performance relative to single processor times):

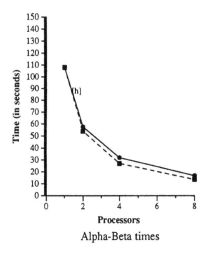

Alpha-Beta times

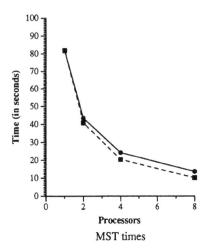

MST times

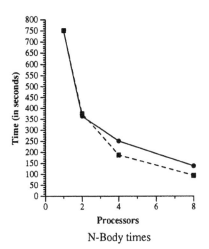

N-Body times

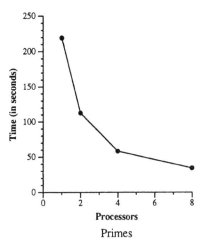

Primes

Fig. 4. Wallclock times on benchmark suite.

1. **Primes:** This program computes the first million primes. It uses a master-slave algorithm; each slave computes all primes within a given range.
2. **Minimum Spanning Tree:** This program implements a geometric minimum spanning tree algorithm using a version of Prim's algorithm. The input data is a fully connected graph of 5000 points with edge lengths determined by Euclidean distance. The input space is divided among a fixed number of threads. To achieve better load balance, each node not in the tree does a parallel sort to find its minimum distance among all nodes in the tree. Although 46,766 threads are created with this input, only 16 TCBs are actually generated. This is again due to Sting's aggressive treatment of storage locality, reuse and context sharing.
3. **Alpha-Beta:** This program defines a parallel implementation of a game tree traversal algorithm. The program does α/β pruning[11, 15] on this tree to minimize the amount of search performed. The program creates a fixed number of threads; each of these threads communicate their α/β values via distributed data structures. The input used consisted of a tree of depth 10 with fanout 8; the depth cutoff for communicating α and β values was 3. To make the program realistic, we introduced a 90% skew in the input that favored finding the best node along the leftmost branch in the tree.
4. **N-body:** This program simulates the evolution of a system of bodies under the influence of gravitational forces. Each body is modeled as a point mass and exerts forces on all other bodies in the system. The simulation proceeds over time-steps, each step computing the net force on every body and thereby updating that body's position and other attributes. This benchmark ran the simulation on 5000 bodies over six time-steps.

Speedup and efficiency ratios for these problems are shown in Fig. 5.

Benchmark	Speedup	Efficiency
Primes	6.4	81
MST	5.9	75
N-Body	5.5	69
Alpha-Beta	6.5	81

Fig. 5. Efficiency and speedup rations on 8 processors.

9 Conclusions

The utility of high-level language abstractions such as first-class procedures and continuations is well-appreciated in sequential symbolic application domains. Sting is an attempt to generalize the purview of these abstractions to systems software. The need for efficiency often compromises generality and elegance in operating system implementations. Our investigations support the hypothesis that with careful engineering and thoughtful design, useful high-level abstractions can be profitably applied to such performance critical domains. The free use of these abstractions also leads to smaller and simpler systems and data structures.

Besides providing an efficient multi-threaded programming system similar to other lightweight thread systems[8, 9, 24], the liberal use of procedural and control abstractions provide added flexibility. First, advanced memory management techniques are facilitated. Second, Sting's pervasive use of continuation objects in implementing thread storage permits a variety of other storage optimizations not available otherwise[18]. Third, first-class procedures provide an expressive basis for implementing flexible message-passing protocols. Finally, Sting is built on an abstract machine intended to support long-lived applications, persistent objects, and multiple address spaces. Thread packages provide none of this functionality since (by definition) they do not define a complete program environment.

In certain respects, Sting resembles other advanced multi-threaded operating systems[3, 23]: for example, it supports user control over interrupts, and exception handling at the virtual processor level. It cleanly separates user-level and kernel-level concerns: physical processors handle (privileged) system operations and operations across virtual machines; virtual processors implement all user-level thread and local address-space functionality. However, because Sting is an extended dialect of Scheme, it provides the functionality and expressivity of a high-level programming language (*e.g.*, first-class procedures, continuations, general exception handling, and rich data abstractions) that typical operating system environments do not offer.

References

1. Harold Abelson and Gerald Sussman. *Structure and Interpretation of Computer Programs.* MIT Press, Cambridge, Mass., 1985.
2. Thomas Anderson, Edward Lazowska, and Henry Levy. The Performance Implications of Thread Management Alternatives for Shared Memory MultiProcessors. *IEEE Transactions on Computers*, 38(12):1631–1644, December 1989.
3. Thomas E. Anderson, Brian N. Bershad, Edward D. Lazowska, and Henry M. Levy. Scheduler activations: effective kernel support for the user-level management of parallelism. In *Proceedings of 13th ACM Symposium on Operating Systems Principles*, pages 95–109. Association for Computing Machinery SIGOPS, October 1991.
4. Andrew Appel. A Runtime System. *Journal of Lisp and Symbolic Computation*, 3(4):343–380, November 1990.
5. John K. Bennett, John B. Carter, and Willy Zwaenepoel. Munin: Distributed Shared Memory Based on Type–Specific Memory Coherence. In *Symposium on Principles and Practice of Parallel Programming*, March 1990.
6. A.D. Birrell, J.V. Guttag, J.J. Horning, and R. Levi. Synchronization Primitives for a Multiprocessor: A Formal Specification. In *Proceedings of the 11th Symposium on Operating Systems Principles*, pages 94–102, November 1987.

7. David L. Black, David B. Golub, Daniel P. Julin, Richard Rashid, Richard P. Draves, Randall W. Dean, Alessandro Forin, Joseph Barrera, Hideyuki Tokuda, Gerald Malan, and David Bohman. Microkernel Operating System Architecture and Mach. In *Workshop Proceedings Micro-Kenels and other Kernel Architectures*, pages 11–30, April 1992.

8. Eric Cooper and Richard Draves. C Threads. Technical Report CMU-CS-88-154, Carnegie-Mellon University, June.

9. Eric Cooper and J.Gregory Morrisett. Adding Threads to Standard ML. Technical Report CMU-CS-90-186, Carnegie-Mellon University, 1990.

10. William Clinger *et. al.* The Revised Revised Revised Report on Scheme or An UnCommon Lisp. Technical Report AI-TM 848, MIT Artificial Intelligence Laboratory, 1985.

11. Raphael Finkel and John Fishburn. Parallelism in Alpha-Beta Search. *Artificial Intelligence*, 19(1):89–106, 1982.

12. David Saks Greenberg. *Full Utilization of Communication Resources*. PhD thesis, Yale University, June 1991.

13. Christopher Haynes and Daniel Friedman. Embedding Continuations in Procedural Objects. *ACM Transactions on Programming Languages and Systems*, 9(4):582–598, 1987.

14. Wilson Hsieh, Paul Wang, and William Weihl. Computation Migration: Enhancing Locality for Distributed-Memory Parallel Systems. In *Fourth ACM SIGPLAN Symposium on Principles and Practice of Parallel Programming*, pages 239–249, May 1993. Also appears as ACM SIGPLAN Notices, Vol. 28, number 7, July, 1993.

15. Feng hsiung Hsu. *Large Scale Parallelization of Alpha-Beta Search: An Algorithmic and Architectural Study with Computer Chess*. PhD thesis, Carnegie-Mellon University, 1990. Published as Technical Report CMU-CS-90-108.

16. Paul Hudak. Para-Functional Programming. *IEEE Computer*, 19(8):60–70, August 1986.

17. Suresh Jagannathan and James Philbin. A Customizable Substrate for Concurrent Languages. In *ACM SIGPLAN '91 Conference on Programming Language Design and Implementation*, June 1992.

18. Suresh Jagannathan and James Philbin. A Foundation for an Efficient Multi-Threaded Scheme System. In *Proceedings of the 1992 Conf. on Lisp and Functional Programming*, June 1992.

19. David Kranz, Robert Halstead, and Eric Mohr. Mul-T: A High Performance Parallel Lisp. In *Proceedings of the ACM Symposium on Programming Language Design and Implementation*, pages 81–91, June 1989.

20. David Kranz, Kirk Johnson, Anant Agarwal, John Kubiatowicz, and Beng-Hon Lim. Integrating Message-Passing and Shared-Memory: Early Experience. In *Proceedings of the Fourth ACM SIGPLAN Symposium on Principles and Practice of Parallel Programming*, pages 54–63, 1993.

21. Peter J. Landin. The Mechanical Evaluation of Languages. *Computer Journal*, 6(4):308–320, January 1964.

22. Kai Li and Paul Hudak. Memory Coherence in Shared Virtual Memory Systems. *ACM Transactions on Computer Systems*, 7(4):321–359, November 1989.

23. Brian D. Marsh, Michael L. Scott, Thomas J. LeBlanc, and Evangelos P. Markatos. First-class user-level threads. In *Proceedings of 13th ACM Symposium on Operating Systems Principles*, pages 110–21. Association for Computing Machinery SIGOPS, October 1991.

24. Sun Microsystems. *Lightweight Processes*, 1990. In SunOS Programming Utilities and Libraries.

25. Robin Milner, Mads Tofte, and Robert Harper. *The Definition of Standard ML*. MIT Press, 1990.

26. James Philbin. *An Operating System for Modern Languages*. PhD thesis, Dept. of Computer Science, Yale University, 1993.

27. M. Rozier, V. Abrossimov, F. Armand, I. Boule, M. Gien, M. Guillemont, F. Herrman, C. Kaiser, S. Langlois, P. Leonard, and W. Newhauser. Overview of the Chorus Distributed Operating System. In *Workshop Proceedings Micro-Kenels and other Kernel Architectures*, pages 39–69, April 1992.

28. Guy Steele Jr. Rabbit: A Compiler for Scheme. Master's thesis, Massachusetts Institute of Technology, 1978.

29. A. Tevanian, R. Rashid, D. Golub, D. Black, E. Cooper, and M. Young. Mach Treads and the UNIX Kernel: The Battle for Control. In *1987 USENIX Summer Conference*, pages 185–197, 1987.

30. David Ungar. Generation Scavenging: A Non-Disruptive High Performance Storage Reclamation Algorithm. In *Proceedings of the ACM SIGSOFT/SIGPLAN Software Engineering Symposium on Practical Software Development Environments*, pages 157–167, 1984.

31. Thorsten Von Eicken, David Culler, Seth Goldstein, and Klaus Schauser. Active Messages: A Mechanism for Integrated Communication and Compuation. In *Proceedings of the 19th International Symposium on Computer Architecture*, 1992.

Language and Architecture Paradigms as Object Classes: A Unified Approach Towards Multiparadigm Programming

Diomidis Spinellis, Sophia Drossopoulou and Susan Eisenbach

Department of Computing
Imperial College of Science, Technology and Medicine
London SW7 2BZ, UK

Abstract. Computer language paradigms offer linguistic abstractions and proof theories for expressing program implementations. Similarly, system architectures offer the hardware abstractions and quantitative theories applicable to the execution of compiled programs. Although the two entities are usually treated independently, object-oriented technology can be used to obtain a unifying framework. Specifically, inheritance can be used to model both programming languages as extensions to the assembly language executed by the target architecture, and system architectures as the root class of those paradigms. We describe how these principles can be used to model, structure and implement real multiparadigm systems in a portable and extendable way.

1 Motivation

Computer language paradigms offer linguistic abstractions and proof theories for expressing program implementations. Similarly, system architectures offer the hardware abstractions and quantitative theories applicable to the execution of compiled programs. It is widely accepted that each paradigm offers a different set of tradeoffs between efficiency, provability, elision, and implementation cost [17]. Thus, in the context of programming languages many believe that the *functional programming paradigm* is suited for applying formal proof theories on programs implemented in it, the *logic programming paradigm* is suited for expressing rule-based systems, and that efficient implementations are best expressed using the *imperative paradigm*. In the area of system architectures the imperative paradigm maps particularly well on the sequential von-Neumann architecture, while the referential transparency of the pure functional and logic languages can be advantageously used by parallel architectures.

Our research goal is the development and subsequent exploitation of a unifying framework for expressing the inter-relationship between system architectures and program paradigms. Such a framework could provide a solid basis for multiparadigm program development [30]. In particular, we are interested in multiparadigm programming environments, that allow a programmer to express a system using the most suitable paradigms. Absolute portability of the resulting implementations is not our goal as efficient implementations are closely tied to the underlying architecture. However, a relative amount of portability can be achieved by using the unifying framework and by abstracting suitable taxonomy characteristics of languages and architectures to identify the optimum coupling points between them.

2 Language and Architecture Paradigms as Classes

The word paradigm (from the Greek word $\pi\alpha\rho\acute{\alpha}\delta\epsilon\iota\gamma\mu\alpha$ which means example) is commonly used to refer to a category of entities sharing a common characteristic. Wittgenstein [29, p. 48] defines a paradigm by examining all the activities we call games. Among those activities there are some which possess some characteristic similarities equivalent to those exhibited by the members of a family. The *notion* "game" can only be defined by creating a list of all these typical cases that we call games, without being able to prescribe specific conditions for labelling an activity as a "game." In other words, we define games by listing some exemplar cases. In order to define an activity as a "game" it must share some common, but unspecified characteristics with those exhibited by the other members of the family; therefore the notion is only vaguely defined.

Kuhn used the notion of a *paradigm* in the scientific process by defining it as the scientist's view of the world and the structure of his or her assumptions and theories[1]. According to Kuhn [11, p. 10] a paradigm has a wider meaning than that of a scientific theory; it encompasses "law, theory, application and instrumentation together." Although Kuhn's examples are drawn from the history of physical science, his paradigm notion has been extended to a number of sciences [8, 6]. Paradigms are the basis of *normal science* which is related to all the activities of the established scientific tradition. Therefore, the formation of a paradigm is a sign of maturity for a given science.

In [24] it is suggested that as programming languages mature, attention is turning from languages to paradigms. An analogous statement can be made in relation to computer architectures where the characteristics of emerging technologies or quantitative theories are grouped into specific architecture paradigms. In trying to define the notion of a *programming* paradigm the most common definition found, is that of a "model or approach in solving a problem" [15], or the system architecture encompassing definition of "way of thinking about computer systems" [30]. A more general definition is given in [25, p. 21], where paradigms are described as rules for determining classes of languages according to some testable conditions. These conditions can be based on a number of abstraction criteria, such as the structure of a program or its state, or the development methodology. Similarly, system architecture classes can be grouped according to the cardinality of processing units, the implicit or explicit control of inter-

---

[1] Our apologies to the many people who are offended by Kuhn's misuse of the word paradigm.

nal state, and the technologies available for mapping languages onto the given architecture.

2.1 Paradigms as Object Classes

Implementation paradigms, whether at the hardware (system architectural) or software (programming language) level, are expressed using an appropriate notation. This notation can resemble the notation used by the machine that will execute the implementation (ranging from logic gate circuit diagrams, to state transition tables, to microcode listings) or it can resemble a more abstract notation suitable for describing implementations in that problem domain. At some point however, the implementation *will* be executed on a real machine and for this reason the semantic gap between the implementation paradigm and the programming paradigm of the target architecture must be bridged. This is usually done by an interpreter, a compiler or a hybrid technique. We regard all these methods as linguistic transforms from the paradigm notation to the target architecture notation. This view, although simple provides us with three insights:

1. A programming paradigm is nothing *magical*. All programming paradigms can be implemented on all architectures. Furthermore, in principle, there is no practical or theoretical reason for not being able to combine different paradigms, since they can all be mapped into the same concrete architecture.
2. The target architecture plays an important role when thinking of programming paradigms. The concept of the target architecture should be an integral part of multiparadigm systems and not an externally imposed specification, or an afterthought.
3. The target architecture naturally suggests a paradigm object hierarchy, with the target architecture forming the root of the hierarchy and other paradigms forming subclasses. Subclassing is used to create new paradigms, and inheritance to combine common features between paradigms.

Under the view outlined, most computer systems are intrinsically "multiparadigm", since they embody a high-level description of some application implementation together with other lower-level elements, such as an operating system, or the underlying architecture microcode. From this point onwards, we will use the term "multiparadigm" to refer to some combination of implementation paradigms regardless of whether these are hardware or software based.

We found that the object metaphor suits the abstraction of a "programming or architecture paradigm", and that by using it a common unifying model can be defined. In the following paragraphs we will examine how important aspects of object-oriented programming can be related to programming paradigms and multiparadigm programming. We will present the elements of the equation [23]:

$$object\text{-}oriented = objects + classes + inheritance$$

and in addition present the definition of class variables, instance variables and methods [14], in the context of programming and architecture paradigms.

In an object-based multiparadigm programming environment every paradigm forms a class, and every module written in that paradigm is an object member of that paradigm's class. Paradigms form the class hierarchy with the target architecture being the root of it. Inheritance is used to bridge the semantic gaps between different paradigms.

Objects An object can be used as the abstraction mechanism for code written in a given programming paradigm or the realisation of a hardware architecture. Such objects have at least three *instance variables* (Fig. 1):

1. *Source code.* The source code contained in an object is the module code provided by the application programmer.
2. *Compiled code.* The compiled code is an internal representation of that specification (generated by the class *compilation method*) that is used by the class *execution method* in order to implement the specification.
3. *Module state.* The module state contains local data, dependent on the paradigm and its *execution method*, that is needed for executing the code of that object.

Every object has at least one *method*:

1. *Instance initialisation method.* The instance initialisation method is called once for every object instance when the object is loaded and before program execution begins. It can be used to initialise the module state variable.

As an example, given the imperative paradigm and its concrete realisation in the form of Modula-2 [28] programs, an object written in the imperative paradigm corresponds to a Modula-2 module. The *source code* variable of that object contains the source code of the module, the *compiled code* variable contains the compiled source, and the *module state* variable contains the values of the global variables. In addition, the *instance initialisation method* is the initialisation code found delimited between BEGIN and END in the module body. In an example closer to the system architecture, the *source code* variable represents the assembly listing of that compiled module, the *object code* variable represents the machine code, while the *module state* is contained in the processor's data memory allocated for that module.

Classes Collections of objects of the same paradigm are members of a class. All classes contain at least one class variable (Fig. 1):

1. *Class_state*: contains global data needed by the *execution method* for all instances of that class.

In addition, paradigm classes define at least four *methods*:

1. *Compilation method.* The compilation method is responsible for transforming, at compile-time, the source code written in that paradigm into the appropriate representation for execution at run-time.

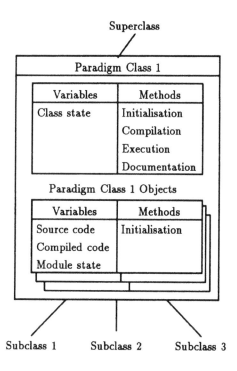

Fig. 1. Programming paradigm classes and objects

2. *Class initialisation method.* The class initialisation method of a paradigm is called on system startup in order to initialise the class variables of that class. It also calls the instance initialisation method for all objects of that class.
3. *Execution method.* The execution method of a class provides the run-time support needed in order to implement a given paradigm.
4. *Documentation method.* The documentation method provides a description of the class functionality. It is used during the building phase of the multi-paradigm environment, in order to create an organised and coherent documentation system.

The compilation and execution methods also contain the machinery needed to implement the import and export call gates described in Sect. 2.3.

Taking as a paradigm class example the logic programming paradigm realised as Prolog compiled into Warren abstract machine instructions [22], the *class state* variable contains the heap, stack and trail needed by the abstract machine. In addition, the *compilation method* is the compiler translating Prolog clauses into abstract instructions, the class initialisation method is the code initialising the abstract machine interpreter, while the execution method is the interpreter itself. The above also holds if the abstract machine is realised as a concrete hardware architecture. In that case the execution method is the processor hardware, or — for microcode-based systems — its microcode.

Inheritance Inheritance is used to bridge the semantic gap between code written in a given paradigm and its execution on a concrete architecture. We regard the programming paradigm of the target architecture as the *root class*. If it is a uniprocessor architecture it has exactly one object instance, otherwise it has as many instances as the number of processors. The *execution method* is implemented by the processor hardware and the *class_state* is contained in the processor's registers. The *compiled code* and *module state* variables are kept in the processor's instruction and data memory respectively.

From the root class we build a hierarchy of paradigms based on their semantic and syntactic relationships. Each subclass inherits the *methods* of its parent class, and can thus use them to implement a more sophisticated paradigm. This is achieved, because each paradigm class creates a higher level of linguistic abstraction, which its subclasses can use.

As an example, most paradigms have a notion of dynamic memory; a class can be created to provide this feature for these paradigms. Two subclasses can be derived from that class, one for programmer-controlled memory allocation and deallocation and another for automatic garbage collection. As another example a simulation paradigm and a communicating sequential processes paradigm could both be subclasses of a coroutine-based paradigm. Subclassing is not only used for the run-time class execution methods. Syntactic (i.e. compile-time) features of paradigms can be captured with it as well. Many constraint logic languages share the syntax of Prolog, thus it is natural to think of a constraint logic paradigm as a subclass of the logic paradigm providing its own solver method, and extension to the Prolog syntax for specifying constraints. A paradigm class tree based around these examples is shown in Fig. 2. It is important to note that Fig. 2 only represents an example based on *one* possible set of abstraction characteristics. Other class hierarchies based on language attributes such as type system or block structure are possible and may be preferable.

2.2 Paradigm Inter-operation Design

Having described the basic structure of a multiparadigm system we must now deal with the problem of paradigm inter-operation. Languages supporting modularisation allow a problem to be decomposed in smaller problems. The entities representing the decomposed problem vary according to the language paradigm as summarised in the Table 1. Each one of them however is based on the basic computational model of input, computation, and output.

When a decomposition entity is invoked (implicitly or explicitly), control is passed to it in conjunction with some input data. After the requisite computation is performed, control passes back to the invoking entity together with the resulting data. In some cases the computation entity may directly interact with input/output devices or modify its internal state. In those cases, input and/or

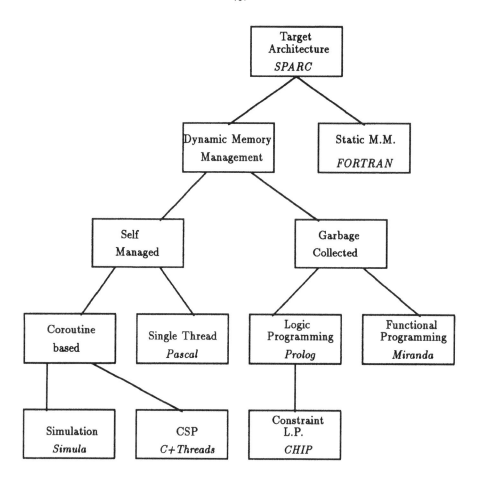

Fig. 2. Paradigm class tree structure example

output data does not have to be provided. Examples of explicit and implicit control transfer are given in Tables 2 and 3 respectively.

Our approach supports explicit control transfer between paradigms, and implicit control transfer within a paradigm. In this way all semantic and implementation issues related to the implicit transfer of control are avoided at the expense of a less expressive system. The restricted version of control transfer is based on the encapsulation properties of the underlying objects: control transfer between the paradigms follows the control transfer conventions of their superclass. This recursive definition is followed until we reach the root class, the target architecture. Using the conventions of the parent paradigm ensures that no unexpected interactions occur. By unexpected interactions we mean implicit and unintended

Table 1. Paradigms and their problem decomposition entities.

Paradigm	Decomposition entity
Imperative	Procedure
Functional	Function
CSP	Process
Logic	Predicate
Object-oriented	Method

Table 2. Explicit control transfer in different paradigms.

Paradigm	Transfer mode
Imperative	Procedure call
Functional	Eager β reduction
Logic	Predicate invocation through its *call* port (c.f. Byrd model [2])

control transfer from one paradigm to the other. The subclasses are coded using the features and caveats of the parent class; this ensures that unexpected interactions will not occur. We will attempt to clarify this statement by four examples:

Non-Preemptive Von-Neumann Target Architecture The basic control transfer mechanism used is the explicit *procedure call*. Implicit calls between other paradigms will be encapsulated within their respective classes and therefore the paradigms remain isolated.

Threads Subclass We assume now the existence of the *thread* programming subclass with primitives that allow the creation of multiple processes. Since this subclass was coded using the mechanisms of the parent paradigm, the threads created will be non-preemptive and therefore all implicit control transfers will happen *within* the threads paradigm. Therefore no implicit control transfers can occur between paradigms, other than subclasses of the

Table 3. Implicit control transfer in different paradigms.

Paradigm	Transfer mode
Imperative	Invocation of an interrupt handler
Functional	β reduction of a suspended expression in a lazy implementation
Logic	Predicate invocation through its *redo* port (backtracking)
CSP	A process is awaken as a result of an external event (e.g. an input or output buffer becomes full)

threads paradigm. These subclasses will of course be coded to anticipate such control transfers.

Event-Driven Target Architecture In an event-driven target architecture, all paradigm compilers have to be able to cope with non-deterministic control transfers. For this reason the paradigms provided will — by definition — be able to cope with such transfers. A straightforward way to implement such a system would be to disable interrupts when paradigms that are unsuitable for such an environment are executed (a higher quality implementation would provide a better solution).

Parallel Target Architecture Here again the target architecture imposes some stringent requirements regarding synchronisation, memory sharing, message passing etc. All these requirements are inherited, and must be handled by all paradigm subclasses.

The data that is exchanged between the computation entities can be divided between data that is commonly available in similar representations[2] across many different paradigms, and data representations that are only supported in a limited number of paradigms. We say that a paradigm supports a given data representation if that form of representation can be

- expressed in the source code of that paradigm,
- dynamically created at run-time, and
- operated upon by the data manipulation primitives or libraries of that paradigm.

Examples of data representation commonly available across different paradigms and representations particular to specific paradigms are listed in Tables 4 and 5.

Table 4. Data representations commonly available across many paradigms.

Data type	Common representation
Integer value	Two, four or 8 bytes
String value	Character vector
Character	One byte (ASCII)
Floating point number	IEEE 488 byte sequence
Boolean	Single byte or bit
A finite predetermined collection of the above (record, structure, term)	Sequence of bytes representing the above

For the sake of simplicity our approach only supports inter-paradigm communication with data representations supported by both paradigms exchanging

---

[2] Some paradigms may *cell* a particular representation in the form of a record usually containing its type and a pointer to the actual value.

Table 5. Data representations particular to specific paradigms.

Data representation	Supporting paradigm
Uninstantiated value	logic programming
Curried function	functional programming
List of infinite length	functional programming
Pointer or reference	imperative programming
Valued data object whose value can be changed	imperative programming
Array	imperative programming

data. When the two paradigm implementations use dissimilar representations, it is straightforward to map one representation to the other. Examples of such transformations include the assembling or disassembling of plain values used in imperative paradigms to the cell-based representations common in declarative paradigms, or the changes between byte orderings for different representations of arithmetic types. This approach again trades limited expressive power for semantic and implementational simplicity.

2.3 Paradigm Inter-operation Implementation Abstraction

Paradigm inter-operation can be designed around an abstraction we name a *call gate*. A call gate is an interfacing point between two paradigms, one of which is a direct subclass of the other. We define two types of call gates, the import gate, and the export gate. In order for a paradigm to use a service provided by another paradigm (this could be a procedure, clause, function, rule, or a port, depending on the other paradigm), that service must pass thought its import gate. Conversely, on the other paradigm the same service must pass through its export gate. The call gates are design abstractions and not concrete implementation models. They can be implemented by the paradigm compiler, the runtime environment, the end user, or a mixture of the three. Each paradigm provides an import and export gate and documents the conventions used and expected. The input of the export gate, and the output of the import gate follow the conventions of the paradigm, while the output of the export gate, and the input of the import gate, follow the conventions of the paradigms' superclass. The target architecture paradigm combines its import and its export gate using the linked code as the sink for its export gate and the source for its import gate. Call gates can make the paradigm inter-operation transparent to the application programmer, and provide global scale inter-operation using only local information.

Figure 3 illustrates an example case. Assume that a module written in *paradigm 2* is using a facility implemented in *paradigm 1.1*. The module written in *paradigm 1.1* will export that facility (using the syntax and semantics appropriate to *paradigm 1.1*) to its superclass (*paradigm 1*) through its export gate, thus converting it to the data types and calling conventions used by paradigm 1.

Paradigm 1 will again pass it through its export gate, converting it to the conventions used by *paradigm 0*, the target architecture. (For example, the calling conventions of the Unix system can include the passing of parameters through a stack frame, and the naming of identifiers with a prepended underscore.) In this form the facility will again be imported from the pool of linked code by *paradigm 1* and made available to its subclasses using its conventions. The facility can then be imported and used by *paradigm 2* which can understand the calling conventions of *paradigm 1*. Although during the path described the facility crossed three paradigm boundaries, in all cases the paradigm just needed to be able to map between its calling conventions and data types and those of its superclass.

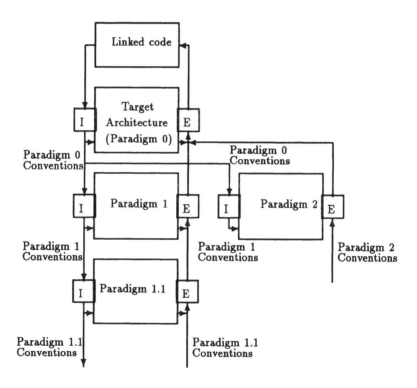

Fig. 3. Paradigm inter-operation using call gates

We must note at this point that the class hierarchy is not visible to the application programmer. The hierarchy is useful for the multiparadigm programming environment implementor, as it provides a structure for building the system, but is irrelevant to the application programmer, who only looks for the most suited paradigm to build his application. This is consistent with the recent trend in object-oriented programming of regarding inheritance as a *producer's mecha-nism* [13], that has little to do with the end-user's use of the classes [4].

3 An Exemplar Multiparadigm Programming Environment

3.1 Design Objectives

Blueprint is an exemplar multiparadigm programming environment, built using the object-based approach. It was implemented in order to prove the viability of the object-based approach to multiparadigm programming and its design was centred around the following objectives:

- realisation of a wide variety of diverse programming paradigms, and imple-mentation methods,
- provision of a non-trivial class hierarchy, including the abstraction of com-mon characteristics in a special superclass,
- incorporation of existing tools, and
- ability to bootstrap the system and implement a non-trivial application in order to test and use it as much as possible.

The *blueprint* name is derived from the acrostical spelling of the paradigms provided, namely:

- BNF grammar descriptions (*bnf*),
- lazy higher order functions (*fun*),
- unification and backtracking (*btrack*),
- regular expressions (*regex*),
- imperative constructs (*imper*) and,
- term handling (*term*).

All paradigms are provided in the form of individual paradigm compilers: tools that convert the code expressed in a given paradigm into object code that can be linked and executed together with code from other paradigms.

3.2 System Structure

The target paradigm of *blueprint* is the imperative paradigm provided by the target architecture, which in our case is that of the Sun SPARC computer. The class structure of the paradigm classes implemented can be seen in Fig. 4. Term expressions are the natural data objects, for both functional and logic languages; the provision of the *term* class is based on this observation and, in addition, provides a practical vehicle for their implementation.

It is important to note that the tree structure is only used in order to design and implement the system. The structure is transparent to a programmer using *blueprint* who is presented with a flat structure of all the paradigms (Fig. 5). In the following paragraphs we briefly describe each *blueprint* paradigm.

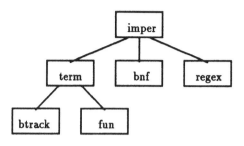

Fig. 4. *Blueprint* class hierarchy

Fig. 5. Programmer's view of *blueprint*

Imperative Paradigm *Imper*, the imperative paradigm, is provided in the form of the C programming language. It is the one closest to the target architecture, and the calling and naming conventions of the language are used as a common interface for the other paradigms.

BNF Grammar and Regular Expression Paradigms The *bnf* (BNF-grammar) and *regex* (regular expression) paradigms are used to encapsulate *yacc* [10] grammar descriptions and *lex* [12] lexical analyser specifications, as objects. The main advantage of this encapsulation is the ability to use more than a single grammar description or lexical analyser specification within the same project. This is achieved by "protecting" the global variable and function names that *yacc* and *lex* define by prepending them their object (module) name. Both paradigms were implemented by using special tools to create multiparadigm environment conformant compilers out of the standard Unix *yacc* and *lex* generators. Inter-operation with the imperative paradigm is achieved by using the standard *lex* and *yacc* interfacing conventions, as modified by the object encapsulation scheme.

Rule-rewrite Paradigm *Term*, the term-based rule rewrite paradigm abstracts the notion of a *term* used by both the functional and logic programming paradigms. Its syntax resembles that of Prolog, but it uses a deterministic rule-rewrite execution model with predefined argument mode declarations, resembling the functionality provided by Strand [7]. *Term* is implemented in *term*, *imper*, *bnf*, and *regex* as a compiler that translates *term* into C. It was bootstrapped using the SB-Prolog compiler [5], and a semi-automatic translation process. Inter-operation with the imperative paradigm is provided by documenting the compiled form of the *term* "predicates" and providing access and constructor functions for the term abstract data type, in its converted form of C *structures*.

Logic Programming Paradigm *Btrack*, the logic programming paradigm provides the backtracking execution model, deep unification and syntax, associated with implementations of the Prolog programming language. It is implemented in *term* as an encoded token interpreter based on a *solve/unify* loop [3, p. 1313]. The *btrack* to *term* token conversion is performed by "compiling" the *btrack* predicates into *term* rules. Inter-operation with *term* is achieved by defining the predicates that are exported using *term* signatures. The *btrack* compiler then creates the necessary interfaces and entry ports.

Functional Programming Paradigm *Fun*, the functional programming paradigm offers lazy higher order functions supporting currying and call-by-name normal-order evaluation. Its syntax resembles that of Miranda [20] omitting the guard and pattern matching constructs. It is implemented in *bnf*, *lex*, and *term* with function evaluation provided by an *eval/apply* interpreter, written in *term*. Inter-operation with *term* is provided by allowing the import of single result *term* rules and exporting functions as *term* rules with a single result. Calls to and from *fun* need to take into account and respect the *fun* data structuring conventions, which are documented as *term* constructors.

3.3 Experience with *Blueprint*

Blueprint was used to bootstrap itself and, in addition, to implement a small algebraic manipulation system. In the implementation of *blueprint* four of the six available paradigms were used, as illustrated in Table 6. Each row of the table shows a breakdown (in lines of code) of the paradigms used to implement the system named in its first column. The class structure of every paradigm was described in a *paradigm description file* (the table column labelled PDF), which was then processed by a special compiler in a way analogous to the one described in [16] to create the requisite translation tools. Some of the paradigms were implemented using existing tools and their implementation consists of just a paradigm description file.

Table 6. *Blueprint* paradigm implementation summary

Paradigm	PDF	imper	bnf	regex	term	fun	Total	%
imper	43						43	1.1
term	70	1192	119	84	666		2131	56.0
btrack	60				316		376	9.9
fun	140		305	59	237	43	784	20.6
bnf	95						95	2.5
regex	379						379	10.0
Total	787	1192	424	143	1219	43	3808	100.0
%		18.5	28.0	10.0	3.4	28.6	1.0	100.0

Table 7. Algebraic manipulation system implementation

Function	Paradigm	Lines
Symbolic integration	btrack	127
Lexical analysis	regex	47
Expression parsing	bnf	76
Numeric integration	fun	75
Interfacing	term	131
Graph creation	imper	51
Total	blueprint	507

The algebraic manipulation system deals numerically and graphically with definite integrals and symbolically with indefinite ones. It was implemented at a fraction of time and effort spent to implement an analogous one in Modula-2 by using all six paradigms available under *blueprint*. The breakdown of the system into the paradigms used is illustrated in Table 7.

4 Related Work

In a survey of multiparadigm programming we have identified more than 100 languages and systems that allow programming in more than one paradigm. A number of multiparadigm projects are described in [9], languages based on distributed system architectures are surveyed in [1], and implementations of different paradigms of parallel computer architectures are examined in [19]. Most of the multiparadigm languages attempt to amalgamate the advantages of their constituent paradigms, either in a pragmatic implementation-targeted way, or by developing an underlying theory. Some typical examples are [18, 26, 21]. Our approach differs from those in that we offer an underlying unifying framework for the combination of arbitrary paradigms instead of examining how specific paradigms can be combined. It is a pragmatic approach weak on its theoretical basis. The *compositional approach* described in [30] can also deal with arbitrary paradigms, but is concerned more with the validation of the resulting system. The intimate relation between the target architecture and the programming language that forms the basis of our approach is examined in [27].

5 Conclusions and Further Work

The abstraction of programming languages and system architectures as object classes provides a unified model for dealing with multiparadigm programming systems. Objects are used as encapsulation entities for modules written in different paradigms, while inheritance is used to bridge the semantic gap between a high level language and the target architecture. Our implementation of an exemplar system based on these principles and its use both as a bootstrapping vehicle and as an implementation platform demonstrated the viability of this approach. We intend to use the same approach on different types of computer architectures in order to test its applicability and limits. One other challenging problem is the development of a theoretical reasoning framework that can be applied to architecture and language systems structured using object classes.

Acknowledgements

We would like to express our thanks to our colleagues at the Distributed Software Engineering group at Imperial College for ideas and discussions during the conduct of this research, and to the anonymous reviewers for their helpful comments. Financial support from the British Science and Engineering Research Council and the DTI under grant ref. IED4/410/36/002 is gratefully acknowledged.

References

1. Bal HE, Steiner JG, Tanenbaum AS (1989) Programming languages for distributed computing systems. *ACM Comput. Surv.*, 21: 261–322.
2. Byrd L (1980) Understanding the control flow of Prolog programs. In *Logic Programming Workshop*, Debrecen.
3. Cohen J (1985) Describing Prolog by its interpretation and compilation. *Commun. ACM*, 28: 1311–1324.
4. Cook WR (1992) Interfaces and specifications for the Smalltalk-80 collection classes. *ACM SIGPLAN Notices*, 27: 1–15. Sevent Annual Conference on Object-Oriented Programming Systems, Languages and Applications, OOPSLA '92 Conference Proceedings, October 18–22, Vancouver, British Columbia, Canada.
5. Debray SK (1988) *The SB-Prolog System, Version 3.0: A User Manual.* University of Arizona, Department of Computer Science, Tucson, AZ 85721, USA.
6. Eckberg DL, Hill , Jr. L (1980) The paradigm concept and sociology: A critical review. In Gutting G, editor, *Paradigms and Revolutions*, pp 117–136. University of Notre Dame Press, Notre Dame, London.
7. Foster I, Taylor S (1990) *Strand: New Concepts in Parallel Programming.* Prentice-Hall.
8. Greene JC (1980) The Kuhnian paradigm and the Darwinian revolution in natural history. In Gutting G, editor, *Paradigms and Revolutions*, pp 297–320. University of Notre Dame Press, Notre Dame, London.
9. Hailpern B (1986) Multiparadigm research: A survey of nine projects. *IEEE Software*, 3: 70–77.
10. Johnson SC (1975) Yacc — yet another compiler-compiler. Computer Science Technical Report 32, Bell Laboratories, Murray Hill, NJ, USA.
11. Kuhn TS (1970) *The Structure of Scientific Revolutions.* University of Chicago Press, Chicago and London, 2nd, enlarged edition. International Encyclopedia of Unified Science. 2:(2).
12. Lesk ME (1975) Lex — a lexical analyzer generator. Computer Science Technical Report 39, Bell Laboratories, Murray Hill, NJ, USA.

13. Meyer B (1990) Lessons from the design of the Eiffel libraries. *Commun. ACM*, 33: 68–88.
14. Nelson ML (1991) An object-oriented tower of Babel. *OOPS Messenger*, 2: 3–11.
15. Shriver BD (1986) Software paradigms. *IEEE Software*, 3: 2.
16. Spinellis D (1993) Implementing Haskell: Language implementation as a tool building exercise. *Structured Programming*, 14: 37–48.
17. Stefik MJ, Bobrow DG, Kahn KM (1986) Integrating access-oriented programming into a multiparadigm environment. *IEEE Software*, 3: 10–18.
18. Takeuchi I, Okuno H, Ohsato N (1986) A list processing language TAO with multiple programming paradigms. *New Generation Computing*, 4: 401–444.
19. Treleaven PC, editor (1990) *Parallel Computers: Object-Oriented Functional, Logic*. John Wiley & Sons.
20. Turner DA (1985) Miranda — a non-strict functional language with polymorphic types. In Jouannaud JP, editor, *Proceedings of the Conference on Functional Programming Languages and Computer Architecture*, pp 1–16, Nancy, France. Springer-Verlag. Lecture Notes in Computer Science 201.
21. Uustalu T (1992) Combining object-oriented and logic paradigms: A modal logic programming approach. In Madsen OL, editor, *ECCOP '92 Europen Conference on Object-Oriented Programming*, pp 98–113, Utrecht, The Netherlands. Springer-Verlag. Lecture Notes in Computer Science 615.
22. Warren DHD (1983) An abstract Prolog instruction set. Technical Note 309, SRI International, Artificial Intelligence Center, Computer Science and Technology Division, 333 Ravenswood Ave., Menlo Park, CA, USA.
23. Wegner P (1987) Dimensions of object-based language design. *ACM SIGPLAN Notices*, 22: 168–182. Special Issue: Object-Oriented Programming Systems, Languages and Applications, OOPSLA '87 Conference Proceedings, October 4–8, Orlando, Florida, USA.
24. Wegner P (1989) Guest editor's introduction to special issue of computing surveys. *ACM Comput. Surv.*, 21: 253–258. Special Issue on Programming Language Paradigms.
25. Wegner P (1990) Concepts and paradigms of object-oriented programming. *OOPS Messenger*, 1: 7–87.
26. Wells M (1989) Multiparadigmatic programming in Modcap. *Journal of Object-Oriented Programming*, 1: 53–60.
27. Wirth N (1985) From programming language design to computer construction. *Commun. ACM*, 28: 159–164.
28. Wirth N (1985) *Programming in Modula-2*. Springer Verlag, third edition.
29. Wittgenstein L (1960) Philophische Untersuchungen. In *Schriften*, volume I. Suhrkamp Verlag, Frakfurt a.M., Germany. In German.
30. Zave P (1989) A compositional approach to multiparadigm programming. *IEEE Software*, 6: 15–25.

Engineering a Programming Language: The Type and Class System of Sather

Clemens Szyperski, Stephen Omohundro, and Stephan Murer

International Computer Science Institute
1947 Center St, Suite 600
Berkeley, CA 94704
Email: {szyperski, om, murer}@icsi.berkeley.edu
Fax: +1 510 643-7684

Abstract. Sather 1.0 is a programming language whose design has resulted from the interplay of many criteria. It attempts to support a powerful object-oriented paradigm without sacrificing either the computational performance of traditional procedural languages or support for safety and correctness checking. Much of the engineering effort went into the design of the class and type system. This paper describes some of these design decisions and relates them to approaches taken in other languages. We particularly focus on issues surrounding inheritance and subtyping and the decision to explicitly separate them in Sather.

1   Introduction

Sather is an object-oriented language developed at the International Computer Science Institute [22]. It has a clean and simple syntax, parameterized classes, object-oriented dispatch, statically-checkable strong typing, multiple subtyping, multiple code inheritance, and garbage collection. It is especially aimed at complex, performance-critical applications. Such applications are in need of both reusable components and high computational efficiency.

Sather was initially based on Eiffel and was developed to correct the poor computational performance of the Eiffel implementations available in 1990. Eiffel introduced a number of important ideas but also made certain design decisions which compromised efficiency. Sather attempts to support a powerful object-oriented paradigm without sacrificing either the computational performance of traditional procedural languages or support for safety and correctness checking.

The initial "0.1" release of the compiler, debugger, class library, and development environment were made available by anonymous FTP[1] in May, 1991 and it was quickly retrieved by several hundred sites. This version achieved our desired efficiency goals [15] and was used for several projects. Our experience with it and feedback from other users has led to the design of Sather 1.0. This improves certain aspects of the initial version and incorporates a number of new language constructs.

---

[1] From ftp.icsi.berkeley.edu, directory pub/sather/

The language design process has been intimately coupled with the design and implementation of libraries and applications. A particularly demanding application is the extensible ICSI connectionist network simulator: ICSIM [24]. The examples in this paper are taken from the actual code and structure of the Sather libraries and applications to make them realistic. The design effort was continually a balance between the needs of applications and constraints on language design, such as simplicity and orthogonality.

One of the most fundamental aspects of the Sather 1.0 design is its type system. Earlier versions of the language were strongly typed, but it was not possible to *statically check* a system for type correctness. Eiffel has the same problem [5], and attempts to solve it by introducing *system-level type-checking* [17]. This is a conservative system-wide global check. A system which satisfies the check will be type safe but many legal programs are rejected. Adding new classes to a system can cause previously correct code to become incorrect.

Sather 1.0 solved this and other problems by completely redesigning the type and class system. This paper describes a number of these issues. We do not describe the whole language here but do include the relevant parts of the grammar in Appendix A. The bulk of the paper is devoted to the interplay between subtyping and subclassing. Section 2 defines these concepts and motivates the decision to explicitly separate them in the language. It describes the Sather version of parameterized classes and object-oriented dispatch. It also describes the three kinds of Sather objects: reference objects, value objects, and bound objects. Bound objects are a particularly clean way of implementing higher-order functions within an object-oriented context. Section 3 describes some of the subtle issues involved in code inheritance. Finally, section 4 describes some more system level issues.

2 Sather Types and Classes

Object-oriented terminology is used in a variety of ways in the programming language literature. A few informal definitions will suffice for the purposes of this paper:

- *Objects* are the building blocks of all Sather data structures. Objects both encapsulate state and support a specified set of operations.
- A *type* represents a set of objects.
- The *signature* of an operation that may be performed on an object consists of its name, a possibly empty tuple of its argument types, and an optional return type. Sather supports both *routines* which perform a single operation and *iters* [20] which encapsulate iteration abstractions[2].
- Each type has an *interface* which consists of the signatures of the operations that may be applied to objects of the type. Sather supports *overloading* which means that an interface may have more than one operation with the same

---

[2] There is some additional information in signatures which are associated with iters which we do not describe here.

name if they differ in the number or types of arguments or in the presence of a return type.

- *Classes* are textual units that define the interface and implementation of types.
- The Sather *type graph* is a directed acyclic graph whose vertices are types and whose edges define the *subtype* relationship between them. We say that a type A *conforms* to a type B if there is a directed path from A to B in the type graph.

2.1 Subtyping and Multiple Subtyping

Every object and every variable in Sather has a uniquely specified type. The fundamental Sather typing rule is: *"A variable can hold an object only if the object's type conforms to the variable's declared type."*.

There are three kinds of object type: *reference*, *value*, and *bound*. We describe these later. Variables can be declared by one of these types, but may also be declared by an *abstract* type. These are types which represent sets of object types and are how Sather describes polymorphism. Abstract types are defined by abstract classes and do not directly correspond to objects.

We say that Sather is *strongly typed* because each variable has a type which specifies exactly which objects it can hold. We say that it is *statically type-safe* because it is impossible for a program which compiles to assign an object of an incorrect type to a variable. The Sather type correctness checking is purely local and does not require a system-wide analysis. It is done by checking calls against the declared signatures in the interface of the type to which the call is applied. Statically-checked strong typing is fundamental to achieving both the performance and the safety goals of Sather.

Type safety is ensured because of a *conformance* requirement on the interfaces of types[3]. If the type A conforms to the type B, then the interface of A is required to conform to the interface of B. This means that for each signature in B's interface there is a conforming signature in A's interface.

Object-oriented dispatch means that the particular implementation for a routine call is made according to the type of the object the call is made on. This object may be thought of as the first argument of the routine. Within the routine, it is referred to as **self**. The type of **self**, denoted as **SAME**, is the type defined by the class that implements the operation.

Under the subtype relation, the **self** parameter is *covariantly* typed. Because of the dispatching, this is typesafe[3]. All other arguments are *contravariantly* typed and the return value is *covariantly* typed. Together, these conformance requirements ensure that if a call is type correct on the declared type of a variable, then it will be type correct when made on all possible objects that may be held by that variable.

---

[3] The language Cecil[4] uses multi-methods to allow multiple covariantly typed parameters in a type-safe way. Some disadvantages of multi-methods are discussed in section 2.7.

Sather allows for *multiple subtyping*: A type can be subtype of more than one type. This is very important for using software types to model types in the world. Real-world types are often subtypes of more than one type. In a system which only supports single subtyping, one is often forced to introduce spurious subtype relations which can destroy the conceptual integrity of a design.

A new feature introduced by Sather is the possibility for a new class to declare itself as a *supertype* of an existing class. Using this facility, it is possible to interpose a new type between two existing types in a hierarchy. This solves an old dilemma of class hierarchy design. On the one hand, for future flexibility one often wanted to introduce many incrementally different types. On the other hand, huge type hierarchies with many similar classes are hard to understand and use. With the ability to insert new types into a hierarchy, intermediate classes can be introduced only when needed.

2.2 Code Inheritance, Subclassing and Multiple Subclassing

Although often confused or combined with subtyping, an entirely different aspect of object-oriented programming is code reuse by means of *code inheritance*, also called *subclassing*. A class A is called a subclass of a class B if A's implementation is based in part on B's implementation. Code reuse in this sense differs from the use of traditional library routines in two important ways. First, the inherited code has direct access to the internal representation of the reusing class. Second, the inherited code may make calls on **self**. Such calls may call other inherited operations or operations explicitly defined in the new class. This intricate tangling of new and old code is powerful but complexity-prone [16].

As with subtyping, Sather allows *multiple subclassing*: A class can be subclass of multiple classes, i.e. it can reuse portions of the implementations of multiple classes. Multiple subclassing introduces many complications that require careful attention. Most languages combine multiple subtyping with multiple subclassing into *multiple inheritance*. The complexity introduced by multiple subclassing has given rise to widespread ambivalent feelings about multiple inheritance. A particularly tricky situation arises when the same code is inherited by a class along multiple paths. The resulting conflicts and Sather's conflict resolution mechanisms are described below in Section 3.

One could imagine introducing a construct for code inheritance which is analogous to the supertyping construct described above (cf. Section 2.1). This would be a form of "code injection" in which classes could add implementation to other classes. This possibility was rejected in the Sather design because it gives rise to many ambiguities and errors which would be hard to find. One would no longer be able to determine the source code of a class by merely looking at the class text and those classes reachable from references in it. When using classes from another system, it would not be clear which source files contributed code to the desired classes. Also, because of the separation between subclassing from subtyping, code can be inherited in the opposite direction from subtyping if desired.

2.3 Separating Subtyping and Subclassing

Traditionally, object-oriented languages are either untyped – e.g. Smalltalk [10] or Self [27] – or tightly bind classes and types – e.g. C++ [8], Eiffel [17] Modula-3 [21], or Oberon-2 [19]. (In contrast to Oberon-2, Oberon [23] keeps the dispatching of implementation variants separate from subtyping issues, essentially by not providing methods at all. Instead, Oberon relies entirely on procedure variables to implement late binding. Nevertheless, Oberon still does not completely separate subtyping from subclassing, cf. Section 2.6.)

The decision to have static type safety caused us to reject the untyped variants. Given that there will be types, one must decide how tight a binding there should be between subtyping and subclassing. The typed object-oriented languages mentioned above bind these notions closely together. Not separating these concepts properly leads to several problems, however.

One approach requires that every subclass relationship obeys the rules of type-safe subtyping. This leads to contravariant typing of routine arguments. It has been argued that this eliminates several important opportunities for code reuse [18, 14].

Another approach introduces subclasses which are subtypes by declaration but not in terms of the interface which is supported. This approach is adopted as a compromise in many languages, including the original version of Sather and Eiffel [17]. This violates the requirement of local type checkability. In the original Eiffel design this was a safety loop-hole [5]. The latest version of Eiffel requires "system-level type checking", which gives up on local type checkability and sometimes rejects dynamically type-safe programs.

Because of these problems, [6] suggested that subtyping should be clearly separated from subclassing. Emerald [11] is one of the few languages that actually implemented this separation. In Emerald, however, the result is a significant burden on the programmer. Often, subtyping and subclassing do go along in parallel, and Emerald requires separate specification even for this common case.

Later language designs, such as Sather 1.0 and Cecil [4], attempt to provide more convenient ways to support the common case. Since Cecil is based on prototype objects, quite similar to Self, its code inheritance is not based on classes. Still, Cecil's counterpart to subclassing has the default behavior of also introducing a subtype. This behavior can be explicitly prevented, however, and it is even possible to have code inheritance and subtyping go in opposite directions. Sather follows a similar path of optimizing the common case. However, instead of introducing defaults, Sather introduces special kinds of classes and an explicit means to implement subclassing and subtyping graphs over these classes.

2.4 Sather Types, Classes, and Variables

As described above, Sather distinguishes between *abstract* and *concrete* types (the names of abstract types are distinguished by a leading "$" to help distinguish them). Abstract classes can have descendants in the type graph, but cannot be instantiated as objects. Concrete classes are always leaf-nodes in the subtype graph, but can be instantiated. This approach is similar to the type system formally defined in [7]. Abstract classes may provide partial implementations to be inherited by subclasses, while concrete classes are required to fully implement their type. Sather code inheritance is explained in Section 3.

All Sather variables are statically typed. If a variable is declared as a concrete type, then only objects of exactly that type can be held by it. As a result, all calls on such variables are monomorphisms, i.e. the actual implementation invoked is statically determined. This is an important source of efficiency for Sather programs. If a variable is declared by an abstract type, then it can hold objects belonging to any of the subtypes of the declared type. Calls made on such "abstract variables" are polymorphisms. This means that the actual implementation invoked is determined at run-time according to the type of the object bound to the variable at the time of the call.

2.5 Examples of Separate Subtyping and Subclassing

Multiple subtyping is important in situations where there is not an obvious hierarchy of object properties. In the Sather library some container classes are internally based on hash tables, others are not. Not every object defines a corresponding hash function, however. We make objects which do provide a hash function be descendants of the abstract class $HASHABLE whose interface defines the single routine "hash":

```
abstract class $HASHABLE is
   hash:INT;
end
```

Note that $HASHABLE doesn't provide an implementation of hash, because there is no generic hash function that works for all types. Abstract classes without implementation information, such as $HASHABLE, only serve for subtyping purposes. The implementation of the required features is left to the descendants. Figure 1 shows a typical inheritance graph for defining an element class to be used as a type parameter in a parameterized, hash table-based set class (cf. Section 2.8. In Figure 1, as well as in the other inheritance graphs, solid and dashed arcs are used to represent subtype and subclass relationships. Class names set in a plain typeface denote abstract types and those in a bold typeface denote concrete types. (cf. Section 2.3).

Multiple subclassing is much less common in Sather programs than multiple subtyping. Nevertheless, there are situations where application programmers prefer to use multiple subclassing. It is used in the *mixin* programming style used extensively in CLOS[2]. ICSIM, the ICSI neural network simulator, uses this style to let the user configure the properties of neuron sites. Sites are subsets of a neuron's connections with identical properties. Sites have connection-oriented properties represented by $PORT descendants and computation-oriented properties represented by $COMPUTATION descendants. $SITE is a subtype of both $PORT and $COMPUTATION.

214

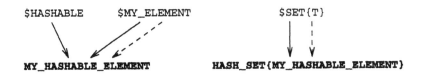

Fig. 1. Typical inheritance graph for $HASHABLE

Typically, ICSIM users do not program their own sites, but instead choose them from built-in classes that provide building blocks for connection and computation code. All types used in ICSIM are also subtypes of $ANY_ICSIM. These relationships are shown in the inheritance graph in Figure 2. It is interesting to note that the MY_SITE class uses multiple subclassing but single subtyping (the opposite of the usual case).

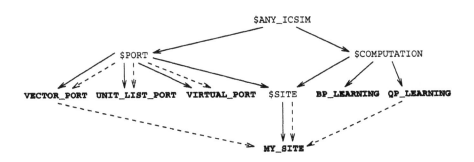

Fig. 2. Multiple Subclassing for Programming by Configuration

2.6 Features of a Class

The features of a Sather class are either *attributes, routines,* or *iters*[4]. Attributes are the analog of record fields in Pascal-like languages. Routines are the equivalent of "methods" in some other object-oriented languages. In particular, routines differ from Pascal-like procedures by having an additional implicit parameter bound to the object that the routine is called on.

Whether a particular operation is implemented as an attribute or as a routine is not visible from the interface of a type. One concrete descendant of an abstract class may define an attribute while another may just provide accessor

---

[4] We do not describe iters here because it would take us too far afield and they have been described elsewhere [20].

and modifier routines. This is a departure from the traditional coincidence of type and structure definition in Pascal-like languages, including Oberon. It is used in some earlier object-oriented languages, however, such as Self. One might argue that the same effect can be achieved by not introducing public attributes, as, for example, can be done in Oberon by not exporting record fields. The price to pay for doing so is the loss of the intuitive and lightweight attribute access notation "x.a" in clients of the class.

Attributes may be declared to be *shared* among all instances of a type. Such shared attributes serve the function of global variables (and the rather difficult to use "once functions" of Eiffel). Shared attributes may specify an initialization expression that is evaluable at compile-time. Similar attributes may be declared *constant*, in which case the binding established by the initializer is permanent.

The features of a Sather class may be declared *private*, allowing only the routines within the class to access them, and only relative to **self**. For attributes, it is possible to declare the accessor and modifier routines individually as private or public. This allows attributes to be read-write, read-only, or invisible from within code external to the class.

2.7 Object-Oriented Dispatch

Variables declared by an abstract type can hold objects of any descendant type. Routine calls made on such variables dispatch on the runtime type of the object to determine the code to execute. This lookup adds a small amount of extra overhead to such calls. By declaring a variable with an abstract or a concrete type, the programmer may decide to pay the price for routine dispatch or to restrict the generality of the code by precisely specifying the object type that the variable can hold.

Some languages support "multi-methods" which can dispatch on all the arguments of a call. Sather does not adopt this approach for both semantic and performance reasons. In Sather routines are grouped into classes according to the type of "**self**". This provides a natural organization principle and is responsible for the encapsulation of functionality into types. The interface of a type encapsulates the abstraction defined by that type. With multi-methods code does not naturally belong to a particular type. Sather deals with multi-method situations by using "**typecase**" statements. These appear in the body of a routine which dispatches on the first argument type and may explicitly dispatch on the second argument type. Unlike a simple "**case**" statement applied to the type, a "**typecase**" statement can branch on abstract types. This means they can be used in the same situations that multi-methods would be helpful though multi-method languages can add new routines without modifying existing code. This approach also makes the performance consequences of a multi-method organization explicit rather than hiding it behind a complex language construct.

Sather routines can also be called directly. A direct call is equivalent to dispatching the routine call on an unbound variable of concrete type (**self** = **void**). Direct-called routines are Sather's version of plain procedures in Pascal, class methods in Smalltalk, and static member functions in C++.

2.8 Parameterized Classes

Sather allows the definition of a family of classes parameterized by types. This is a similar mechanism to the generic packages of Ada [28] and templates in the newer versions of C++. Sather type parameters have associated type constraints. The values specified for the type parameters are required to be subtypes of these constraint types. The supertyping feature introduced in Section 2.1 is quite useful for defining such constraints. A constraint type representing an arbitrary union of types can be introduced by forming an appropriate supertype.

A second form of genericity in Sather is related to the typing of arguments and return values in inherited code. Sather allows such types to be declared as SAME, similar to Eiffel's like-current. If a class A inherits code which refers to the type SAME, it behaves as if the type were replaced by A. For a subclass to be also a subtype, however, this replacement has to follow the subtyping rules stated in Section 2.1. This is very different from Eiffel's like-current, where a subclass formed in this way is automatically considered a subtype, even though it might well have introduced a conformance conflict.

2.9 Reference and Value Classes

Sather distinguishes between *reference, value,* and *bound* objects. Most user-defined objects are reference objects. These are passed by reference as routine arguments and may be aliased. The fundamental types representing boolean values, integers, characters, floating point values, etc. are called *value* objects. These are always passed by value and it is not possible to alias them (i.e. to reference the same object under two names). More pure object-oriented languages such as Smalltalk and Self try to unify these notions.

Languages that only operate over values are typically called *functional languages,* and operations defined only over value types are side-effect free and therefore *referentially transparent.* On the other hand, reference objects are best used to model entities that have an identity plus a current state. The idea of an object identity bound to a modifiable state introduces referential opaqueness and allows for side-effects.

Sather distinguishes between these at the level of types. Instances of value types have value semantics: Once created they never change, and there is no such thing as a "reference" to a value object. Reference objects have an identity and the state of a reference object can be modified by writing to its attributes. A variable of abstract type can be used to store either value or reference objects.

The special properties of value objects make them especially amenable to compiler optimization techniques. Most important, a value object can be copied freely without the possibility of aliasing conflicts. Logically, when value objects are passed as arguments, their value is first copied and then the operation is invoked on the copy (*call by value* semantics). Of course, the compiler is free to eliminate this copying whenever it can deduce that the invoked operation cannot modify the object.

The introduction of separate value and reference classes imposes certain restrictions on subclassing: An abstract class can only be a subclass of other abstract classes; a value class can only be a subclass of abstract classes and other value classes; a reference class can only be a subclass of abstract classes and other reference classes.

2.10 Bound Routines

A controversial feature of non-functional programming languages are closures or higher-order functions. While expressive and powerful, certain formulations are difficult to implement efficiently. Hence, many non-functional programming languages provide more lightweight but much less powerful facilities.

Pascal [12] introduced procedure parameters, but no procedure variables. This allowed implementations to strictly adhere to a stack discipline, but prevented the use of procedures as first-class values in data structures. In Modula-2 [29] this was changed to allow for procedure variables, but the restriction was added that only global procedures can be assigned. C [13] has function pointers with a similar semantics. While this allows first-class procedure values, it restricts such procedures to operate on the global state only, while in Pascal it was possible to pass a nested procedure that in turn could operate on the current bindings of local variables of the passing procedure.

Sather, as in C, has no nested routines, hence the C / Modula-2 solution would work without any constraints. However, the fact that routines bound to a variable can only operate on the global state is disallows constructs that many applications need. For example, to implement a routine which produces the complement of a boolean argument routine or the composition of two argument routines there must be internal state associated with the routine.

The Sather solution is to introduce *bound routines*[5] to express higher-order functions and closure-like constructs. The key idea is that the parameters of a routine, including the implicit **self**, can be bound to objects. The resulting bound routine can then be assigned to a routine variable of the appropriate type. For example, it is possible to take a routine with two integer parameters, bind one of these to an integer value, and then assign the resulting bound routine to a variable that asks for a routine with a single integer parameter. Bound types describe the resulting signature of a bound routine. Conformance is defined as contravariant conformance of the type signature.

3 Code Inheritance

3.1 The Textual Inclusion Model for Code Inheritance

The semantics of *code inheritance* in Sather is defined by textual inclusion of the inherited code. So-called "include" clauses are used to incorporate source code from a specified class. The choice of the keyword "include" was made

---

[5] Sather also introduces *bound iters*.

218

to indicate the textual semantics for the inheritance model. References to the type "SAME" in the inherited code represent the type of the inheriting class. Newly defined features in a class override inherited features with a conforming signature (as defined in Section 2.1). This approach differs from that used in Smalltalk and most other object-oriented languages, in which a call conceptually climbs up in the class hierarchy until a corresponding method is found. For most common cases, the two approaches produce identical results. In complex situations, however, the textual inclusion approach seems easier to understand and to reason about.

It is sometimes convenient for a new version of a feature to call the old version that it overrides. Smalltalk solves this problem by providing the "super"-call, which bypasses any matching implementations in the object's class and passes the call directly to the superclass. We found that in Sather, this approach would be confusing in certain circumstances. The problem arises when code which makes a super call is itself inherited. The ambiguity for programmers was whether the inherited "super" call refers to the "super" class of the original defining class or of the inheriting class.

To eliminate this problem, Sather replaces the "super"-call approach with a general "renaming" facility in the include clauses which define code inheritance. The include clause comes in two forms: one is used to include and possibly rename a single feature from another class and the other includes an entire class but may cause features to be undefined or renamed. Renaming is shallow, i.e. renaming affects only the definition of the specified feature but not calls on that feature[6]. Appendix A.10 includes the syntax of the construct. Figure 3 uses the example of extending a simple unit (neuron) in ICSIM to a unit with back-propagation learning to show how the "super"-call problem is solved in Sather. The routine accumulated_input inherited from SIMPLE_UNIT is renamed as the private routine: "SIMPLE_UNIT_accumulated_input" in SIMPLE_BP_UNIT.

One may argue that the renaming solution for "super"-calls unnecessarily clutters the name space. Our experience shows that we use this style of programming infrequently, and if we need it we make the renamed version of the old routine private in order not to affect the external interface of a class. Because every routine has a specified name, the approach eliminates any ambiguity in the interpretation of code. As shown in the next section, the renaming approach is also more general than the "super"-call approach.

3.2 Multiple Subclassing and Conflict Resolution

Sather supports *multiple subclassing* (multiple code inheritance) by allowing multiple include clauses per class. Since more than one of the superclasses may provide a feature with the same signature, multiple subclassing leads to *inheritance conflicts*. Two routines or iters are said to *conflict* if they have the same

---

[6] Thus, renaming or undefining a feature may break inherited code. If this is the case the compiler signals a "subclassing error" associated with the corresponding include clause.

```
class SIMPLE_UNIT is
  ...
  accumulated_input: REAL is
    -- Compute the dot product of input values * weights
    input_port.get_outputs_into_vec(input_values);
    res := input_values.dot_v(weights)
  end;
  ...
end;

class SIMPLE_BP_UNIT is
  include SIMPLE_UNIT
    accumulated_input -> private SIMPLE_UNIT_accumulated_input ... ;
  ...
  accumulated_input: REAL is
    -- Compute the dot product of weights * inputs + the bias value
    res := SIMPLE_UNIT_accumulated_input + bias.val
  end;
  ...
end
```

Fig. 3. Using renaming instead of a "super"-call.

name, the same number and types of arguments, and both either have or do not have a return value. Reference [9] describes four ways to cope with inheritance conflicts:

1. *Disallow conflicts*: Signal an error in the case of a conflict.
2. *Resolve conflicts by explicit selection*: Require the user to make a selection in case of a conflict. This is Sather's approach, as described below.
3. *Form disjoint union of features*: Create a separate feature for each conflicting feature. This is the approach of C++ where feature names of sub- and super-classes are in different scopes. The user selects between conflicting features using the scope resolution operator "::".
4. *Form composite union of features*: Create one single feature for each conflicting feature by algorithmically resolving the conflict. CLOS [1] follows this approach by linearizing the class hierarchy.

1. to 3. are *explicit* conflict resolution methods, 4. is an *implicit* method. Cecil [4] takes an intermediate stance between 3. and 4. by imposing only a partial ordering on classes, and requiring any remaining conflicts to be resolved explicitly by the programmer. We agree with [25] that CLOS-style linearization of the inheritance graphs may lead to unexpected method lookups, and result in faulty and hard to debug programs.

Sather, therefore, adopts an explicit conflict resolution scheme in which the programmer has to explicitly choose in case of conflicts. A class may not explicitly define two conflicting routines or iters. A class may not define a routine which conflicts with the reader or writer routine of any of its attributes (whether explicitly defined or included from other classes). If a routine or iter is explicitly defined in a class, it overrides all conflicting routines or iters from included classes. The reader and writer routines of a class's attributes also override any included routines and must not conflict with each other. If an included routine or iter is not overridden, then it must not conflict with another included routine or iter. Renaming or undefining in include clauses is used to resolve these conflicts.

Any language which supports code inheritance must deal with the problem of the same code inherited along two different paths. Some languages introduce complex mechanisms to deal with this case, but these tend to be confusing to programmers and rarely do exactly what is desired. Sather's solution is implied by the rules given above. Sather does not consider the origin of code and resolves inheritance solely based on the body of the class itself and the bodies of the classes it includes (*after* their own code inheritance has been resolved). This behaves like the non-virtual inheritance of C++ for diamond-shaped inheritance graphs, i.e. features from a common superclass are included along each edge. This sometimes necessitates explicitly choosing a single version of a routine inherited along multiple paths, but it eliminates complex rules which depend on the structure of the code inheritance graph.

Our experience with the Sather libraries is that we use multiple subclassing only rarely. We therefore felt that these special cases were too weak a justification to introduce a complex graph-based subclassing scheme or a strategy based on structural equality of feature definitions.

3.3 Separate Compilation

Sather has no explicit notion of structural units comprising multiple classes. The Sather programming environment is intended to manage and maintain the source code of multiple classes. In particular, when compiling a new class it is often required that the Sather compiler has access to the source code (or at least the type interface and dependency information) of all classes referred to by it.

For example, the compiler automatically inlines short routines to improve efficiency. There tend to be many short routines in object-oriented programming because a routine which is needed only for the purposes of an abstract interface often just calls another routine. In addition to eliminating an extra routine call, inlining allows much more optimization to be done within a routine with inlined code. On the one hand, compiler-controlled inlining requires that the code to be inlined is available to the compiler and to the compiler's analysis process, i.e. that the source is at hand. On the other hand, inlining introduces hidden dependencies between implementations.

For large systems, there are arguments for introducing another level of modularity. In some cases, one doesn't want to require that all source code be available or allow arbitrary dependencies between compiled units. Such large systems are usually composed of subsystems. For a limited subsystem the global analysis is acceptable. For a composed system, however, it should be possible to define the subsystems in a way that global analysis is not required.

For Sather, it is possible to form subsystems with strict boundaries in terms of compiler analysis. Such a subsystem must be limited by an interface presenting only types, i.e. empty abstract classes, to subtyping clients, and allowing for direct calls to routines (c.f. Section 2.7) defined by classes within the subsystem.

The most prominent mechanism that cannot be allowed to cross subsystems is code inheritance, which of course is a direct consequence of specifying the semantics of code inheritance based upon the actually inherited source text. Also to be excluded from a subsystem's interface are parameterized classes: The current Sather compiler cannot completely check a parameterized class before its parameters actually get specified. This defect in the checkability of Sather's parameterized classes is unfortunate and an issue of ongoing research. However, this problem is not specific to Sather, the same holds for C++, where such errors might be detected as late as at link-time(!), Eiffel, and Ada. Possible solutions tend to either restrict the usefulness of parameterized classes, or to introduce a complicated apparatus to specify sufficiently strong bounds on the parameters.

Explicit support for expressing subsystem boundaries, such as modules [26], might be a useful extension to Sather. In particular, a module construct would help to package helper classes, to explicitly treat subsystem invariants, to reduce the probability of conflicts in the global class name space, and allow limitations to be placed and how much of a source needs to be revealed for purposes of compilation. The best form for such a construct is not yet clear, however, and so the current version of Sather will address these issues at the level of the development environment rather than in the language.

4 Foundations

The type system described so far needs to be grounded in explicit built-in classes. A class can do no more than define attributes of types introduced by itself or other classes, or define routines operating over its own or shared attributes, or invoking other operations. What is missing are the foundational entities to start with. Such foundation entities are present in all languages[7] in the form of built-in types or operations with predefined semantics. In Sather, certain predefined classes serve this purpose.

A language that claims to be "general-purpose" also has to be able to express interfaces to the outside world. For example, a Sather program should be able to call non-Sather libraries, including functions of the underlying operating system and graphical user interface. It is not reasonable to expect that a fixed set of built-in classes will ever suffice to serve this purpose in full generality. For these purposes Sather has *external classes*. Predefined and external classes are described in the next two subsections.

---

[7] In theory, the λ-calculus, e.g. with syntax $E ::= x \mid EE \mid \lambda x.E$, is sufficient. Such languages tend to have efficiency problems, though!

4.1 Built-in Classes

Most classes are defined by explicit code in a Sather program, but there are several classes which are automatically constructed by the compiler. These classes have certain built-in features that may be defined in an implementation dependent way. In each case, the choices made by the implementation are described by constants which may be accessed by a program. This section provides a short description of some of the most important built-in classes. The complete and detailed semantics and precise interface is specified in the Sather class library documentation.

- **$OB** is automatically an ancestor of every class. Variables declared by this type may hold any object.
- **BOOL** defines value objects which represent boolean values.
- **CHAR** defines value objects which represent characters.
- **STR** defines reference objects which represent strings.
- **INT** defines value objects which represent machine-dependent integers. The size is implementation dependent. Classes representing fixed-sized integers with a different number bits may be defined by inheriting from **INT** and redefining the constant "**bsize**". All the routines work with an arbitrary "**bsize**".
- **INTINF** defines reference objects which represent infinite precision integers. They support arithmetic operations but do not support bit operations.
- **FLT**, **FLTD**, **FLTE**, and **FLTDE** define value objects which represent floating point values according to the single, double, extended, and double extended representations defined by the IEEE-754-1985 standard.
- **ARR{T}** is a reference class defining dynamically-sized arrays of elements of type **T**. Classes which inherit from this are called *array classes*. They allocate space for the array and the attribute **asize:INT** whose value is the number of elements in the array.
- **TYPE** defines the value objects returned by the **type** routine.

4.2 Interfacing to External Code

Sather provides a few special built-in classes to interface to external code, as listed below. Additionally, Sather's external classes can be used to interface with code from other languages. External classes are not classes in the traditional sense. They can neither be instantiated, nor can they be in a subclass or subtype relationship with any other class. It is merely for the sake of uniformity of the language that external routine interfaces are grouped into external "classes".

Each external class is typically associated with an object file compiled from a language like C or Fortran. External classes may only contain routines with distinct names (overloading is not allowed in external classes). The external object file must provide a conforming function definition with the same name as each routine which doesn't have an implementation in the external class. Sather code may call these external routines using a class call expression of the form

`EXT_CLASS::ext_rout(5)`. Similarly, the external code may call one of the non-abstract Sather routines[8] defined in the class by using a name consisting of the class name, an underscore, and the routine name (eg. `EXT_CLASS_sather_rout`).

- **BITS** may be inherited by value classes which represent a single field of data. The descendant may define the two constants `bsize:INT` and `balign:INT` to specify the size in bits of the object and its alignment requirements.
- **$EXTOB** is used to refer to "foreign pointers". These might be used, for example, to hold references to C structures. Such pointers are never followed by Sather and are treated essentially as integers which disallow arithmetic operations. They may be passed to external routines.

5 Conclusions

The design of Sather 1.0 involved trading off an interesting set of constraints regarding efficiency, clarity, reusability and safety. We have described several important aspects of the type and class system and compared them with the solutions chosen by other object-oriented languages. These give rise to a language with a unique combination of conceptual clarity, safety and support for high performance.

Acknowledgements

Many people were involved in the Sather 1.0 design discussions. Jerry Feldman, Ben Gomes, Ari Huttunen, Chu-Cheow Lim, Heinz Schmidt, and David Stoutamire made suggestions which were especially relevant to the topics discussed in this paper.

A Syntax of the Sather Class and Type System

The following sections give examples of actual Sather code fragments together with the corresponding grammar rules. The grammar rules are presented in a variant of Backus-Naur form. Non-terminal symbols are represented by strings of letters and underscores in italic typeface and begin with a letter. The nonterminal symbol on the lefthand side of a grammar rule is followed by an arrow " $\Rightarrow$ " and right-hand side of the rule. The terminal symbols consist of Sather keywords and special symbols and are typeset in the `typewriter` font. Italic parentheses "(...)" are used for grouping, italic square brackets "[...]" enclose optional clauses, vertical bars "... | ... " separate alternatives, italic asterisks "... *" follow clauses which may appear zero or more times, and italic plus signs "... +" follow clauses which may appear one or more times.

---

[8] The calling conventions and the layout of objects are described in the implementation manual of individual versions.

A.1 Class definition lists

```
class A is ... end; class B is ... end
```
class_def_list ⇒ *[class_def]* | *class_def_list* ; *[class_def]*

A.2 Class definitions

```
class A{S,T:=INT,U<B} is ... end
value class B < $C,$D is ... end
abstract class $E > G,H is ... end
```
class_def ⇒ *[value* | **abstract** | **external]** **class** *class_name*
  [{ param_dec (, param_dec)}]* *class_inheritance* **is** *class_elt_list* **end**

param_dec ⇒ *ident [< type_spec] [:= type_spec]*

class_inheritance ⇒ *[< type_spec (, type_spec)*]*
  [> type_spec (, type_spec)]*

A.3 Type specifiers

```
A{B,C{$D}}
ROUT{A,B,C}:D
ITER{A,B!,C}
```
type_spec ⇒ *[class_name] [{ type_spec_list }]* |
  ROUT *[{ type_spec_list }] [: type_spec]* |
  ITER *[{ type_spec [!] (, type_spec [!])*}] [: type_spec]*

type_spec_list ⇒ *type_spec (, type_spec)**

A.4 Features

class_elt_list ⇒ *[class_elt]* | *class_elt_list* ; *[class_elt]*

class_elt ⇒ *const_def* | *shared_def* | *attr_def* | *rout_def* | *iter_def* | *include_clause*

A.5 Constant attribute definitions

```
const r:FLT:=45.6
private const a,b,c
```
const_def ⇒ *[private]* **const** *ident (: type_spec := expr* | *[:= expr] [,ident_list])*

ident_list ⇒ *ident (, ident)**

A.6 Shared attribute definitions

```
private shared i,j:INT
shared s:STR:="name"
readonly shared c:CHAR:='x'
```
shared_def ⇒ *[private* | **readonly]** **shared**
  (ident : type_spec := expr | *ident_list : type_spec)*

A.7 Object attribute definitions

```
attr a,b,c:INT
private attr c:CHAR:='a'
readonly attr s:STR:="a string"
```

$attr\_def \Rightarrow$ [private | readonly] attr
  $(ident : type\_spec := expr \mid ident\_list : type\_spec)$

A.8 Routine definitions

```
a(FLT):FLT pre arg>1.2 post res<4.3 is ... end
b is ... end
private d:INT is ... end
c(s1,s2,s3:STR)
```

$rout\_def \Rightarrow$ [private] $ident$ [(arg_dec (, $arg.dec)^*)$] [: $type\_spec$] [pre $expr$]
  [post $expr$] [is $stmt\_list$ end]

$arg\_dec \Rightarrow$ [ident_list :] $type\_spec$

A.9 Iter definitions

```
elts!(i:INT, x:FLT!):T is ... end
```

$iter\_def \Rightarrow$ [private] $iter\_name$ [($iter\_arg\_dec$ (, $iter\_arg\_dec)^*)$] [: $type\_spec$]
  [pre $expr$] [post $expr$] [is $stmt\_list$ end]

$iter\_name \Rightarrow ident!$

$iter\_arg\_dec \Rightarrow$ [ident_list :] $type\_spec$ [!]

A.10 include clauses

```
include A a:INT->b, c(INT)->, d:FLT->private d;
private include D e:STR->readonly f;
include A::a(INT)->b;
```

$include\_clause \Rightarrow$ include $type\_spec$ :: elt_mod |
  [private] include $type\_spec$ [elt_mod (, elt_mod)*]

$elt\_mod \Rightarrow ident$ [($type\_spec\_list$)] [: $type\_spec$] ->
  [[private | readonly] $ident$]

References

1. D. G. Bobrow, L. G. DeMichiel, R. P. Gabriel, S. Keene, G. Kiczales, and D. A. Moon. The common lisp object system specification. Technical Report 88-002R, X3J13, June 1988. Also in special issue of SIGPLAN Notices 23 (Sep. 1988) and Lisp and Symbolic Computation (Jan. 1989).

2. Gilad Bracha and William R. Cook. Mixin-based inheritance. In *Proceedings of the Conference on Object-Oriented Programming, Systems, and Applications and European Conferance on Object-Oriented Programming (OOPSLA/ECOOP'90)*, Ottawa, Canada, October 1990. Also in SIGPLAN Notices, 25:10, Oct. 1990.

3. Luca Cardelli. Typeful programming. Technical report, DEC Systems Research Center, Palo Alto, CA, May 1989.

4. Craig Chambers. The Cecil language - specification and rationale. Technical Report 93-03-05, Department of Computer Science, University of Washington, Seattle, WA, March 1993.

5. William R. Cook. A proposal for making eiffel type safe. In *Proceedings of the Third European Conference on Object-Oriented Programming (ECOOP'89)*, pages 57–70, Nottingham, England, 1989. Cambridge University Press.

6. William R. Cook, Walter L. Hill, and Peter S. Canning. Inheritance is not subtyping. In *Proceedings of the ACM Conference on Principles of Programming Languages (POPL'90)*, pages 125–135. ACM Press. Addison-Wesley, 1990.

7. Mahesh Dodani and Chung-Sin Tsai. ACTS: A type system for object-oriented programming based on abstract and concrete classes. In *Proceedings of the Sixth European Conference on Object-Oriented Programming (ECOOP'92)*, pages 309–328, Utrecht, Netherlands, 1992.

8. Margaret A. Ellis and Bjarne Stroustrup. *The Annotated C++ Reference Manual.* Addison-Wesley, 1990.

9. Richard P. Gabriel, Jon L White, and Daniel G. Bobrow. CLOS: Integrating object-oriented and functional programming. *Communications of the ACM*, 34(9):29–38, September 1991.

10. Adele Goldberg and David Robson. *Smalltalk-80, The Language and its Implementation.* Addison-Wesley, 1985.

11. Norman Hutchinson. *Emerald: An Object-Oriented Language for Distributed Programming.* PhD thesis, Department of Computer Science and Engineering, University of Washington, Seattle, WA, January 1987.

12. Kathleen Jensen and Niklaus Wirth. *PASCAL: User Manual and Report.* Springer-Verlag, 2d ed. corr. print edition, 1978.

13. Brian W. Kernighan and Dennis M. Ritchie. *The C Programming Language.* Prentice-Hall, 1978.

14. B.B. Kristensen, O.L. Madsen, B. Moeller-Pedersen, and Kristen Nygaard. The BETA programming language. In B.D. Shriver and P. Wegner, editors, *Research Directions in Object-Oriented Programming*. MIT Press, 1987.

15. Chu-Cheow Lim and Andreas Stolcke. Sather language design and performance evaluation. Technical Report TR-91-034, International Computer Science Institute, May 1991.

16. Boris Magnusson. Code reuse considered harmful. *Journal of Object Oriented Programming*, 4(3), November 1991.

17. Bertrand Meyer. *Eiffel - The Language.* Prentice-Hall, 1988.

18. Bertrand Meyer. *Object-oriented Software Construction.* Prentice-Hall, 1988.

19. Hanspeter Mössenböck and Niklaus Wirth. The programming language oberon-2. *Structured Programming*, 12(4), 1991.

20. Stephan Murer, Stephen Omohundro, and Clemens A. Szyperski. Sather iters: Object-oriented iteration abstraction. Technical Report TR-92-xxx, International Computer Science Institute, 1993.

21. Greg Nelson, editor. *Systems Programming with Modula-3.* Prentice Hall, 1991.

22. Stephen Omohundro. Sather provides nonproprietary access to object-oriented programming. *Computers in Physics*, 6(5):444–449, 1992.
23. Martin Reiser and Niklaus Wirth. *Programming in Oberon. Steps Beyond Pascal and Modula.* Addison-Wesley, 1992.
24. Heinz W. Schmidt and Benedict Gomes. ICSIM: An object-oriented connectionist simulator. Technical Report TR-91-048, International Computer Science Institute, November 1991.
25. Alan Snyder. Encapsulation and Inheritance in object-oriented programming languages. In *Proceedings of the First ACM Conference on Object-Oriented Programming, Systems, and Applications (OOPSLA'86)*, pages 38–45, Portland, OR, November 1986. Also in SIGPLAN Notices, 21:11, Nov. 1986.
26. Clemens A. Szyperski. Import is not Inheritance – why we need both: Modules and Classes. In *Proceedings of the Sixth European Conference on Object-Oriented Programming (ECOOP'92)*, Utrecht, The Netherlands, June 1992.
27. David Ungar and Randall B. Smith. Self: The power of simplicity. In *Proceedings of the Second ACM Conference on Object-Oriented Programming, Systems, and Applications (OOPSLA'87)*, Orlando, FL, October 1987. Also in SIGPLAN Notices, 22:12, Dec. 1987.
28. U.S. Department of Defence. *Ada Reference Manual: Proposed Standard Document*, July 1980.
29. Niklaus Wirth. *Programming in Modula-2.* Springer-Verlag, 1982.

OPAL: **Design And Implementation of an Algebraic Programming Language**[*]

Klaus Didrich[1], Andreas Fett[2], Carola Gerke[1], Wolfgang Grieskamp[1], Peter Pepper[1]

[1] Technische Universität Berlin, Fachbereich Informatik, Institut für angewandte Informatik, Franklinstr. 28/29, 10587 Berlin
[2] now at Daimler-Benz AG, Forschung Systemtechnik, Alt-Moabit 91b, 10559 Berlin

1 Introduction

The algebraic programming language OPAL has been designed as a testbed for experiments with the specification and development of functional programs. The language shall in particular foster the (formal) development of production-quality software that is written in a purely functional style. As a consequence, OPAL molds concepts from *Algebraic Specification* and *Functional Programming* into a unified framework.

The amalgamation of these two concepts — which are both based on a sound theoretical foundation — constitutes an ideal setting for the construction of formally verified "safe" software. Moreover, the high level of abstraction and the conceptual clarity of the functional style can increase the productivity of software development significantly.

Unfortunately, the acceptance of the functional style suffers from the widespread prejudice that its advantages are paid for by a high penalty in run-time and storage consumption. Yet, this is not true.

Due to their semantic elegance and clarity, functional programs are amenable to a wide variety of powerful optimizations, such that the programmer need not be concerned with low-level considerations about machine pecularities. This way, she can concentrate on finding good algorithmic solutions instead.

The potential for automatic optimizations is highly increased, when functional programs are combined with algebraic specifications. Then, algebraic properties of the functional programs can be used to guide optimizations which go far beyond traditional optimization techniques.

In this paper we illustrate some principles of algebraic programming. Moreover, we introduce the language OPAL, sketch its compilation strategies, and point out some challenges for further research in the area of algebraic programming languages.

[*] This work was partially sponsored by the German Ministry of Research and Technology (BMFT) as part of the project "KORSO – Korrekte Software".

2 The Functional Core of OPAL

The core of OPAL is a strongly typed, higher-order, *strict* functional language, which belongs to the tradition of HOPE and ML ([10, 7]). Due to our orientation towards algebra-based programming environments, the core language already has a distinctive algebraic flavour in the tradition of languages like CIP-L, ACT ONE, OBJ, and others ([1, 3, 2, 5]). This flavour shows up in the syntactical appearance of OPAL, in the preference of parameterization to polymorphism, and last but not least, in the semantics of OPAL. In the following, we first consider some examples, then point out the specialities of OPAL, and finally sketch the semantics.

2.1 Items

The basic syntactical parts of OPAL are *items*. The following OPAL text consists of four items, which declare and define the data type of sequences and the sequence concatenation:

```
DATA seq   == <>                       -- empty sequence
              ::(ft: data, rt: seq) -- first element and rest sequence
FUN ++ : seq ** seq -> seq     -- declaration of concatenation function

DEF (Ft :: Rt) ++ S == Ft :: (Rt ++ S)
DEF <>          ++ S == S
```

- The DATA item is in fact a shortcut for the following items:

```
SORT seq
FUN <>      :  seq                      -- constructors
FUN ::      :  data ** seq  -> seq
FUN <>? ::? :  seq          -> bool  -- discriminators
FUN ft      :  seq          -> data  -- selectors
FUN rt      :  seq          -> seq
```

It does not matter whether we add these declarations explicitly to the text or not since declarations can be repeated; collections of declarations collapse to sets.
- The DATA item does not only induce the above declarations, but also a *free algebraic type*, which ensures for example the following properties:

$$\forall ft_1, ft_2, rt_1, rt_2. \, ft_1 \not\equiv ft_2 \vee rt_1 \not\equiv rt_2 \Rightarrow ft_1 :: rt_1 \not\equiv ft_2 :: rt_2$$
$$\forall s. \, s \equiv <> \vee (\exists d, s'. \, s \equiv d :: s')$$

These axioms, together with the generation principle (all elements of seq can be denoted by the constructors), allow for the pattern matching used in the left-hand sides of the definitions of the concatenation function ++. They also contain enough information to induce the definitions for the automatically declared functions.

2.2 Structures

In OPAL, a program is a collection of "structures" which are connected to each other by import relations. Motivated by principles from software engineering, as well as by compilation aspects, we have split every structure into two parts:

structure = (visible) signature + implementation.

The signature part provides the *externally visible view* of the structure, and the implementation part provides the *hidden internal view*. That is, the signature is the syntactic interface of the software module, whereas the implementation constitutes its constructive realization.

An interface to a structure which implements edge-labeled directed graphs might look as shown in Table 1.

Table 1. A Signature of Edge-Labeled Graphs

```
SIGNATURE Graph[label]
   SORT label                          -- the parameter

   IMPORT Seq[node] ONLY seq           -- selective import
          Seq[edge] ONLY seq           -- only seq is visible here

   SORT graph edge node

   FUN empty   : graph                 -- empty graph

   FUN nodes   : graph -> seq[node]    -- nodes of graph
   FUN =       : node ** node -> bool   -- equal nodes
   FUN outs    : graph ** node -> seq[edge] -- outgoing edges
   FUN source                          -- source of edge
       destin  : edge  -> node          -- destination of edge
   FUN label   : edge  -> label         -- label of edge

      -- and many more functions on graphs
```

Note that the structure Graph is *parameterized* by the sort label. On the basis of this interface, we may define a structure for the calculation of *minimal distances* in a directed edge-labeled graph. The signature is shown in Table 2.

We have *actualized* the graph interface by the sort dist, which is a parameter itself. Actualizing can be thought of as textually replacing all occurences of the identifier label by the identifier dist; for a full description see e.g. [3]. Note that parameterization in contrast to polymorphism allows not only the abstraction from sorts, but also from operations.

A prototypical implementation of MinDist is given by the implementation part in Table 3.

Table 2. Signature for Calculating Minimal Distances

```
SIGNATURE MinDist[dist,+,min,infinity]
  SORT dist                 -- the abstract sort of distances
  FUN  +        : dist ** dist -> dist    -- accumulation
  FUN  min      : dist ** dist -> dist    -- choose
  FUN  infinity : dist                    -- worst case


  IMPORT Graph[dist] ONLY graph node


  FUN distance : graph -> node ** node -> dist
  -- calculate the minimal distance between two nodes
```

Table 3. Implementation Part of `MinDist`

```
IMPLEMENTATION MinDist
  IMPORT Seq              COMPLETELY
         SeqMapReduce     COMPLETELY
         Graph[dist]      COMPLETELY


  DEF distance(G)(x,y) == distance(G,nodes(G))(x,y)


  FUN distance : graph ** seq[node] -> node ** node -> dist
  DEF distance(G,cand :: Rest)(x,y) ==
      -- choose between indirect path via cand and other paths
      min(distance(G,Rest)(x,cand) + distance(G,Rest)(cand,y),
          distance(G,Rest)(x,y) )
  DEF  distance(G,<>)(x,y) ==
      -- construct the sequence of weights for outgoing edges,
      -- and choose the best one
      (min,infinity) / (weight * outs(G,x))
      WHERE weight == \\edge.IF destin(edge) = y
                             THEN label(edge)
                             ELSE infinity  FI
```

A few remarks on this implementation:

- Parameterized structures can be imported omitting the actualization. Actualizations are then inferred from the context of applications of objects of the structure, a process quite similar to the instantiation of principal typings in polymorphic languages.
- In the second equation of **distance** we have made some use of the benefits of "programming with functions". The symbol \\ is the ASCII representation of the λ symbol. The λ-construct introduces a local function which returns for a given edge its weight, if it is connected to the destination. We apply

two functions from the parameterized structure **SeqMapReduce**: firstly, the map function *, which applies a function to each element of a sequence, is used to produce a sequence of weights; secondly, the reduce function /, which collapses a sequence to a single value by "accumulating" the sequence elements, chooses the best weight.

2.3 Virtual Constructors

OPAL provides the TYPE item to declare an algebraic data type which is implemented later on by the DATA item. But the declaration of an algebraic type need not be identical to its implementation. Thus, it is possible to hide constructors or components of constructors, which are used for implementation-dependent formulations of exported functions. One may even use completely different sets of constructors in the interface and in the implementation. An application of this feature is given by the example in Table 4, which is taken from the OPAL standard library.

Table 4. Natural Numbers in OPAL

```
SIGNATURE Nat
  TYPE nat == 0   succ(pred: nat)
  -- and many more declarations

IMPLEMENTATION Nat
  IMPORT BUILTIN     COMPLETELY
  DATA nat           == abs(rep: NUM)
  DEF 0              == abs(zeroNum)
  DEF 0?(abs(x))     == eqZeroNum(x)
  DEF succ(abs(x))   == abs(succNum(x))
  DEF succ?(abs(x))  == ~(eqZeroNum(x))
  DEF pred(abs(x))   == IF succ?(abs(x)) THEN abs(predNum(x)) FI
  -- and many more definitions
```

This implementation uses the built-in integral numbers to implement natural numbers. In the interface, the TYPE item ensures the abstract properties of an algebraic free type and enables the use of pattern matching. Internally, natural numbers are as efficient as integral numbers, since simple sort embeddings as given by the DATA item are eliminated by the optimizer.

2.4 What Is Special About OPAL?

We consider it essential when writing easily readable programs to have access to powerful and flexible syntactic concepts that are realized in an orthogonal manner. For this reason OPAL has no built-in "syntactic sugar" for coping with special situations, but a generalization of these features which is accessible to the user:

- Identifiers in OPAL consist of sequences of either alphanumeric or graphical characters. Thus, there is nothing special about the identifiers **0**, **++**, **::**, or **<>** used in the examples above.
- Function names may be placed arbitrarily before, between, or after their arguments in applications.
- Collections of objects are only separated by commas, when the order is important. Hence, parameters in function calls are seperated by commas, but declarations are not.
- OPAL supports overloading in a very general way by means of OPAL's *name concept*. Every object is identified by a *name* which has three components besides its *identifier*: the *origin identifier* is the name of the structure in which the object is declared, the *instance* is the list of the actualizations of that structure, and the *kind* distinguishes sorts and operations, reflecting the operation's functionality based on the names of the used sorts. The append function, for example, as applied in the implementation part of the structure `MinDist`, has the name `::'Seq[node]:node**seq'Seq[node]->seq'Seq[node]` [1].
  Of course, all components of a name except its identifier can be omitted, if the context allows the unique derivation of the missing parts. (In order to perform this resolution, all available context information is used, including mutual dependencies between names.) *Note:* On the level of *names* there is actually *no* overloading; but the programmer normally uses only the *identifiers* and then has the effect of overloading.

Another important point in the design of OPAL is the incorporation of software engineering aspects:

- Structures have a distinct export interface, called signature part (see Section 2.2), from which other structures can selectively import objects. To enable pattern matching over exported data types without violating the principle of information hiding, OPAL incorporates the concept of virtual constructors (see Section 2.3).
- Parameterization provides a high degree of abstraction and reusability of structures (see Section 2.2). It likewise allows the abstraction from sorts as well as from operations and gives a stronger typing. The inference of actualizations of parameterized structures fits nicely into OPAL's name scheme. Omitting the actualization of an import of a parameterized structure is explained as *importing the (infinite) set of possible actualizations*. The application of a partial name from such an import must contain the actualization explicitly, or it must be deducible from the context as described above.

The following feature of OPAL is motivated by the desire to compile to efficient code, but it also affects the programming of functions in OPAL:

- OPAL is a *strict* language. Hence, infinite lists and streams cannot be expressed adequately. However, this disadvantage is not that important in the

---
[1] `node` is used here as short-cut for `node'Graph[dist]:SORT`, and so forth.

everyday use of functional languages. The advantage of being able to produce time- and space-efficient code is considered more valuable.

2.5 Semantics

The algebraic flavour of OPAL is apparent in the semantic definition of the functional language. We presume that the reader is familiar with the basic concepts of algebraic specification languages (see e.g. [3]), and only sketch the essentials of the semantics here.

The signature and the implementation part are both assigned a specification $\mathcal{S}_e^F$ and $\mathcal{S}_i^F$ respectively. The respective signatures Σ_e^F and Σ_i^F of these specifications are related by $\Sigma_e^F \subseteq \Sigma_i^F$; the validity of this relation is ensured by OPAL.

The *internal functional semantics IntSem^F* of an OPAL structure $\mathcal{S}$ is the class of all $\mathcal{S}_i^F$-algebras. These algebras are defined on the basis of cpo's in the usual way. (Note: We employ a loose semantics here due to some constrained nondeterminism in guarded expressions and pattern-based definitions.)

The *external functional semantics ExtSem^F* of $\mathcal{S}$ is a class of $\mathcal{S}_e^F$-algebras that is derived from the internal functional semantics by a forget-restrict scheme: Let $A \in IntSem^F(\mathcal{S})$ be an internal model. Then we obtain the corresponding external model as follows: first, we form the Σ_e^F-reduct $A' = reduct_{\Sigma_e^F}(A)$, then we extract the subalgebra $A'' = restrict_{\Sigma_e^F}(A')$ of reachable elements w.r.t. Σ_e^F-operations. A'' then belongs to $ExtSem^F(\mathcal{S})$.

This distinction of the external semantics by a forget-restrict process is a feature we have borrowed from the area of algebraic specification. It has turned out to be mandatory for any reasonable development and verification methodology. Whether an additional *identify* operator should be used for the external functional semantics is a question we consider worthy of further research.

3 The Property Language

As mentioned earlier, we aim at an integration of specification concepts and functional concepts. "Pure" specification languages focus on expressing algebraic properties of one or several stages of the software development process. Hence, the need for executable specifications is only a minor point in the concepts of these languages and efficiency is not taken into consideration at all.

In contrast, the concept of OPAL is just the other way round: we concentrate on the production of executable, efficient software, but are therefore more restrictive in our specification features. To emphasize this difference in motivation, we use the term OPAL "property language" rather than OPAL "specification language".

3.1 Laws

Properties are expressed by first-order predicate-logic formulas. The primitive predicates are congruence (===) and definedness (DFD) of functional expressions.

Assume we have a function `connected` for graphs. We can specify the behavior of this function as follows:

```
FUN connected: graph -> node ** node -> bool
LAW ALL G x y.connected(G)(x,y) <==>
        Y in (destin * outs(G,x)) OR
        (EX z. z in (destin * outs(G,x)) AND connected(G)(z,y))
```

We would like to point out the following:

- We distinguish boolean values in the functional language from truth values in the logical language. This is necessary because the former express *computable* values, whereas the truth values are not necessarily computable. Nevertheless, we have introduced an abbreviation feature for the denotation of boolean expressions in places where formulas are expected. Actually, the above formula reads: ... `connected(G)(x,y) === true <==>` ...

- Note the difference between the congruence symbol `===` and the definition symbol `==`. The former denotes strong equality (which yields true for two undefined values, and is in general not computable), whereas the latter denotes *fixpoint* equality. Hence, the definition DEF `f(x) == f(x)` sets `f` to the least fixpoint which satisfies the equation (i.e. is the function which is totally undefined), while LAW ALL `x. f(x) === f(x)` is a tautology and therefore does not say anything about `f` at all.

3.2 Structures

Following the same software engineering principles as for the functional sublanguage, we again distinguish two parts of properties:

properties   =   external properties   +   internal properties .

As is to be expected, the external properties are the only ones that are available to the environment, whereas the internal properties express facts about implementation details and therefore must be hidden from the environment.

For example, in the structure **MinDist** we might have an external property part which simply expresses the fact that the function **distance** is total:

```
EXTERNAL PROPERTIES MinDist
  LAW ALL G x y. DFD distance(G)(x,y)
```

Of course, it is possible to also express the input-output behavior of the function **distance**. But in general it is probably quite sufficient in our framework to formally specify only those properties of functions which interest us because of special safety requirements the produced software has to obey.

The internal property part of a structure may be used to express certain facts about the implementation of a structure. This can be used to express facts about auxiliary functions not visible in the external view of a structure, or to hide some facts which are not essential for the external view. For example, we might express the internal property:

```
INTERNAL PROPERTIES MinDist
  LAW ALL G x y. NOT connected(G)(x,y) ==>
                            distance(G)(x,y) === infinity
```

This property may be used for example by an optimizer to enhance applications of the function **distance** in contexts where preconditions ensure that **x** and **y** are not connected.

Note that due to the semantics given in Section 2.5, the implementation does not necessarily satisfy the properties given in the external property part. That is, some properties given in the interface rely on the fact that the forget-restrict operation has been performed.

For example, the sort **graph** might be implemented by adjacency matrices, using a DATA declaration like DATA **graph == graph(adj:matrix[bool])**. While the functions in the interface guarantee that the adjacency matrix is always quadratic, the internal constructor **adj** does not enforce this. So the definedness property of **distance** stated above is not valid in the implementation part, since the application of the function **distance** to a graph with a nonquadratic adjacency matrix is not defined.

One might wonder why property parts are conceptually distinguished parts of a structure, and are not merged with the signature or implementation part. This is motivated as follows:

- The external as well as the internal property parts require auxiliary functions — just as the implementation often requires auxiliary functions. But for specifying properties, other auxiliary functions may be needed than in the implementation part. There might even be a necessity to specify functions which are not intended to be implemented. Therefore, it makes sense to separate these functions from the implemented auxiliary functions in the implementation part.
- We would like to enable the user to view the functional part of a structure without its property parts.

3.3 Semantics

Now we have a situation where there is a specification as well as a functional program, and this necessitates certain compatibility requirements. Therefore we will first sketch the semantics of the property parts (which is a standard loose semantics) and then consider the compatibility criteria.

The relationship between the declarations of the several parts of an OPAL structure is documented by the following diagram:

Signature $\mathcal{S}_e^F$	External Properties $\mathcal{S}_e^P$	visible
Implementation $\mathcal{S}_i^F$	Internal Properties $\mathcal{S}_i^P$	hidden

OPAL automatically includes the signatures from the visible into the corresponding hidden parts and from the functional parts into the corresponding property parts. Thus, the signatures are related by $\Sigma_e^F \subseteq \Sigma_i^F$, $\Sigma_e^F \subseteq \Sigma_e^P$ and $\Sigma_i^F \subseteq \Sigma_i^P$, $\Sigma_e^P \subseteq \Sigma_i^P$.

The *external property semantics* $ExtSem^P$ of an OPAL structure $\mathcal{S}$ is constructed from the (loose) class of $\mathcal{S}_e^P$-algebras. On this class of algebras we perform the reduct operation with respect to the signature of the signature part.

The *internal property semantics* $IntSem^P$ is analogously defined as the Σ_i^F-reduct of the class of $\mathcal{S}_i^P$-algebras.

The notion of *model correctness* states how the property semantics of an OPAL structure is related to the functional semantics:

- *Internal model correctness:* $\emptyset \neq IntSem^F(\mathcal{S}) \subseteq IntSem^P(\mathcal{S})$
- *External model correctness:* $\emptyset \neq ExtSem^F(\mathcal{S}) \subseteq ExtSem^P(\mathcal{S})$

Note that for model correctness, the connection between the internal and external property part is only established implicitly through their common relationship to the implementation. However, correctness can only be determined if a complete implementation of the structure is present.

In order to enable the evolutionary development of implementations by correctness-preserving transformations, we envisage an additional notion of correctness which allows for an implementation being only partially present. The situation here is complicated by the constrained nondeterminism of the functional language OPAL: we cannot simply amalgamate properties with definitions from the implementation, nor can we just construct the intersection using the loose class of algebras in $IntSem^F(\mathcal{S})$ and the loose class of algebras in $IntSem^P(\mathcal{S})$, since the first one stands for "don't know which model is chosen", whereas the second stands for "don't care which model is chosen". This subject is currently a topic of ongoing research.

4 Compiling Algebraic Programming Languages

For algebraic programming to become feasible in practice, it is of fundamental importance that the resulting programs are executed efficiently enough to be competitive with programs written in more machine-oriented languages such as C or Pascal. The *basic* compilation of functional languages has been a topic of intensive research during the past decade. The results are encouraging, although not totally satisfactory. The *extended* compilation of algebraic languages is a topic of ongoing research in which we invest considerable effort. The idea is, basically, to exploit algebraic properties for advanced compilation schemes.

4.1 Basic Compilation

Under the catchword "basic compilation" we understand the mapping of functional programs to von-Neumann-like machine architectures *without* ambitious structural transformations of the functional source. The problems in the basic compilation are the treatment of recursion, functional values, lazy evaluation, and recursive data types, particularly with respect to storage reclamation ([6, 11, 12]).

For OPAL we take the approach of compiling to the "high-level assembly language" C instead of some concrete machine architecture. This has several advantages: the resulting code is highly portable, and we can benefit from many of the optimization techniques for imperative languages nowadays performed by C compilers. Moreover, we have a well-defined interface for interlanguage working, which allows us to access the huge amount of existing standard software available in C.

On the way from OPAL to C code several analysis and transformation phases are invoked. We merely sketch some of them:

- The *Import Analysis* collects information for system global optimizations, such as definitions and properties of the implementations of imported structures. This phase is usually only activated for ready-to-ship systems, since it creates recompilation dependencies between structure implementations.
- The *Unfold Analysis* examines which functions have to be unfolded and in which order. Unfolding is a prerequisite for common subexpression elimination and simplification.
- The *Common Subexpression Elimination* phase detects all common subexpressions local to functions. Note that the potential for common subexpressions is very large in functional languages, because of the absence of side-effects.
- The *Simplification* phase performs fusion and tupling of function local computations.
- The *Translation* phase translates to intermediate imperative code, mainly establishing the notion of *memory* and reference. This is achieved by using a reference-counting scheme.
- The *Selective Update / Reusage* analysis introduces the immediate reclamation of released memory (instead of returning it to the free memory pool), and in particular cases, the *selective updating* of data objects. In our framework, a selective update is performed if a released memory cell is immediately reused for constructing a new cell, and some of its components should be simply copied to the new cell – then we can omit copying (see example below). The reuse analysis is based on static and dynamic reference-counting information.
- The *C Code Generation* phase finally produces the C target code. This is also the place where several forms of recursion (tail recursion, tail recursion modulo constructor application) are mapped to iteration (see example below).

We include a short example of the generated C code, which mainly illustrates the effect of the selective update / reusage optimization, the compilation of recursion, and the handling of functional values. The example is the filter function on sequences as shown in Table 5. The compiler generates the (hand-

Table 5. The Filter Function on Sequences

```
FUN filter : (data -> bool) -> seq[data] -> seq[data]
DEF filter(P)(ft::Rt) == IF P(ft) THEN ft :: filter(P)(Rt)
                            ELSE        filter(P)(Rt)  FI
DEF filter(P)(<>)      == <>
```

formatted and commented) C code given in Table 6. We would like to point out the following properties of the generated C code:

- **RELEASE, FIELD** etc. are of course C macros which expand to C statements.
- The function **filter**, although not tail-recursive in nature, is compiled to an iterative loop. This is achieved by introducing the "result pointer" **resp**, which always points to the place where the result of the next iteration shall be stored.
- In case that the processed sequence is exclusive – i.e. each cell in the linked list representation has a reference count of one – no new list elements are allocated, and only the link to the next cell is occasionally updated. Hence, the only overhead in this case compared to an imperative program is the test of the reference counter, which is compiled to a single machine instruction.

Just to give a brief impression of the attainable efficiency, we mention two classical benchmarks, viz. the functions *Tak* and *Tree* (originally suggested in [14]). *Tak* tests the behaviour of recursion, while *Tree* is a typical program which manipulates recursive data structures. We compare the OPAL compiler with Standard ML of New Jersey, Version 0.65, and also include benchmarks of versions of the algorithms written in (moderately hand-optimized) C [2], compiled with the GNU C-Compiler, Version 1.4, on a SUN-4/75 (32MB):

	C	SML	OPAL
Tak	3.9	35.910	4.415
Tree	2.045	6.110	1.228

For further details about our storage optimization and garbage collection techniques, we refer to [12] and [11].

---

[2] The C version of the *Tree* benchmark uses the memory allocation routines from the standard library. It make uses of imperative destructive update.

Table 6. Generated C Code for the Filter Function on Sequences

```
OBJ filter(OBJ P,OBJ S)
{OBJ res; OBJ *resp=&res;        /* resp points to the location where to
                                    store the result of the next iteration. */
 for(;;){
  if(IS_PRIMITIVE(S)){           /* S is a primitive value -- in our context
                                    this must be the empty list. */

   RELEASE(P,1);                 /* release 1 reference to closure P. We do not
                                    need to release S since it is primitive. */

   *resp=_Slg;                   /* '_Slg' is the alphanumerical */
   break;                        /* representation of '<>' */
  }else{
   {OBJ Ft = FIELD(S,1);         /* select ft and rt components */
    OBJ Rt = FIELD(S,2);
    int Excl_S = IS_EXCL(S));     /* Check if we are the only reference to the */
    OBJ T1,T2;                   /* cell S. */
    if (Excl_S){
       RESERVE(Ft,1);            /* We plan to consume 2 references to Ft and */
    } else {                     /* 1 to Rt (see below). But in the exclusive */
       RESERVE(Ft,2);            /* case we can 'borrow' some from S. */
       RESERVE(Rt,1);
    }
    RESERVE(P,1);
    T1 = METHOD(P,1)(P,Ft);      /* Each closure carries an array of methods
                                    describing how to evaluate it with N
                                    parameters. Here, N=1.
                                    Now we have consumed Ft the 1st time! */
    if (IS_TAGGED(T1,1)){        /* Test if the predicate yields true */
     if (Excl_S){
        *resp = S;               /* We can reuse S. And moreover, Ft is   */
     } else {                    /* already at its place. */
        RELEASE(S,1);
        *resp = CREATE(2);       /* Create cell of size 2, and initialize it. */
        FIELD(*resp,1) = Ft;     /* This is where Ft is consumed the 2nd time.*/
     }
     S = Rt;                     /* Prepare next iterate, consuming Rt. */
     resp = &FIELD(*resp,2);     /* Setup location to store result of  */
     continue;                   /* next iteration. */
    } else
     if (Excl_S){
        DISPOSE(S,1);            /* Dispose the cell S refers to. Note that */
                                 /* the components are already released */
     } else {                    /* indirectly (we have consumed them). */
        RELEASE(S,1);            /* Release the reference S. */
     }
     S = Rt;                     /* resp still points to the place where
                                    the result of the next iteration
     continue;                     shall be stored. */
   }
  }
  return res;                    /* return front of the filtered list. */
}
```

4.2 Extended Compilation

Even though the basic compilation performs already quite nicely, our ambitions
go further. We want to apply more elaborated optimization tactics on the source
level, such as extended recursion removal, function combination, function composition, partial evaluation, finite differencing, and so forth ([8]). None of the
existing compilers for functional languages applies these techniques. The main
reason is that these tactics usually require information about the program which
is not deducible from the program text alone — at least not automatically by

means of a compiler. Hence, the additional information has to be incorporated into the compilation by the programmer as an "advice" to the compiler. This approach has been called *extended compilation* ([4]).

In our setting additional information can be naturally expressed by algebraic properties. Moreover, optimization tactics themselves can be expressed by *algorithm theories* [13], leading to a knowledge-based extendable compilation system. We illustrate the principal ideas of this approach by an example.

The following function computes the chromatic number of a graph G.

```
DEF chromaticNumber(G) ==
  IF complete?(G) THEN cardinality(nodes(G))
                  ELSE min(chromaticNumber(amalgamate(G)),
                           chromaticNumber(connect(G)))   FI
```

Here, the functions **amalgamate** and **connect** choose the next pair of unconnected vertices in the graph, and identify or connect them, respectively. Clearly, one of the efficiency problems of this algorithm is that each recursion level creates modified versions of the graph– this is in particular painful if the graph is represented by a monolithic data structure such as an adjacency matrix. Since the graph is shared between the two recursive incarnations, the compilation techniques for selective updating of data structures cannot be applied. A better solution would be based on a backtracking algorithm, which undoes the modifications at each recursion level and uses only one "single-threaded" instance of the graph.

The compiler would be capable of generating this algorithm automatically, if it is provided with a suitable theory of backtracking. One possible backtracking theory matching our problem is given in Table 7. It is expressed as an ordinary parameterized OPAL structure with properties. *Applying* the theory **BackTrack** to our problem means finding a semantically correct instantiation of the parameter, such that function **f** is instantiated with function **chromaticNumber**. Once having found this instantiation, it is a simple matter of term rewriting to replace applications of **chromaticNumber** by the function **bt** (using the LAW **bt_law**), and to apply specialization techniques to simplify the implementation of the definition of **bt** according to the concrete instantiation.

The least information the compiler requires for finding an instantiation automatically is the following:

```
FUN unconnect    : graph -> graph -> graph
FUN unamalgamate : graph -> graph -> graph
LAW ALL G . (unconnect(G)    o connect   )(G) === G
LAW ALL G . (unamalgamate(G) o amalgamate)(G) === G
```

Here, the function **unamalgamate** extracts the necessary information from a graph to construct a function, which undoes an amalgamation on the given graph; the function **unconnect** behaves similar. Of course, these functions must be implemented by the programmer such that they behave as specified.

Table 7. A Backtracking Theory

```
SIGNATURE BackTrack [f,A,B,b,t,h,d1,d2,inv1,inv2]
  SORT A B                        -- parameter
  FUN f t    : A -> B   b : A -> bool   h : B ** B -> B
      d1 d2 : A -> A    inv1 inv2 : A -> A -> A
  FUN bt : (A -> A) ** A -> A ** B   -- introduced function


EXTERNAL PROPERTIES BackTrack
  -- properties of the parameter
  DEF f(x) == IF b(x) THEN t(x) ELSE h(f(d1(x)),f(d2(x))) FI
  LAW inv1 == ALL x . (inv1(x) o d1)(x) === x
  LAW inv2 == ALL x . (inv2(x) o d2)(x) === x
  -- properties of introduced function
  LAW bt_law == ALL x . f(x) === y WHERE (_,y) == bt(\\z.z,x)


IMPLEMENTATION BackTrack
  DEF bt(inv,x) ==
      IF b(x) THEN (inv(x),t(x))
      ELSE bt(inv1(x),d1(x))   ; (\\x1,y1.
           bt(inv2(x1),d2(x1)) ; (\\x2,y2.
           (inv(x2),h(y1,y2))                )) 
             WHERE ; == \\a,b,C. C(a,b)          FI
```

A sketch of a possible implementation looks as follows:

```
DEF unconnect(G) ==
    LET (u,v) == unconnectedVertexPair(G)
    IN \\NewG.removeEdge(NewG,u,v)
DEF unamalgamate(G) ==
    LET (u,v) == unconnectedVertexPair(G)
        ModifiedEdges == << all edges touching u or v >>
    IN \\NewG.restoreEdges(NewG,ModifiedEdges)
```

Note the use of higher-order functions to "store" the information on how to undo a modification of the graph[3]. Clearly, the function unamalgamate as given here is in the worst case as expensive as copying a graph; depending on the underlying implementation of graphs, more efficient versions are possible.

Given these functions and properties, the compiler can now syntactically prove that the following is a correct instantiation of the backtracking theory:

---

[3] Technically, when calling e.g. unamalgamate(G) this information is stored in a closure as an implicit parameter to the anonymous function returned.

243

```
IMPORT BackTrack [chromaticNumber,graph,nat,complete?,
                  \\G.cardinality(nodes(G)),min,
                  amalgamate,connect,unamalgamate,unconnect]
```

The proof is performed by syntactic unification techniques. More generally, for each function definition the compiler tries to instantiate from a set of given algorithm theories. This operation is still very expensive, and there are several approaches under investigation to guide and improve this search. A promising one is the use of *property theories*. Applied to our example, the properties of an inversion function would be put into a distinguished structure:

```
THEORY LocalInversion[A,f,i]
  SORT A   FUN f : A -> A   i: A -> A -> A
  LAW ALL x. (i(x) o f)(x) === x
```

Now, instead of the axioms inv1 and inv2 this theory would be instantiated in **BackTrack**, and instead of giving the properties of **unconnect** and **unamalgamate** explicitly, the programmer would instantiate this theory as well[4]. A syntactic proof that the inversion properties hold now simplifies to a check whether a particular theory has been instantiated. The goal of this approach is to find a small set of property theories such as **LocalInversion**, **Monoid**, **Lattice** etc. which is shared by a large amount of algorithm theories.

There are several challenges for future research in the area of extended compilation, as it is applied to algebraic programming. On a conceptual level, algorithm and property theories suitable for this approach have to be recovered, collected and systematized. On an engineering level, the implementation technology has to be worked out by combining methods from term rewriting, higher-order unification, and classical compiler construction.

5   Conclusion

We feel that there is great leverage to be gained from making maximum use of the amalgamation of functional and algebraic paradigms. It is by now a well-known thesis that functional programs can be formally derived from algebraic specifications. The research project KORSO sets out to convert this thesis into practically applicable methodology, including tool support. Whereas other languages used in this project mainly focus on the aspect of developing specifications, OPAL fosters the transition from specification to executable (functional) code.

Currently, OPAL is used in teaching Compiler Construction and Software Engineering Principles at our university[5]. Furthermore, it is employed in a joint

---

[4] The integration of these kinds of theories into OPAL, which are used only for bracketing algebraic properties, but do not represent an executable piece of the program, is currently under investigation.

[5] The OPAL compiler is available by anonymous ftp from ftp.cs.tu-berlin.de, directory pub/local/uebb/ocs. For comments and requests please contact us at the e-mail address opal@cs.tu-berlin.de.

project with Daimler-Benz AG in implementing and verifying a compiler for a programmable controller language.

But there is more to be gained from the fusion of algebra and functional programming. An elaborate usage of algebraic properties can generate optimizations that possibly go far beyond the capabilities of classical compilation. And this may help to take functional and algebraic approaches from academic prototype-building to the practical software production stage.

Acknowledgement

We thank our colleagues from the OPAL project and from the KORSO project for many stimulating discussions. This also applies to our former colleagues Gottfried Egger, Michael Jatzeck and Wolfram Schulte. Niamh Warde and Manuela Weitkamp-Smith provided valuable assistance in formulating the paper.

References

1. F. L. Bauer, R. Berghammer, M. Broy, W. Dosch, F. Geiselbrechtinger, R. Gantz, E. Hangel, W. Hesse, B. Krieg-Brückner, A. Laut, T. Matzner, B. Möller, F. Nickl, H. Partsch, P. Pepper, K. Semelson, M. Wirsing, and H. Wössner. *The Munich Project Cip*, volume 1. LNCF Springer, Berlin, 1985.
2. I. Classen. Semantik der revidierten Version der algebraischen Spezifikationssprache ACT ONE. Technical Report 88/24, TU Berlin, 1988.
3. H. Ehrig and B. Mahr. *Fundamentals of Algebraic Specifications I, Equations and Initial Semantics*. EATCS Monographs on Theoretical Computer Science 6, Springer, Berlin, 1985.
4. M.S. Feather. A survey and classification of some program transformation approaches and techniques. In L.G.L.T. Meertens, editor, *Program Specification and Transformation*. North-Holland, 1987.
5. K. Futatsugi, J. A. Goguen, J.-P. Jouannaud, and J. Meseguer. Principles of OBJ2. In *Proc. POPL*, 1985.
6. Simon L. Peyton Jones. *Implementing Functional Languages*. Prentice Hall, 1992.
7. R. Milner, M. Tofte, and R. Harper. *The Definition of Standard ML*. MIT Press, 1990.
8. H. Partsch. *Specification and Transformation of Programs - a Formal Approach to Software Development*. Springer–Verlag, Berlin, 1990.
9. P. Pepper. The Programming Language OPAL. Technical Report 91-10, TU Berlin, June 1991.
10. N. Perry. Hope+. Internal report IC/FPR/LANG/2.51/7, Dept. of Computing Imperial College London, 1988.
11. W. Schulte and W. Grieskamp. Generating efficient portable code for a strict applicative language. In J. Darlington and R. Dietrich, editors, *Declarative Programming*. Springer Verlag, 1992.
12. Wolfram Schulte. *Effiziente und korrekte Übersetzung strikter applikativer Programmiersprachen*. PhD thesis, Technische Universität Berlin, 1992.
13. Douglas R. Smith and Michael R. Lowry. Algorithm theories and design tactics. *Science of Computer Programming*, 14:305–321, 1990.
14. R. Stansifer. Imperative versus functional. *SIGPLAN Notices*, 25, 1990.

Architectural Issues
in
Spreadsheet Languages

Alan G. Yoder
David L. Cohn
University of Notre Dame

{agy, dlc}@cse.nd.edu

Abstract. We have recently begun to develop a programming language model for concurrent computation based upon the popular spreadsheet metaphor. We call it the *Generalized Spreadsheet Model* (GSM); it provides an easy-to-use, geometrically appealing interface to a concurrent computational facility. This interface directly captures the Dataflow relationships inherent in many important problems and makes them available for determining the order of computations. GSM's hierarchical approach to problem modeling also allows us to control granularity. In this paper we use a simple neural net example program to explore some architectural implications of our approach to concurrent computation; we conclude that some reasonably simple hardware assistance would yield significant performance gains.

1. Introduction

The Generalized Spreadsheet Model (GSM) is the result of an effort to formalize the contributions of the popular spreadsheet metaphor and apply them to the domain of concurrent computing. GSM is not a language; rather, it is a framework upon which one can build a concurrent language. As in spreadsheets, a GSM language represents data objects as *cells*, which are grouped into *blocks*. GSM extends the metaphor by allowing blocks to be *n*-dimensional and nested to an arbitrary depth.

In this paper we examine the execution of a program in a hypothetical GSM language from the point of view of a processor in a processor pool. Section 2 provides a brief introduction to GSM. Section 3 introduces the processor's view of a program as a society of mobile cells communicating via message-passing. Section 4 evaluates the opportunities for hardware support implicit in the GSM paradigm. Section 5 concludes.

2. GSM Basics

In an abstract sense, cells are simply lists of *properties*. *Names*, *values*, *types*, *methods* and *data dependencies* are examples of properties. GSM extends the spreadsheet metaphor by defining *structures* and *functions*. It does this by defining *blocks* of cells to be first-class objects, so that a cell may contain a block of other cells. Structures then are just nested arrays of cells. Functions are cells (possibly containing blocks of other cells) which export message handlers.

2.1 Example Structure

Consider the specification of a *Rectangle* structure, defined as two corner *Points* and a *baseline angle*. It looks like this (where indentation denotes containment):

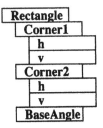

Unless specifically constrained, cells in such a structure can be evaluated concurrently. One moves the Rectangle by adding an offset to the corner Points. This requires a method for adding two Points **A** and **B** together. Normally, such a method amounts to:

$$\textbf{A.h} + \textbf{B.h} \quad \| \quad \textbf{A.v} + \textbf{B.v};$$

i.e. concurrent evaluation of the two sums.

2.2 Evaluating Cells

While users need to be aware that they are manipulating a concurrent computational resource, they needn't usually concern themselves with the details. Computation takes place in any processor in a processor pool, but the pool appears to the user as one large virtual machine. Cells are the unit of computation, so code in a single cell is guaranteed to be executed sequentially. We normally expect this code to be on the order of a typical C++ object method; GSM languages may (or may not) guarantee that it will also be executed indivisibly on a given processor.

Cells can contain both data and functions, so they strongly resemble OOP objects. Communication between cells is by means of message-passing, so most cell methods are used to respond to messages (we term the reception of a message an *event* and often call sending a message *posting an event*).

2.3 Dependencies and Migration

When cell A uses one of cell B's properties to compute or set one of its own properties, a *dependency* from B to A is established. This dependency is stored as one of B's properties; in this way a fully distributed representation of the global dependency graph is available at all times.

Work can be moved from one processor to another by means of *cell migration*. At runtime, a cell consists of its properties (name, value, methods etc.) and any pending messages. When a processor becomes overloaded, it ships off a subset of the cells it currently "owns" to some other processor. This cell migration is automatic and invisible to the user.

Cell migration also necessitates a copy of the inverse of the dependency graph. Suppose cell A, mentioned above, migrates to a new location. When B changes and attempts to notify A of the change, its "pointer" to A will no longer be valid. For this reason **A** must also maintain an *anti-dependency* property[1] containing a link to **B**; when **A** moves it sends **B** a message containing its new location.

Inter-cell dependencies generate asynchronous parallelism in a Dataflow fashion. Synchronous parallelism is provided by the map() primitive. Cell evaluation can also be constrained to be sequential by use of the mapseq() primitive.

2.4 Displaying Cells

Cells connect to the outside world by means of *reflective links*. A reflective link to an interactive display device allows the user to change the linked cell, while any automatic changes as a result of ongoing calculations show up on the display. This in effect introduces a cycle in the dependency graph, which is illegal by any other means.

Cells may display themselves in arbitrary ways; as icons, pictures, text and so on. A *defaults* mechanism provides for dynamic alteration of the runtime environment, which includes the display and interaction mechanisms.

2.5 Ease of Use

It is easy for the details to obscure the main point of the GSM model. We know that commercial spreadsheet programs have been used for everything from finite-element analysis to financial planning to parts databases. Clearly, users are comfortable with a metaphor that places data objects in a familiar row-by-column geometric relationship. GSM's goal is to maintain this level of comfort and intuition, but within a framework that provides sophisticated programming tools and a facility for automatic code parallelization. For more information on the GSM model, see [YC 93].

3. Processor's View of the World

A processor in a GSM environment is one member of a continuously running processor pool. The model does not assume either heterogeneous or homogeneous processors; this decision is left for the designer of a GSM language system. We also do not assume that processors are "real"; a Unix process on each node of a workstation cluster could easily serve as a virtual GSM processor, as could each node of an Intel Paragon or any other multiprocessor machine.

3.1 An Example System

Our discussion will assume the existence of a system based on a toy GSM language called *gsm* with the following properties:

- Processors in a pool are connected by a network able to establish point-to-point links between arbitrary processor pairs.
- Each processor has a continuously running copy of the compute server for *gsm*.
- Within each cell, *gsm* allows local variables and the standard blocking, control and iteration statements *begin-end*, *if*, *if-else*, *for* and *while*. Recursion within cells and private functions within cells are not implemented. Functions are defined in individual cells and structures are defined with nested arrays (blocks of cells).
- Processors are reliable.
- The underlying message system implemented by *gsm* hides heterogeneity, so that processors all look alike.

---

1. We are in the market for a short and intuitive name for this.

- Message transmission is reliable.
- The *gsm* language allows processors to store information about cells onto cell properties reserved for use by themselves and other processors.
- When users "start up" a *gsm* program, the cells in the program file are loaded into a single processor and the top-level cell is sent the EXEC message.

3.2 An Example Program

In GSM, a *program* is a block of cells; examples of valid programs are:
- a matrix filled with data
- a block of cells which queries the user for a number and then computes its factors
- a set of cells simulating fluid flow through a flow network
- a single cell which displays the time of day
- a set of cells implementing a digital neural network.

We will look at the neural network example, a program called *nnet*. We choose this somewhat complex example because it contains most of the elements of an interesting computation. It would be easy to program this problem using "doall" loops in a conventional parallel language. But such a solution would not give the continuous visual feedback possible in *gsm* (it's not hard to imagine a library that would represent cell weights as colors).

Our simple neural network will be a group of *n* *neuron* cells $N_1..N_n$ whose values are fed through an $n \times n$ weighting matrix where each matrix element w_{ij} represents the weighting factor cell N_j uses on the value of cell N_i when computing its own value. In concept then, the network is a fully connected graph of N nodes, with weights w_{ij}, $i \neq j$ on each arc of the graph. The values of the neurons N_i are computed by the recurrence relation

$$N_i(n+1) = f\left(\sum_k w_{ik} N(n) - \Theta_i\right)$$

where f is a sigmoid thresholding function and Q_i is N_i's threshold. Pictorially, a 3-neuron network looks like this:

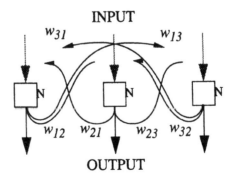

Being a typical GSM language, *gsm* requires the dependency graph to be a dag. We therefore model this graph as a vertical array of *n* neuron cells N_i, each with a row of *n* weighting cells extending from it, thus forming the weighting matrix. A horizontal array of n summing cells calculate the total of each column of the weighting matrix. To avoid introducing a cycle in the dependency graph, we will cause each summing cell S_i to send a message to its feedback cell N_i whenever it changes value.

N_1		w_{12}	w_{13}
N_2	w_{21}		w_{23}
N_3	w_{31}	w_{32}	
	S_1	S_2	S_3

The neuron cells N_i can decide (using the sigmoid function f and their thresholds Q_i) whether or not to act on messages from the weighted sums S_i and update their values or not[2]; updated values cause RECALC messages to propagate to the weighting cells. When all neurons finish acting upon messages by changing their values, the network is in equilibrium and computation ceases until a new input is presented.

Program *nnet*'s top-level cell, upon activation by the EXEC message, does the following:

- Build a set of display cells for getting input to the net from the user.
- Build a set of display cells for displaying the state of the network.
- Link the neuron cells to the user input cells.
- Link the output display cells to the neuron and weighting cells.
- Set up a cell as an "**OK**" button on the display. When the user presses this button it sends an INPUT message to each of the neuron cells, which then reads the corresponding input cell, causing computation to begin cascading through the weight cells.

3.3 Running the Example

Suppose *nnet* has been loaded into processor P_i. At the time the user presses the OK button, the following process occurs for each neuron cell C_i, $i = 1 .. n$:

- C_i receives a message from its corresponding display cell D_i containing a new input pattern value. It sets its own value to this input value.
- C_i sends a RECALC message to each of its dependents $w_{i1} .. w_{in}$.
- C_i sends a RECALC message to the display cell which is linked to it. The display cell updates its displayed value.
- Each cell w_{ij}, $j = 1 .. n$ calculates an output value by multiplying the value of C_i by its own internal weights, possibly adjusting that weight in the process. If the new output value is different from the old one, w_i in turn sends a RECALC message to the summing cell S_j, and a RECALC message to the display cell linked to it, which updates its displayed value.

---

2. Such decisions are necessary if the network is Lyapunov-stable, but not absolutely stable.

3.4 The Processor's Job

To understand what all this looks like at the processor level, we need to examine the way cells are represented and stored in our hypothetical *gsm* language. To the processor, a cell is a list of property name:value pairs. Let there be seven properties of a cell which are visible at the cell level:

- globally unique id
- name
- value
- message-handlers (methods which may incorporate automatic and static local variables)
- dependents
- anti-dependents
- reflective links

Let there be four additional properties used by the processor:

- computational "mass" of the message-handlers
- pending messages
- total mass of the pending messages
- recalculation priority level of the cell

The first of these additional properties is a list of heuristically derived (history-based) processing times for the message-handlers defined in the cell. The second is a queue of messages waiting to be processed. The third is a running total of the weights of the queued messages. The fourth determines the order in which pending messages to cells are serviced.

There are a number of ways to store the set of cells inhabiting a given processor. Our little *gsm* language settles on indexing them two ways:

- by id
- by recalculation priority level

The first is necessary for quick lookup at the time a message is received, so the message can be added to the target cell's pending queue; *gsm* also then stores a pointer to the cell in a priority queue which uses the cell's recalculation priority as the sort ordering.

Recalculation priority works as follows: do a topological sort of the global dependency graph. Let a cell having a position of a leaf node in this sorted graph be assigned priority level 1; let its dependents be assigned level 2, and so on (this analysis can be done "statically" at cell creation time). At least in concept, we arrive at a dependency tree which looks like this:

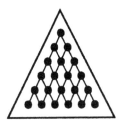

Processor P_i decides that it is overloaded when the combined weight of the event queues of all the cells currently in it exceeds some tunable bound. Suppose that this happens soon after an input pattern is presented to our network program. P_i then attempts to find another processor Pj with available capacity (we do not discuss the mechanism for this in this paper). If successful, it needs to select a set of cells which can be migrated to Pj.

We need this selection process to do at least two things:

- Minimize the number of dependencies which cross processor boundaries, thereby keeping down message traffic.
- Maximize the height of the priority tree in each processor; since priorities have the effect of serializing computations, maximizing the height of the priority tree has the effect of minimizing the number of potentially parallel computations which must be performed by an individual processor.

An easy heuristic partitioning scheme simply divides the dependency tree like this:

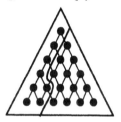

Intuitively, we want a minimum cut of maximum height through the dependency tree. More formally, for some selected height h, we want to partition the priority tree such that at least one node of every priority less than or equal to h is included in each partition, and the number of edges crossing from one partition to the other is minimized. Our little *gsm* language uses a greedy strategy (with lookahead for resolving ties) as a heuristic. This strategy is probably not optimal; still, we choose it as an inexpensive way of doing the partition. The reason it is being done, after all, is that the processor is overloaded with work.

When a processor sends a group of cells to another processor, it sends all the properties of each cell, including any pending messages. Once this is done, references to them in other cells are no longer valid. Accordingly, the processor must send all the dependents and anti-dependents of these cells messages informing them of the new location of the cells. These messages are of higher priority than RECALC messages, and so are serviced immediately at the host processors of the affected cells.

Still, some messages may be sent to old incorrect addresses in the time between the start of the move and the arrival of the new address messages. Our *gsm* language provides for this by using a protocol described in [YC 93].

Briefly, the sending processor P_s and receiving processor P_r synchronize, and after this point the sender retains all messages for the sent cells in a special holding area. After the cells are sent, the receiver notifies dependents and anti-dependents of the move. Each processor receiving these notification messages is able to determine when it has finished processing a given batch of them (because no more such messages remain to be serviced), and notifies P_r of this. When all the processors which have been involved in the move have acknowledged the move, P_r notifies P_s which then forwards all the messages in its holding area.

This solution does not guarantee serialization of messages, nor does the underlying asynchronous message-passing system. Serialized messages are gotten by means of blocking synchronous message calls.

3.5 Cell Addressing

Cell addressing in *gsm* works as follows. Cells have a 64-bit unique ID, of which 16 bits denote the originating processor, 32 bits are an ID number assigned by that processor, and 16 bits denote the current host processor. (Much of the overhead of this ID, and other often-duplicated or serial cell properties as well, can be eliminated in practice by clever use of *cell sequences* at the processor level, but we do not discuss that here). Processors keep cells in indexed arrays in memory. When a message arrives for a cell, the processor must look it up, attach the message to the cell, and store a pointer to the cell in a *to-do* priority queue which tells it which cell to service next. The main processor loop looks something like this:

```
loop
   if (msg pending)
      look up target cell
      store msg in cell
      enqueue ptr to msg
   else
      dequeue msg ptr
      service msg
```

Assume now that our sample program *nnet* is running with a neural net of 1000 neurons. This results in a million *gsm* cells; let's suppose they are spread out over 100 processors. What sort of problems might occur?

Observe first that each summand cell S_i depends on 999 weight cells $w_{1,i} .. w_{999,i}$. Each of these will send a RECALC message to S_i whenever it changes, resulting in a possible 999-fold increase in the number of calculations performed by S_i in the course of one iteration of the recurrence formula.

This problem is already mostly solved by the automatic assignment of recalculation priorities that happens at cell "compilation" time. This has the effect of giving the summing cells S_i a lower priority than any of the corresponding weighting cells w_{ji} upon which they depend. So long as there are weighting cells awaiting recalculation, processors will not evaluate the summing cells. Unless communication latencies are extreme, computation will proceed in a reasonably synchronized fashion.

4. Architectural Implications

We designed the GSM model without an overt architectural preference. In this section we compare message-passing and shared memory as implementation platforms. We next remark on the effect current communications protocols have on implementation efforts. Finally, we examine two possibilities for direct hardware enhancements to a GSM system.

4.1 Caching and Message Passing

To deliver messages to cells, the runtime system must implement a cell lookup mechanism. In our toy *gsm* language, intercell messages are keyed by the absolute ID of the target cell, whether that cell is local or remote. This means target cells must be looked up by some indexing scheme.

Consider *nnet*; its recurrence equation can be divided into three parts, and the number of messages for each iteration per neuron totals 1999:

(a) 999 recalculation messages from N_i to the weight cells w_{ij}, $j = 1..1000$, $j \neq i$

(b) 999 recalculation messages from weight cells w_{ji}, $j = 1..1000$, $j \neq i$ to summing cell S_i.

(c) One simple message from S_i to N_i.

About 50% of the messages (those of type b) are for the summing cells. Thus, a small cell lookup cache should hit on nearly all messages for these cells. This means that approximately half the messages will be delivered at cache hit speeds; the rest will cost an indexed cell lookup.

4.2 Shared Memory

A shared-memory system would obviate the need for lookup entirely; our little *gsm* language would be built differently, because cells would be defined as master pointers into the global shared memory. In fact, even the anti-dependencies would be unnecessary, since cell master pointers never need to change. To compare this approach to message-passing, we need to know the relative overhead of the shared memory system compared to raw message-passing.

Problem Mapping

We will assume that all processors cache every reference they make to the shared memory area. Then every write to the shared memory area will involve as many messages as there are readers of the written datum. We will further assume that the cell population in *nnet* is distributed optimally and uniformly across 100 processors, i.e. one neuron N_i, one summing cell S_i and 999 weighting cells w_{ij}, $j = 1..1000$, $j \neq i$ inhabit each processor P_i.

Message Patterns

Each of the horizontal rows of the weighting matrix lives in a single processor along with the neuron of the same index. This means all the writers and readers of each neuron occupy the same host, so memory accesses for the type (a) messages are local and occur in constant time. The messages of type (b) are all remote; as we have modeled shared memory there will be 990 messages notifying remote processors of a write to a cached memory location.

Suppose the shared memory is implemented using a virtual file system paging mechanism [BKT 93]. At *best*, each of the 100 processors will have all 10 of the corresponding w_{ji} cells on a single page. Each neuron will then cause 100 page faults per iteration of the recurrence relation. We do not know of a way to measure the cost of this analytically. It will depend upon the size of individual cells, on the mapping of cells to shared memory pages, on the relative arrival times of messages from various neuron cells and on the relative cost of an inter-processor page fault over the raw messaging time. Intuitively, we fear the hit will be large.

Trade-offs and Question Marks

It appears that the number of costly, non-cached messages is approximately the same for message-passing and shared-memory. The increased system overhead involved in the shared-memory model buys a decreased memory requirement and management responsibility on the part of the language implementor.

This must be offset against the fact that either the user or an intelligent agent in the runtime system must properly map the problem to shared memory. In our minds, this problem is nearly as thorny as discovering parallelism in sequential code—a problem that GSM was specifically designed to avoid.

Although empirical experiments would be of interest, it seems that distributed shared memory is unlikely to be of much assistance in the development of GSM languages.

4.3 Communications Protocols

Message-passing implementations, particularly those intended for networks or clusters of processors, will clearly be heavily dependent upon the underlying messaging protocols. Many sites today (ours among them) are committed to the UDP/IP datagram services and TCP/IP connection-oriented protocols.

In our *gsm* language, most messages are quite small–much smaller than the UDP packet size limit of 8K bytes (in fact they are smaller than the IP packet size of 576 bytes). Furthermore, since messages from each processor to every other processor are possible (and often inevitable, as in our example), the message traffic reduces in the abstract to a fully connected graph. This poses a problem for implementations which use TCP; although there is no theoretical reason why a processor should not be able to support 1000 TCP connections, existing implementations that we are familiar with begin running out of kernel resources long before that number is reached [Aue 93].

The obvious solution is to use datagram protocols for ordinary message passing, augmented with temporary TCP connections for large messages and transfers of cells between processors. But the widely available UDP and IP protocols do not guarantee in-order delivery of packets; they do not even guarantee delivery at all. This means that users of the protocols which desire these extra features must implement them in software. We are not able to tackle the thorny problems inherent in a specification for an RDP protocol (*Reliable* Datagram Protocol) here. We only note that many of the problems stem from a natural desire to fully generalize the solution. For our purposes a "parameterized protocol" would suffice nicely–we are willing to do without pack-etizing and reassembly of long messages if we can get hardware-based in-order guar-anteed delivery of packets with notification of failure to deliver or receive after some number of tries.

4.4 Hardware Assistance

There are two other options for hardware assistance in the GSM metaphor. These are *cell lookup* and *calculation queues*. We have already mentioned caching as a necessary strategy for cell lookup. Here we propose moving it into the hardware. Calculation queues are simply priority queues maintained by the runtime system to determine which cell should be given the processing resource next.

Cell Lookup

Here we argue that hardware support for cell look-up is possible, easy and valuable. First, we consider how messages are handled. Our little *gsm* language handles incoming messages like this:

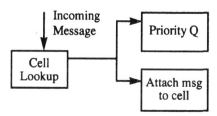

Outgoing messages are dispatched to the host processor named in the ID of the cell to which the message is intended.

Implementing the cell lookup cache in hardware is an obvious enhancement, but such a cache cannot be invisible to the runtime system. A hardware cache must be able to communicate with the runtime system in order to convert 64-bit cell ID's into local memory addresses and invalidate entries based on those same ID's. Arrays of simple cells can often be stored economically as *sequences* of fixed-size cells. Thus, a more sophisticated cache can do a better job with less memory if it is able to deal with these.

A hardware cache protocol therefore needs to include the following atomic interactions with the runtime system software:

- VALIDATE (*bid, ncells, csize*)
- INVALIDATE (*bid, ncells, csize*)

    bid = beginning cell ID
    ncells = number of cells in sequence
    csize = size in bytes of each cell

- IDADDR (*id, addr, seqaddr*)

    id = cell ID to look up (in)
    addr = local address, or `nil` (out)
    seqaddr = address of enclosing sequence, or `nil` (out)

All three of these are commands from the runtime system to the cache. VALIDATE causes the cache to store information about a sequence, flushing older information if necessary. INVALIDATE informs the cache that a set of cells is no longer resident at the previously validated location. IDADDR queries the cache for the local address of a given cell ID. If the cache hits, it returns the address in *addr* and the address of the enclosing sequence (if any) in *seqaddr*. If it does not hit it returns `nil` in both *addr* and *seqaddr*.

Calculation Queues

Once a cell's address is looked up in *gsm*, two things need to happen to complete the message reception process:

- Attach the message to the target cell's message queue.
- Add a pointer to the cell to a priority queue (if one is not already there) so the processor can know when to deliver the processing resource to the cell.

The first of these is best done by the processor (which also takes care of cell memory allocation). The second can be done in hardware.

A protocol for this follows the standard queuing forms, but may also include a call for flushing a given cell from the hardware queue, for use when cells move.

Performance Gains

In estimating what we stand to gain from these, we have only to remember that lookup and queuing are both $O(n \lg n)$ operations. The cost of lookup grows with the number of cells in the system, while the costs of queueing and dequeuing grow with the number of pending messages.

Adding the message to the pending message list of a cell is relatively inexpensive once the address of the cell is known. Interprocessor latency accounts for the remainder of the communication cost. Therefore, the total cost of communications is bounded precisely by the items we propose to put in hardware. We may accordingly expect the usual order-of-magnitude performance increases from the hardware assists we have outlined.

4.5 Other New Architectures

There are architectures today which completely separate the communications and computational aspects of the computing system. The MIT Jellybean Machine is a good example [Dal 92]. There may be some promise here, since the separation between messaging and computing engines is equally clean in GSM.

Of course, efficiency concerns cause problems. There is no place to conveniently add the hardware lookup or queueing mechanisms we have just described to such a system. To make such an architecture work, therefore, we would need to add two pieces to each processing element:

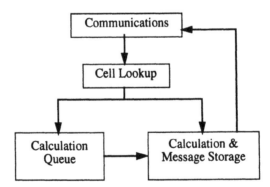

Obviously, hardware systems allowing the interposition of customized hardware between the communications and computation modules would be greeted with a measure of joy. Our desire for such flexibility must be balanced against the performance gains made possible by single-chip integration.

Conclusion

We have examined the GSM metaphor in the light of a representative problem, and found that two simple hardware extensions have the potential to dramatically improve the communications performance of such a system.

References & Additional Reading

[Agh 91]   Gul Agha. The Structure and Semantics of Actor Languages, Dept. of Computer Science, UI Urbana.

[Aue 93]   Joshua Auerbach. IBM T.J. Watson Research Center, personal communication.

[AU 91]   Arvind, L. Bic and T. Ungerer. Evolution of Data-Flow Computers. In *Advanced Topics in Data-Flow Computing*. Prentice-Hall, 1991.

[BKT 93]   Arindam Banerji, Dinesh Kulkarni, John Tracey, Paul Greenawalt and David Cohn. High-Performance Distributed Shared Memory Substrate for Workstation Clusters. HPDC-93, Spokane, Washington, IEEE, 1993.

[CCA 88a]   Jo Ann.C. Carland, James W. Carland, and Carroll D. Aby, Jr.. Spreadsheets: Placebos or Panacea? In *Journal of Research on Computing in Education* 21(1): 112-19. Fall 1988.

[Car 88b]   Sven A. Carlsson. A Longitudinal Study of Spreadsheet Program Use. In *Journal of Management Information Systems* 5(1): 82-100. Summer 1988.

[Cas 92]   Rommert J. Casimir. Real Programmers Don't Use Spreadsheets. In *ACM SIGPLAN Notices* 27(6): 10-16. June 1992.

[CG 87]   Robert L Chew and Rajoo Goel. Transaction Processing Using Lotus 1-2-3. In *Journal of Systems Management* 38(1): 30-37. Jan 1987.

[Cli 81]   W. D. Clinger. *Foundations of Actor Semantics*. AI-TR-633, MIT Artificial Intelligence Lab, May 1981.

[Dal 92]   William J. Dally et al, The Message-Driven Processor: A Multicomputer Processing Node with Efficient Mechanisms. In *IEEE Micro*, pp 23-39, April 1992.

[Den 91]   The Evolution of "Static" Data-Flow Architecture, Jack Dennis; in *Advanced Topics in Data-Flow Computing*, J.L. Gaudiot, L. Bic, eds. (35-91). Prentice-Hall, 1991.

[Fra 86]   N. Francez. *Fairness*. Springer-Verlag, NY, 1986.

[Hew 77]   C. Hewitt. Viewing Control Structures as Patterns of Passing Messages. *Journal of Artificial Intelligence*, 8(3):323-64, June 1977.

[HM 90]   Charles E. Hughes and J. Michael Moshell. Action Graphics; A Spreadsheet-based Language for Animated Simulation. In *Visual Languages and Applications*, Ichikawa, Jungert and Korfhage, eds. Plenum Press, 1990.

[Jag 88]   Suresh Jagannathan. *A Programming Language Supporting First-Class, Parallel Environments*. Technical Report LCS-TR 434, MIT, Dec. 1988.

[Jag 90]   Suresh Jagannathan. Coercion as a Metaphor for Computation. In *Proceedings of the 1990 International Conference on Computer Languages*. IEEE Computer Society Press, 1990.

[JA 92]   Suresh Jagannathan and Gul Agha, *A Reflective Model of Inheritance*, Dept of Computer Science Technical Report, UI Urbana 1990 ("to appear in ECOOP '92 proceedings, Springer-Verlag LNCS 615").

[JNZ 93]   Jeff Johnson, Bonnie Nardi, Craig Zarmer, James Miller. ACE: Building Interactive Graphical Applications. In *CACM* 36(4): 41-55. April 1993.

[KM 88]   Ken Kahn and Mark Miller. Language Design and Open Systems. In *The Ecology of Computation*, North Holland, 1988.

[Kam 90]   Samuel Kamin. *Programming Languages; An Interpreter-Based Approach*. Addison Wesley, 1990.

[Lie 86]   Henry Liebermann. Using Prototypical Objects to Implement Shared Behavior in Object-Oriented Systems. In *OOPSLA '86 Conference Proceedings*, pp 214-223, 1986. Published as SIGPLAN Notice 21(11), November 1986.

[Lit 90]   C. Litecky. Spreadsheet Macro Programming: a Critique with Emphasis on Lotus 1-2-3. In *Journal of Systems and Software* 13(3): 197-200. Nov 1990.

[Mae 87]   Pattie Maes. Concepts and Experiments in Computational Reflection. *Proceedings of OOPSLA '87*, pp. 147-155. Oct 4-8, 1987.

[Mas 89]   D. Mason. An Empirical Analysis of Spreadsheet Usage: A Solution Storing Up Problems. In *Journal of Information Technology* 6(4): 159-63. Sep 1989.

[Mye 91]   Brad A. Myers. Graphical Techniques in a Spreadsheet for Specifying User Interfaces. In *CHI '91 Proceedings*. (New Orleans, LA April 27-May 2, 1991) ACM/Addison Wesley, 1991.

[NLB 89]   H. Albert Napier, David M. Lane, Richard R. Batsell, and Norman S. Guadango. Impact of a Restricted Natural Language Interface on Ease of Learning and Productivity. In *CACM* 32(10): 1190-98, Oct 1989.

[Nie 93]   Jakob Nielsen. Noncommand User Interfaces. In *CACM* 36(4):83-99.

[RCM 93]   George Robertson, Stuart K. Card and Jock Mackinlay. Information Visualization using 3D Interactive Animation. In *CACM* 36(4): 57-71. April 1993.

[SP 90]   J. Sajaniemi and J. Pekkanen. An Empirical Analysis of Spreadsheet Calculation. In *Software: Practice and Experience* 20(11): 1097-1114. Nov 1990.

[WL 90]   Nicholas Wilde and Clayton Lewis. Spreadsheet-based Interactive Graphics: from Prototype to Tool. In *CHI '90 Proceedings*. ACM Press 1990.

[YC 93]   Alan G. Yoder and David L. Cohn. *Making Concurrent Programming Easy*. Univ. of Notre Dame, Tech Report 93-8: Postscript version by anon. ftp from invaders.dcrl.nd.edu, /pub/TechReports/1993/tr-93-8.ps

Technological Steps toward a Software Component Industry

Michael Franz

Institut für Computersysteme, ETH Zürich
CH-8092 Zürich, Switzerland

Abstract. A *machine-independent abstract program representation* is presented that is twice as compact as machine code for a CISC processor. It forms the basis of an implementation, in which the process of code generation is deferred until the time of loading. Separate compilation of program modules with type-safe interfaces, and dynamic loading (with code generation) on a per-module basis are both supported.

To users of the implemented system, working with modules in the abstract representation is as convenient as working with native object-files, although it leads to several new capabilities. The combination of portability with practicality denotes a step toward a *software component industry*.

1. Introduction

The rapid evolution of hardware technology is constantly influencing software development, for better as well as for worse. On the downside, faster hardware can conceal the complexity and cost of badly-designed programs; Reiser [Rei89] is not far off the mark in observing that sometimes "software gets slower more quickly than hardware gets faster". On the other hand, improvements in hardware, such as larger memories and faster processors, have also provided the means for better software tools.

Often enough, methodical breakthroughs have been indirect consequences of better hardware. For instance, software-engineering techniques such as *information hiding* and *abstract data types* could only be developed after computers became powerful enough to support compilers for modular programming languages. Mechanisms such as *separate compilation* place certain demands on the underlying hardware and so had a chance of proliferation only after sufficiently capable computers were commonplace.

This paper presents another example of a systematic technological advance that owes its viability to improved hardware. The first part of the paper describes a technique for representing programs abstractly in a format that is twice as compact as object code for a CISC processor. Combined with the speed of current processors and abundance of main storage, the use of this intermediate representation makes it possible to accelerate the process of code generation to such a degree that it can be performed on-the-fly during program loading on ordinary desktop computers, even with a high resulting code quality.

A system has been implemented that permits the convenient use of program modules in this machine-independent intermediate representation, as if they had been compiled to native code. The second part of the paper discusses some of the possibilities that arise from this scenario, reviews related work, and concludes in an outlook to future developments.

2. Abstract Program Representation

2.1. Intermediate Languages

Ever since the late 1950's, there have been attempts to design a *universal computer-oriented language* (UNCOL) that would be powerful enough so that programs originating in any problem-oriented language could be translated into it, and code for any processor architecture could be generated from it [SWT58]. Having such a language would enable us to construct compilers for *m* languages to be run on *n* machines by writing only *m+n* programs (*m* front-ends and *n* code generators), instead of *m✶n*.

To this date, no proposed UNCOL has been met with general agreement. However, many compilers have employed some form of *intermediate language* (IL) on the path from source code to final executable code image. Such an IL is distinguished from the transient data structures found in memory between different passes of a compiler by providing a stand-alone program representation that can be stored on a data file.

A straightforward method for obtaining an IL representation of a program is to express it as an instruction sequence for some *fictitious computer*, also called an *abstract machine* [PW69]. Most ILs, including the very first UNCOL ever proposed [Con58], follow this pattern, and many implementations, such as *BCPL* [Ric71], *SNOBOL4* [Gri72], and *Pascal-P* [NAJ76], have been based on abstract machines. Besides abstract machines, ILs have also been based on linearized parse-trees [GF84, DRA93a]. Furthermore, there are compilers that compile via ordinary high-level programming languages by way of source-to-source program translation [ADH89].

This paper introduces an IL approach that results in a highly compact intermediate representation. I call it *semantic-dictionary encoding* (SDE). SDE preserves the full semantic context of source programs while *adding further information* that can be used for accelerating the speed of code generation. SDE forms the basis of an implementation, in which a code generator has been combined with the module loader, forming a *code-generating loader*.

2.2. An Overview of Semantic-Dictionary Encoding

SDE is a dense representation. It encodes a syntactically correct source program by a succession of indices into a *semantic dictionary* (SD), which in turn contains the information necessary for generating native code. The dictionary itself is not part of the SDE representation, but is constructed dynamically during the translation of a source program to SDE form, and reconstructed before (or during) code generation. This method bears some resemblance to commonly used data compression schemes [Wel84].

With the exception of wholly constant expressions, SDE preserves all of the information that is available at the level of the source language. Hence, unlike abstract-machine representations, transformation to SDE preserves the *block structure* of programs, as well as the *type* of every expression. Moreover, when used as the input for code-generation, SDE in certain cases provides for *short-cuts* that can increase translation efficiency.

A program in SDE form consists of a *symbol table* (in a compact format, as explained in [Fra93b]) and a series of *dictionary indices*. The symbol table describes the names and the internal structure of various entities that are referenced within the program, such as *variables*, *procedures*, and *data types*. It is used in an initialization phase, in the course of which several *initial entries* are placed into the SD. The encoding of a program's actions consists of references to these initial dictionary entries, as well as to other entries added later in the encoding process.

Dictionary entries represent nodes in a directed acyclic graph that describes the semantic actions of a program abstractly. In its most elementary form, an SD is simply such an *abstract syntax-tree* in tabular shape, in which the references between nodes have been replaced by table indices. Each dictionary entry consists of a *class* attribute denoting the semantic action that the entry stands for (e.g. *assignment, addition,* etc.), and possibly some references to objects in the symbol table and to other dictionary entries.

What differentiates a tabular abstract syntax-tree from an SD is that the latter can describe also *generic characteristics* of *potential nodes* that might appear in such a tree. In addition to *complete entries* that directly correspond to nodes of an abstract syntax tree, semantic dictionaries contain also *generic,* or *template entries.* These templates have the same structure as complete entries, but at least one of their attributes is missing, as recorded in a *status* flag. In SDE, complex program statements can be represented not only by complete entries, but also by templates in combination with other entries. For example, the statement a + b can be represented by the index of an "addition" template followed by the indices of two variable-reference entries.

Now suppose that a template existed in the SD for every construct of the source language (*assignment, addition,* etc.), and that furthermore the SD were initialized in such a way that it contained at least one entry (possibly a template) for every potential use of every object in the symbol table. For example, an object describing a *procedure* in the programming language *Oberon* [Wir88] requires a minimum of four entries in the dictionary, relating in turn to a *call* of the procedure, *entry* of the procedure, *return* from the procedure, and *addressing* the procedure, which is used in the assignment of the procedure to procedure variables. There are no other operations involving procedures in Oberon.

These preconditions can be fulfilled by initializing the SD in a suitable way. They enable us to represent any program by only a symbol table and a succession of dictionary indices. As an example, consider the following module M in the programming language Oberon:

```
MODULE M;

    VAR i, j, k: INTEGER;

    PROCEDURE P(x: INTEGER): INTEGER;
    BEGIN ...
    END P;

BEGIN
    i:=P(i); i:=i+j; j:=i+k; k:=i+j; i:=i+j
END M.
```

In order to encode this program by the method of SDE, we first need to initialize the SD. Initialization depends on the source language (Oberon in this case) and on the objects in the symbol table. The symbol table for the example program contains three integer variables (*i, j,* and *k*) and a function procedure (*P*). After initialization, the corresponding SD might look like the following (the individual entries' indices are represented by symbolic names so that they can be referenced further along in this paper, and missing attributes of templates are denoted by a dot).

Index	Class	Meaning	Status
asgn	assignment	. := .	both arguments missing
plus	addition	. + .	both arguments missing
...	...	...	...
vi	variable	i	complete
vj	variable	j	complete
vk	variable	k	complete
refp	address	P	complete
callp	function call	P(.)	argument missing
entp	entry	P-BEGIN	complete
retp	return	P-RETURN	argument missing
...	...	...	...

The instruction sequence that constitutes the body of module *M* may then be represented by the following sequence of 24 dictionary indices:

```
asgn   vi   callp vi          i := P(i)
asgn   vi   plus vi vj         i := i + j
asgn   vj   plus vi vk         j := i + k
asgn   vk   plus vi vj         k := i + j
asgn   vi   plus vi vj         i := i + j
```

This is where the second major idea of SDE comes in. What if we were to keep on *adding* entries to the dictionary during the encoding process, based on the expressions being encoded, in the hope that a *similar expression* would occur again later in the encoding process? For example, once we have encoded the assignment i := P(i) we might add the following three entries to the dictionary:

Class	Meaning	Status
function call	P(i)	complete
assignment	i := .	right-side argument missing
assignment	i := P(i)	complete

Thereafter, if the same assignment i := P(i) occurs again in the source text, we can represent it by a single dictionary index. If another assignment of a different expression to the variable *i* is come across, this may be represented using the template "assign to *i*", resulting in a shorter encoding. In this manner, the body of module M can be encoded by only 16 dictionary indices, instead of the previous 24. This shorter encoding is shown below, along with the entries added to the SD in the process (assuming that, after initialization, the dictionary comprised *n 1* entries).

```
asgn   vi   callp vi          i := P(i)
n+1         plus vi vj         i := i + j
asgn   vj   n+3 vk             j := i + k
asgn   vk   n+5               k := i + j
n+6                           i := i + j
```

Index	Class	Meaning	Status
...	...	...	...
n	function call	P(i)	complete
n+1	assignment	i := .	right-side argument missing
n+2	assignment	i := P(i)	complete
n+3	addition	i + .	right-side argument missing
n+4	addition	. + j	left-side argument missing
n+5	addition	i + j	complete
n+6	assignment	i := i + j	complete
n+7	addition	. + k	left-side argument missing
n+8	addition	i + k	complete
n+9	assignment	j := .	right-side argument missing
n+10	assignment	j := i + k	complete
n+11	assignment	k := .	right-side argument missing
n+12	assignment	k := i + j	complete
...	...	...	...

2.3. Increasing the Speed of Code Generation

The decoding of a program in SDE form is similar to the encoding operation. At first, the dictionary is initialized to a state identical to that at the onset of encoding. Since the symbol table is part of the SDE file-representation, the decoder has all required information available to perform this task. Thereafter, the decoder repeatedly reads dictionary indices from the file, looking up each corresponding entry in the SD. Whenever a complete entry is found in this manner, its meaning is encoded directly in the dictionary and the decoder can proceed to process the next index on the input stream. If a template is retrieved instead, it is copied, further entries corresponding to its undefined attributes are input in turn, and the modified copy is added to the dictionary as a new complete entry. Moreover, additional templates are sometimes added to the SD according to some fixed heuristics, in the hope that a corresponding branch of the program's syntax tree will show up further along.

The interesting fact now is that SDE actually provides *more explicit information* than the source text, namely about *multiple textual occurrences of identical subexpressions* (including, incidentally, *common designators*). This can sometimes be exploited during code generation, resulting in an increase in the speed of code output. Consider again the (second, more compact) SDE representation of module M above. Now suppose that we want to generate object code for a simple stack machine directly from this SDE form. Let us assume that we have processed the first two statements of M's body already, yielding the following instruction sequence starting at address a:

```
a     LOAD    i      load i onto stack
a+1   BRANCH  P      call procedure P with argument i
a+2   STORE   i      assign result to i
a+3   LOAD    i      load i
a+4   LOAD    j      load j
a+5   ADD            add i to j
a+6   STORE   i      assign their sum to i
```

Let us further assume that we have kept a note in the decoder's SD, describing which of the generated instructions correspond to what dictionary entry. For example, we might simply have recorded the program counter value twice for every entry in the SD, both before and after generating code for the entry:

Index	Class	Meaning	Begin	End	Invar
...	...	...	...	...	...
n	function call	P(i)	a	a+1	No
n+1	assignment	i := P(i)	a	a+2	No
...	...	...	...	...	...
n+5	addition	i + j	a+3	a+5	Yes
n+6	assignment	i := i + j	a+3	a+6	Yes
...	...	...	...	...	...

A setup such as this allows us to bypass the usual code generation process under certain circumstances, replacing it with a simple *copy operation* of instructions already generated. For example, when we encounter the second reference to entry $n+6$ in the semantic-dictionary encoding of module M, we know that we have already compiled the corresponding statement i := i + j earlier on. In this case, we may simply re-issue the sequence of instructions generated at that earlier time, which can be found in the object code between the addresses $a+3$ and $a+6$, as recorded in the dictionary.

Of course, this method cannot be applied in every case and on all kinds of machine. First of all, code-copying is possible only for *position invariant* instruction sequences. For example, a subexpression that includes a call of a local function by way of a *relative branch* is not position invariant. It happens that information about position invariance can also be recorded conveniently in the dictionary, as shown in the example above.

Secondly, the instruction sequences obtained by code-copying may not be optimal for more complex processors with many registers and multistage instruction-pipelines. For these machines, it may be necessary to employ specific optimization techniques. However, since SDE preserves all information that is available in the source text, arbitrary optimization levels are possible at the time of code generation, albeit without the shortcuts provided by code-copying. One then simply treats the semantic dictionary as an abstract syntax-tree in tabular form.

The interesting part is that code-copying is beneficial especially on machines that are otherwise slow, i.e., CISC processors with few registers. Consider module M again, in which the subexpression i := i + j appears three times. On a simple machine that has only a single accumulator or an expression stack, the identical instruction sequence will have to be used in each occurrence of the subexpression. Code-copying can accelerate the code-generation process in this case. The optimal solution for a modern RISC processor, on the other hand, might well consist of three distinct instruction sequences for the three instances of the subexpression, due to the use of register variables and the effects of instruction pipelining. However, such processors will generally be much faster, counterbalancing the increased code-generation effort, so that an acceptable speed of loading can still be delivered to interactive users relying on dynamic code-generation.

3. Implementation

3.1. Basic Architecture

A *code-generating loader* (CGLoader) has been implemented for an experimental version of MacOberon [Fra90, Fra93a]. MacOberon is a full implementation of the Oberon System [WG89] for Apple Macintosh II [App85] computers and supports *separate compilation* with static interface checking of programs written in the programming language Oberon [Wir88]. It also offers *dynamic loading* of individual modules, meaning that separately compiled modules can be linked into an executing computing session at any time, provided that their interfaces are consistent with the interfaces of the modules that have been loaded already. In MacOberon, dynamic loading is accomplished by a *linking loader*, which modifies the code image retrieved from an object file in such a way that all references to other modules are replaced by absolute addresses.

The new implementation adds a *second loader* to MacOberon, namely a CGLoader processing semantic-dictionary-encoded program files (SDE-files). It has been possible to integrate this CGLoader transparently into the existing MacOberon environment, in the sense that SDE-files and native object-files are completely interchangeable in the implemented system and can, therefore, import from each other arbitrarily. Depending on a *tag* in the file (first two bytes), either the native linking loader or the CGLoader is used to set up the module in memory and prepare it for execution.

The CGLoader has not been incorporated into the core of MacOberon, but constitutes a self-contained module package at the application level. The existing module loader was modified slightly, introducing a *procedure variable* into which an *alternate load procedure* can be installed. Once initialized, this procedure variable is up-called whenever an abnormal tag is detected in an object file, and so initiates the passage of control, from the linking loader that is part of the system core, to the CGLoader external to it. The installation of such an alternate load procedure is considered a privileged operation, analogous to the modification of an interrupt vector. Besides the obvious advantages during the design and testing phases, this architecture permits us to view machine-independence quite naturally as an *enhancement of an existing system*. It also hints at the possibility of adding successive external CGLoaders as time progresses and standards for machine-independent object-file formats emerge. Several such formats could in fact be supported simultaneously, among which one would distinguish by different file-tags.

3.2. Interchangeability of Object Files

SDE-files and their native-object-file counterparts are completely interchangeable in the implemented system. This flexibility doesn't come as readily as it may seem. Not only does it require that the CGLoader is able to interpret *native symbol-files* and the corresponding entry-tables in memory, but also that information can be communicated from the new load architecture to the old one, in formats that are *backward-compatible* with the native compiler and loader.

In the present case, the more difficult part of this problem had an almost trivial solution, since the two varieties of "object" files in the implemented system share a *common representation of symbol-table information*. As documented in an earlier paper [Fra93b], a *portable symbol-file format* had been introduced into MacOberon some time ago, leading to improvements in performance completely unrelated to portability. But now, there were further benefits from the previous investment into machine-independence: SDE-files use exactly the same symbol-table encoding as the symbol files of the native MacOberon compiler, and the encoded information starts at the same position in both types of file. Consequently, the native MacOberon compiler cannot distinguish SDE-files from its regular symbol-files. It is able to compile modules that import libraries stored in SDE form, as it can extract the required interface information directly from SDE-files.

Apart from the native compiler, the *native loader* need also be able to handle modules that import machine-independent libraries. This is achieved by enabling the CGLoader to construct on-the-fly not only object code, but also all of the remaining data structures that are usually part of a native object-file, such as entry tables. Once that a module has been loaded from an SDE-file, it loses all aspects of portability, and can henceforth be maintained (enumerated, unloaded, etc.) by the regular MacOberon module manager.

3.3. Execution Frequency Hierarchy Considerations

Programs are usually *compiled* far less frequently than they are *executed*. Therefore, it is worthwhile to invest some effort into compilation, because the benefit will be repeated. The same holds for program *loading* versus *execution*. While a program is normally loaded more often than it is compiled, the individual machine instructions are executed even more often, many times on average per load operation. Unfortunately, we need to shift workload unfavourably when we employ a loader that performs code generation. This can be justified only if important advantages are gained in return, or if the resulting code quality is increased. After all, it is execution speed that matters ultimately.

From the outset, three desired properties were therefore put forward that a system incorporating a CGLoader should possess in comparison to a system employing a traditional compiler and linking loader:

1. run-time performance should be at least as good
2. loading time should not be much worse
3. source-to-IL-translation time should be tolerable

These requirements could be met. Not only was it possible to construct a fast CGLoader for *Motorola 68020* processors [Mot87], but the native code produced by it is of high quality. Indeed, although its method of code-generation is quite straightforward, the CGLoader is sometimes able to yield object code that would require a much more sophisticated code-generation strategy in a regular compiler. This is mainly due to the fact that the absolute addresses of all imported objects are known at loading time, which enables the CGLoader to optimize access to them; for example, by using short displacements instead of long ones.

The time required by the CGLoader for creating native code on-the-fly turned out to be quite acceptable, too. Loading with code-generation takes only slightly longer than loading of a native object-file with linking; the difference is barely noticeable in practice. Moreover, the Oberon system has a modular structure, in which many functions are shared between different application packages. In traditional systems, these common functions would be replicated in many applications and statically linked to each of these applications. As a consequence of *modular design*, each new application adds just little code to an already running system. Hence, at most times during normal operation, the CGLoader need only process the moderate number of modules that are unique to an application, while other modules that are also required will already be in memory from past activations of other applications.

4. Results

Four different Oberon application-packages have been used in the benchmarks in this section. *Filler* is a program that draws the Hilbert and Sierpinski varieties of space-filling curves onto the screen. *Hex* is a byte-level file editor of medium sophistication. *Draw* is an extensible object-oriented graphics editor [WG92], of which the three main modules and three extension modules are included in the survey. *Edit* is a relatively sophisticated extensible document processing system [Szy92], five modules of which are studied here.

4.1. Memory and File-Store Requirements

Table 1 gives an impression of the compactness of SDE-files, the influence of the presence of run-time integrity checks on code size, and the memory requirements of the CGLoader. Its first three columns compare the sizes (in bytes) of native MacOberon object-files with those of SDE-files. Two different values are given for the former, reflecting different levels of run-time integrity checking. SDE-files contain enough information to generate code with full integrity checking, but the user need only decide at loading time whether or not he requires code that includes these run-time checks, and which ones should be included.

The column labelled *"Native −"* in the table below displays the size of native object-files that include reference information for the MacOberon post-mortem debugger, but no extra code for nil-checking, index-checking, type-checking, nor for the initialization of local pointers to NIL. Conversely, the column labelled *"Native +"* gives the sizes of native object-files that include not only reference information, but also code for the aforementioned run-time checks. Neither kind of object file includes symbolic information for interactive debugging, which can be added by enabling a further option in the MacOberon compiler.

The fourth column of Table 1 shows the maximum size to which the semantic dictionary grows during compilation and code-generation. The CGLoader requires about 80 times this number of bytes as temporary storage while loading a module.

Module	Native −	Native +	SDE	Dictionary
Filler	2900	3236	1232	1003
Hex	10810	12678	4480	1308
Graphics	12692	15935	5281	1382
GraphicFrames	10045	12186	4273	1549
Draw	6033	7133	2193	1498
Rectangles	3255	4297	1378	1271
Curves	5810	7340	2168	1287
Splines	4611	5659	1955	1566
TextFrames	30687	37491	11667	1892
TextPrinter	11503	12537	4494	1486
ParcElems	15259	17540	5273	1437
Edit	11670	13431	5157	1659
EditTools	18756	20898	7397	1488
	144031	170361	56948	

Table 1: Object-File Size (Bytes) and Dictionary Size (Entries)

This data shows that on average, SDE-files are about 2.5 times more compact than native MacOberon object-files not containing run-time integrity checks, and about 3 times more compact than native files incorporating these checks. It is also noteworthy that the maximum size, to which the semantic dictionary grows during encoding and decoding, is not proportional to the overall size of a module, but only roughly proportional to the length of the longest procedure in a module. This seems to level out at a relatively small value in typical modules, much smaller than anticipated originally.

Table 2 indicates how much information is output into memory by the CGLoader. The first column repeats the sizes of SDE-files from the previous table. The second and third columns give two different values for the size of the object code generated on-the-fly, depending on whether or not run-time integrity checks (in the same combinations as before) are emitted. The remaining columns list the sizes of dynamically-generated constant data (including type descriptors), reference data for the Oberon post-mortem debugger, and link data. The latter comprises entry tables generated on-the-fly, so that native client modules can later be connected to dynamically generated library modules.

Module	SDE-File	Code –	Code +	Const	Ref	Link
Filler	1232	2352	2682	139	244	8
Hex	4480	9224	11020	267	1094	12
Graphics	5281	9036	12140	799	1452	204
GraphicFrames	4273	8268	10354	433	824	76
Draw	2193	4754	5752	236	451	60
Rectangles	1378	2644	3596	104	292	20
Curves	2168	5042	6462	100	439	24
Splines	1955	3812	4806	137	437	20
TextFrames	11667	25878	32536	629	2843	156
TextPrinter	4494	9176	10180	478	1367	40
ParcElems	5273	12944	15180	493	1085	52
Edit	5157	9368	10940	555	1098	72
EditTools	7397	14844	16886	723	2009	116
	56948	117342	142534	5093	13635	860

Table 2: Sizes of SDE-Files and of Dynamically-Generated Data (Bytes)

It is notable that SDE-files encode programs more than twice as densely as object code for the MC68020 architecture [Mot87]. If code that includes run-time-checking is considered instead, the factor becomes even 2.5. This is in spite of the fact that SDE-files can additionally also serve as symbol files and contain reference information as well.

4.2. Performance

Unless otherwise noted, the following benchmarks were all carried out under *MacOberon Version 4.03* on a *Macintosh Quadra 840AV* computer (*MC68040* Processor running at 40MHz), using version *7.1.1* of the Macintosh System Software. All results are given in real time, i.e. the actual delay that a user experiences sitting in front of a computer, with a resolution of 1/60th of a second. Each benchmark was executed after a cold start, so that no files were left in the operating system's file cache between a compilation and subsequent loading, and the best of three measurements is given in each case. Every attempt has been made to present the system as a regular user would experience it in everyday use.

Table 3 presents the times (in milliseconds) required for compilation and for module loading. The first two columns compare the native compilation times of the MacOberon compiler with the times that the semantic-dictionary encoder requires for generating an SDE-file out of an Oberon source program. The remaining two columns report the loading times for both varieties of object file, including disk access, and, in the case of SDE-files, also including on-the-fly generation of native code. On average, 10% of the object code emitted by the CGLoader can be generated by code-copying. All timings apply to the situation in which no run-time integrity checking is used.

Module	Compile	Encode	Ld Native	Ld SDE
Filler	583	633	83	100
Hex	666	900	100	116
Graphics	766	1083	116	166
GraphicFrames	650	950	66	116
Draw	516	683	66	100
Rectangles	400	500	50	83
Curves	450	650	50	116
Splines	433	583	66	83
TextFrames	1416	2183	216	316
TextPrinter	666	950	100	133
ParcElems	750	1150	116	200
Edit	716	1166	100	166
EditTools	900	1483	150	233
	8912	12914	1279	1928

Table 3: Compilation versus SDE-Encoding and Module Loading Times (ms)

As can be seen from these timings, semantic-dictionary encoding on average takes about 1.5 times as long as normal compilation. I think that there is still room for improvement, as the speed of the encoder was of no major concern during its development. Conversely, I aimed at optimal speed of loading with code generation. This currently takes about 1.5 times as long as normal loading. However the times spent directly on module loading tell only half the story.

Table 4 presents a more realistic measure of loading time, as it takes into account not only the time required for loading an application, but also the duration of *loading and displaying a typical document*. The following timing values indicate how long (in milliseconds) a user must wait after activating a document-opening command before he can execute the next operation. Commands take longer the first time that they are issued, because the modules of the corresponding application have to be loaded as well. Subsequent activations then take up much less time.

Command	Native –	SDE –	Difference
first Draw.Open Counters.Graph	1333	1450	+ 9%
subsequent activations	700	700	
first Edit.Open OberonReport.Text	1400	1766	+ 26 %
subsequent activations	883	883	

Table 4: Command Execution Times with Checks Disabled (ms)

Table 5 demonstrates the effect of run-time integrity-checking on loading time. The measurements of Table 4 have been repeated, but with native object-files that incorporate run-time integrity checks and are therefore larger, and with on-the-fly emission of the same checks in the CGLoader.

Command	Native +	SDE +	Difference
first Draw.Open Counters.Graph	1366	1466	+ 7%
subsequent activations	700	700	
first Edit.Open OberonReport.Text	1550	1800	+ 16%
subsequent activations	916	916	

Table 5: Command Execution Times with Checks Enabled (ms)

The timings in Tables 4 and 5 show that, in practice, on-the-fly code-generation is already almost competitive to dynamic loading of pre-compiled native code from regular object-files. This applies to current state-of-the-art CISC hardware, and is in part due to the fact that disk reading is relatively slow. The compactness of the SDE representation speeds up the disk-access component of program loading considerably, and the time gained thereby counterbalances most of the additional processing necessary for on-the-fly code generation. This argument is supported by a comparison of Tables 4 and 5. It seems to be more efficient to generate run-time checks on the fly than to inflate the size of object files by including them there.

A noteworthy trend in hardware technology today is that processor power is rising more rapidly than disk access times and transfer rates. This trend is likely to continue in the future, which means that hardware technology is evolving in favor of the ideas proposed here. Consider Tables 6 and 7, which repeat the timings of Table 5 for different processors of the MC680x0 family, using the identical external disk drive.

Machine	Native +	SDE +	Difference
Macintosh II (16MHz MC68020)	5683	7966	+ 40%
Macintosh IIx (16MHz MC68030)	4300	5933	+ 38%
Macintosh IIfx (40MHz MC68030)	1833	2283	+ 25%
Macintosh Quadra 840AV (40MHz MC68040)	1366	1466	+ 7%

Table 6: Time for *Draw.Open Counters.Graph* on Different Machines (ms)

Machine	Native +	SDE +	Difference
Macintosh II (16MHz MC68020)	4733	8850	+ 87%
Macintosh IIx (16MHz MC68030)	3600	6750	+ 88%
Macintosh IIfx (40MHz MC68030)	1550	2400	+ 55%
Macintosh Quadra 840AV (40MHz MC68040)	1550	1800	+ 16%

Table 7: Time for *Edit.Open OberonReport.Text* on Different Machines (ms)

Extrapolating from these results, there is reason to believe that on-the-fly code-generation from small SDE-files may eventually become faster than dynamic loading of larger native object-files, unless of course secondary storage universally migrates to a much faster technology. One way or the other, on-the-fly code-generation will definitely become faster in absolute terms as clock speeds increase further, so that the relative speed in comparison to native loading should not much longer be of importance anyway. Ultimately, the speed of loading needs to be tolerable for an interactive user – that is all that matters.

4.3. Code Quality

The last point that needs to be addressed concerns the quality of code that is generated dynamically. Table 8, by way of a popular benchmark, compares the quality of code generated on-the-fly with that generated by the *Apple MPW C compiler for the Macintosh (Version 3.2.4)* with all possible optimizations for speed enabled (*"-m -mc68020 -mc68881 -opt full -opt speed"*). The generation of integrity checks was disabled in the CGLoader because no equivalent concept is present in *C*.

The table lists execution times in milliseconds (less is better). Due to processor cache effects, these timings can vary by as much as 15% when executed repeatedly; the figures give the best of three executions. The *C* version of the *Treesort* benchmark exceeded all meaningful time bounds, due to apparent limitations of the standard Macintosh operating system storage-allocator. The CGLoader is not bound by these limitations, as it generates calls to the storage-allocator of *MacOberon*, which includes an independent memory-management subsystem.

Benchmark	SDE –	MPW C
Permutation	**83**	113
Towers of Hanoi	**83**	121
Eight Queens	50	**43**
Integer Matrix Multiplication	**150**	173
Real Matrix Multiplication	**133**	171
Puzzle	**800**	**800**
Quicksort	66	**61**
Bubblesort	117	**88**
Treesort	**83**	> 1000
Fast Fourier Transform	133	**123**

Table 8: Benchmark Execution Times (ms)

From this data, it can be inferred that the CGLoader emits native code of high quality that can compete with optimizing *C* compilers. In some cases, it surpasses even the output of the official optimizing compiler recommended by the manufacturer of the target machine, which is orders of magnitude slower in compilation. It should be possible to improve the remaining cases in which the CGLoader is currently inferior, without sacrificing much of the speed of on-the-fly code generation. Since the efficiency of the CGLoader is rooted in the compactness of the abstract program representation, rather than in any machine-specific details, it should be possible also to duplicate it for other architectures.

The *Apple MPW C* compiler requires about 6.9 seconds for compiling the C source program for the benchmark, and a similar time additionally for linking, compared to 1.1 seconds that are needed for encoding the corresponding Oberon program into an SDE-file and 233 milliseconds for loading it with on-the-fly code generation (including file access after a cold start).

5. Applications

5.1. Software Components

At the 1968 NATO conference, McIlroy [McI68] argued that "software production today appears in the scale of industrialization somewhere below the more backward construction industries" and attributed this to the absence of a *software component industry*. Yet more than twenty-five years later, not much has changed in the way we construct software. Reuse of software at best takes place within commercial organizations, but not between them. An industrial programmer still cannot just open a catalog of standard software parts and order a module from it that will execute an algorithm according to some given specifications. At best, he can hope that the algorithm he requires is built into the operating system. Strangely enough, today we witness an ongoing standardization of software functions by incorporation into operating systems and related libraries, instead of a separate industry developing software components.

Why, then, has no independent market for platform-independent standard software components developed over the years? Why have operating systems and their supporting libraries instead grown to an awesome complexity, encompassing functions as diverse as user-interface management and data-base support? The answer may simply be that there is currently no commercial incentive to develop *plug-in software components* because the maintenance costs would be prohibitive in today's marketplace. An independent software-component vendor would have to provide his products either in a multitude of link and object formats, which is costly, or in source form, which requires complex legal arrangements to protect the intellectual property of the authors and might cost even more. On the other hand, there is a direct commercial advantage for an operating-system vendor when he adds functionality to his product. As a consequence, we see a proliferation of operating-system-level enhancements instead of an intermediate industry for operating-system-independent support libraries.

The work presented in this paper contributes to lowering the cost of providing drop-in software components. It demonstrates the feasibility of a platform-independent software distribution format that enables portable modules to be used right out-of-the-box, without any off-line steps of compilation or linking. The *on-line* aspect is important because it means that even end-users can migrate libraries in their possession to new hardware platforms. It also suggests that different component-vendors could offer competing implementations of the same library, which an end-user would install or replace simply by *plugging in*. This is analogous to the situation in the personal computer hardware market, in which end-users are expected to buy and install themselves certain parts such as floating-point coprocessors.

5.2. Run-Time Integrity Checks

Apart from simplifying the distribution of software modules that are to be used on different target architectures, the proposed scheme eliminates also many of the cases that traditionally require *several versions* of a module to coexist side-by-side on a single target machine. During development, we often require variants of standard library modules because the software being developed might not be robust enough to guarantee that all constraints of the library are fulfilled. Accordingly, a *development version* includes additional checks that validate the arguments passed to library routines. For reasons of efficiency, however, one would not want to perform these validations during regular operation when all clients of the library have been thoroughly tested and can be trusted. Consequently, the tests are usually removed from the *production version*.

For example, consider a procedure *ReadBytes(f: File; VAR b: ARRAY OF CHAR; n: LONGINT)* in module *Files*. It is essential that the procedure never reads beyond the end of the input buffer, i.e. the precondition $n <= LEN(b)$ must hold. In some cases, the compiler alone may be able to verify that this condition is satisfied, given that there is a suitable mechanism for specifying such context requirements. However, there will always be cases in which the precondition can only be verified at run-time. Therefore, a corresponding validation needs to be present in the development version of module *Files*, but not necessarily in the production version. Unfortunately, a serious management problem arises from having to maintain more than one version of the same module, and having to keep track of which one is which. One constantly has to make sure that changes in a source text are propagated to all variants of the module that may exist concurrently, and use an elaborate naming scheme to differentiate between compiled variants of the same module.

On-the-fly code generation does away with module variants and the associated management overhead. Only at the time of loading do we need to indicate whether we require a *development* or a *production* version of a module. The implemented system supports several independent run-time integrity guards that can be enabled selectively (Table 9). Instead of the individual object file, it is then the run-time-environment that dictates whether or not these guards will be generated. Likewise, symbolic reference information for a debugger can be created as needed, and need not take up disk and memory space in configurations that are used solely for running finished applications. For example, the current implementation provides for the optional insertion of routine names in the code, in a format acceptable to the standard debugger of the host machine.

Switch	Effect if Enabled
debug	annotate code for symbolic debugging
clear	initialize local pointer variables to NIL at procedure entry
nil	test pointers for NIL prior to dereferencing
index	check that array subscripts lie within bounds
type	perform run-time type tests (used with type extension)
assert	generate code for ASSERT function

Table 9: Code Generation Switches

The key to *argument validation* in the implemented system lies in the standard function *ASSERT* of the programming language Oberon [Wir88]. *ASSERT* accepts as its arguments a Boolean expression and an Integer constant. It has the effect of a run-time test to check if the Boolean expression yields *TRUE*. If it doesn't, a run-time trap to the exception vector indicated by the second argument is taken. The main point about *ASSERT* is that it can be turned off by a compiler option, so that no code will be generated at all. Allowing the user to decide at run-time whether or not assertions should be verified eliminates the need for a separate development environment.

5.3. Management of Changes

Having available a system in which native code is generated only at the time of loading reduces also the organizational overhead required to keep a modular application consistent. Each time that a module is loaded, the implemented system performs a recompilation of the module's implementation, but in a manner that is completely transparent to the user. Consequently, the effects of certain changes of library modules can remain invisible, in contrast to other systems in which source-level recompilations of client modules are unavoidable.

A module needs to be recompiled whenever its own implementation is changed, or whenever it is invalidated by a change in one of the modules it depends on. Deciding which clients are invalidated by a change in a library is the difficult part. The easiest solution, implemented in tools such as the *Make* utility [Fel79] of the UNIX operating system [TR74], is to invalidate *all clients* each time that a library is changed. However, this introduces many recompilations that could be avoided in principle.

Tichy and Baker [TB85, Tic86] have introduced the notion of *smart recompilation*. This technique is founded on a detailed analysis of import/export relationships, considering not only the involved modules as a whole, but also the individual features that are exported from one module and imported by another. The results of this analysis are maintained in a database, along with a module dependency graph. One can then mechanize the decision which modules need to be recompiled by comparing the *changed feature set* of a modified library with the *referenced feature sets* of clients. In some cases, such as the addition of further procedures to a library, an interface change need not invalidate all existing clients then.

The proposed method opens a path to even further reduce the number of (source-text) recompilations. A large proportion of changes that typically occur during the software life-cycle has no effect on the *behavior* of a program but nevertheless on the machine code being generated, because native code contains addresses, sizes, and relative offsets *literally*, and addressing modes often depend on particular address values. An example for a change that has consequences only in the code generator is the *insertion of an additional data field* in an exported record type. Apart from the possible error that may occur if the corresponding identifier is used already within the scope of the record, a condition that is detected easily, this addition preserves the semantics of all client modules. Unfortunately, however, the addition alters the size of the record type, and may change the relative offsets of some of the existing record fields, so that new code needs to be generated for all modules that use the record type. In a system such as the implemented one, code generation happens transparently and need not be of concern to users, because all client modules will be updated automatically when they are loaded the next time.

In principle, therefore, in a system in which code generation occurs at loading time, recompilation (of source code) is *not required to propagate changes*. There are, of course, certain changes that invalidate the source code of some client modules completely, such as removing a routine from a library module or changing the result type of a library function, but these require some *source-level re-coding* of all affected clients and cannot simply be dealt with by simple recompilation. Such situations can be detected in advance using the techniques described by Tichy and Baker, or will in any event be flagged when the CGLoader senses an interface mismatch, resulting in a load error.

Accordingly, in a system offering on-the-fly code generation at load time, the only factor that determines whether (source-text) recompilation of clients is necessary in reaction to changes in libraries, is the strategy that is used for describing the inter-module links in the symbol table. The current implementation uses *per-module fingerprints* to ensure interface consistency, so that more recompilations are needed than would be necessary if *per-object fingerprints* were used. In the latter case, no recompilations of clients would be necessary ever. However, this aspect has not been the main focus of the current work and, therefore, not been pursued.

5.4. Improving Code Quality by Targeted Optimizations

The fact that object code is generated anew each time that a module is loaded could also be exploited for increasing the *overall performance* of the whole system, although this has currently not been implemented. Borrowing from ideas discussed by Morris [Mor91] and Wall [Wal91], an *execution profile* obtained in a previous run of the system could guide the level of optimization applied by the CGLoader in the creation of the next executable version.

While fine-grained profiling might be useful as a basis for specifically-targeted optimizations, run-time profiling, which is associated with an overhead, might not even be necessary in a modular system. Instead, one might simply use the *import counter*, which indicates how many clients a module has, as an estimate for the "relative importance" of a module. The more important a module is, the greater the potential benefits of optimization.

The idea of targeted optimization may be developed even further, taking into account that object code can be re-created from SDE-files at any time. The system might expend its *idle time* on the recompilation of whole subtrees of the loaded-module graph, employing a higher optimization level than the currently loaded version. After recompiling such a subtree, it may then attempt to unload the old version, and if unloading is successful adjust the global module graph to include the new, optimized version. Note that the unloading step may fail if further clients are added to the originally loaded modules while re-generation is underway, or if installed procedures from the original modules remain active in the system.

5.5. Further Applications

A machine-independent abstract program representation from which high-quality code can efficiently be generated on-the-fly might also prove to be valuable in the context of heterogeneous distributed systems consisting of several different hardware platforms.

For example, the proposed technique could form the basis for a very general *remote procedure-call* mechanism [Nel81]. Instead of compiling a separate stub for each procedure to be called remotely and installing it as a process on the target machine, one might send a complete instruction sequence in a machine-independent format, to be processed by a single *code-generating stub* on the side of the receiver. Cryptological authentication measures could be applied to prevent misuse in an open network.

Consistency problems between program segments for different machines in a distributed application could be avoided trivially by sending a *consistent version* across the network before the start of the distributed computation, thereby guaranteeing that identical code executes on all machines taking part in the computation.

Last but not least, designers of object-oriented systems could use an even broader definition of *object persistence*. A persistent object might contain its own code in an abstract format. Such an object could then migrate over a network or be transported on some storage medium. At a destination site, first the code to handle the object would be generated dynamically. Thereafter, the object's data would be read in.

6. Comparison with Related Work

The project described here was started in 1990, and first results were published in September of 1991 [FL91]. At that time, it seemed almost exotic to attempt any revival of the old *UNCOL* idea. Today, the topic again seems "hot", and many researchers are working on related projects. The most notable of these other projects is the *architecture neutral distribution format (ANDF)* initiative by the *Open Software Foundation (OSF)*.

6.1. The OSF ANDF Project

The Open Software Foundation (OSF) has recently adopted a technology called *TDF* [OSF91, DRA93a, DRA93b], designed and implemented by the United Kingdom Defence Research Agency (DRA), to serve as the basis of an *architecture neutral software distribution format* (ANDF). TDF has many characteristics in common with semantic-dictionary encoding. Just like SDE, and unlike previous UNCOL attempts, TDF is not based on an abstract machine, but on a tree-structured intermediate language in conjunction with an embedded symbol table.

TDF has been designed to be both source-language and target-architecture independent, although as of June 1993, the only existing compiler front-end for TDF was for the programming language *C* [KR78]. TDF is claimed to be useful for source languages other than C, and compilers translating into the TDF representation from other languages are being developed. However, although TDF is extensible to accommodate specific features of future programming languages, it is not guaranteed that this can be done in an upward-compatible manner [DRA93c].

In contrast, the method of SDE is not a program representation in its own right, but a *meta-technique* for encoding programs abstractly. It is parametrized by the initial configuration of the dictionary and the heuristics used for dictionary management. So far, SDE has been applied only to programs originally written in the programming language *Oberon* [Wir88]. However, since semantic meaning is instilled into each SDE-file solely by the *initial configuration* of a dictionary, it is possible to support easily future language requirements, even without invalidating existing SDE-files in an old format. All that is necessary is a key in the SDE-file that uniquely identifies the initial configuration of the dictionary that has to be used for decoding. This might, for example, be simply the name of a file in which the configuration is stored.

Hence, SDE allows us to *evolve the set of encodeable language constructs* independently of the actual file-formats. In fact, by using configuration files, individual SDE-decoders could be made completely independent of the file formats they need to process. It would require a standardization only of the *meanings* of different meta-language constructs. Software developers would then be able to choose freely which of these meta-language constructs to use in the encoding of their programs, and at which positions of the initial dictionary they would place these constructs. The smaller the set of meta-language constructs, the more difficult it will be to reverse-engineer the encoded program, although the use of fewer constructs could also affect compactness and optimizeability adversely.

The question of reverse-engineerability is important for software developers wishing to offer their products in a portable format. SDE and TDF both preserve the abstract structure of programs and can, therefore, be reverse-engineered to produce a "shrouded" source program, i.e. one that contains no meaningful internal identifiers [Mac93]. However, with current technology, reverse-engineering to a similar degree is possible also from binary code. Many of the algorithms that have been developed for object-code-level optimization [DF84] are useful for these purposes. Moreover, the statement [DRA93c] is probably correct that portable formats are such attractive targets to reverse-engineer that suitable tools will become available anyway, regardless of how difficult it is to produce such tools. It would, therefore, not make much sense to jeopardize the advantages of SDE in an attempt to make reverse-engineering more difficult.

SDE has some further advantages over TDF. To start with, it is more compact. Although in [OSF91] the OSF recognized that *"it is important that the ANDF file size be as small as possible"*, TDF is in fact less compact than object code. TDF is quoted [DRA93c] as being "around twice the size of the binary of CISC machines", while SDE is at most half the size of MC68020 binary code. Because of the lack of a common operating platform on which both mechanisms have been implemented, a direct performance comparison of SDE versus TDF is currently not possible. However, DRA [DRA93c] state that native object-file generation, which is an off-line process in their implementation, takes between 32% and 83% of native C compile time. Considering that Oberon compilers usually outperform C compilers by a factor of more than 15 in compilation speed [BCF92], and that loading of SDE-files is more than four times faster than regular Oberon compilation, it seems reasonable to claim that, at least on CISC processors and without taking code quality into account, SDE should provide for much faster code generation than TDF.

6.2. Dynamic Code-Generation

The concept of *dynamic translation* of programs from one representation into another has been around for some time. Early implementations, such as one by Brown [Bro76], were developed with the aim of balancing execution speed and memory requirements under the extreme hardware constraints that were then the norm. These early implementations applied only to programming languages without block-structure and performed the generation of native object-code on a statement-by-statement basis.

For a long time, the main reason for implementing dynamic translation remained the fact that it allowed an elegant trade-off between execution efficiency and memory consumption. A paper by Rau [Rau78] classifies program representations into three categories, namely *high-level*, *directly interpretable*, and *directly executable*, and discusses the use of dynamic translation between these categories as a means for achieving speed and compactness simultaneously.

Then came Deutsch and Schiffmann's [DS84] landmark paper on the efficient implementation of the programming language *Smalltalk-80* [GR83]. They used dynamic translation for increasing execution speed while *retaining virtual-machine code-compatibility* with existing implementations. The latter was necessary because the Smalltalk-80 virtual machine is actually visible to user programs, and much of the system code depends on it. In Deutsch and Schiffmann's implementation, native code for the actual target machine is generated on-the-fly and cached until it is invalidated by changes in the source program, or until it is overwritten in the code-cache due to lack of space.

A more recent application of dynamic translation comes from the implementation [CUL89, CU89] of the programming language *Self* [US87], a dynamically-typed language based on prototypes. Just as the Smalltalk-80 system, it can benefit enormously from on-the-fly code generation, because type information, although unavailable statically, is available at run-time. By allowing the dynamic generation of several variants of an expression, optimized for different run-time types of the component variables, the efficiency of such systems can be multiplied, but still cannot compete with statically-typed programming languages.

In contrast, the implemented system attempts to deliver run-time performance comparable to traditional compilers and offers on-the-fly code generation primarily as a means for increased user convenience, not code quality. It ties dynamic translation intimately to the two concepts of *separate compilation* and *dynamic module loading*. The information contained in SDE-files is equivalent to the source description, so that there is no fundamental limit to the obtainable code quality. Hence, the most effective optimizing code generators could potentially be built into a CGLoader operating on SDE.

It is also true that *modules* are much better suited as the unit of code generation than procedures. A module is a collection of data types, variables, and procedures that are loaded together always, and, almost equally important, unloaded together always. Furthermore, a module can be loaded only after all of its servers (imported library modules) have been loaded successfully, and unloaded only after all of its clients (importing modules) have been unloaded. Consequently, code is generated from the bottom upwards, and the addresses of all callees are known when compiling a caller. Likewise, there are never any clients that need to be invalidated explicitly when a module is unloaded.

In systems in which the unit of code generation is the *procedure*, such as Smalltalk-80 and Self, program execution is interspersed with code-generation. Each procedure call may potentially fault, at which point native code needs to be generated dynamically before the call can be completed. Unfortunately, this may sometimes generate formidable amounts of such faults in succession, each associated with a re-load of the instruction cache, causing disruptive delays for interactive users.

6.3. Dynamic Binary-To-Binary Object-Code Translation

A technology that has emerged only recently is *binary translation* of object code [SCK93]. It enables programs for one architecture to be executed directly on another via true translation of whole object programs into the native instruction set of the new target machine. The main rationale behind binary translation is the need to protect previous investments into software when migrating to a new hardware architecture. Rather than constituting a *portability technique* in the spirit of "software components", binary translation represents a *capitulation* before the fact that much of the software in existence is not portable, and cannot be ported by ordinary means; for example, because the source texts and the original design documents are no longer available.

Binary translation is a complex technique. Since it is put to use mainly in circumstances in which little is known about the programs that serve as its input, the translation mechanism needs to be suitable for any program that could possibly have executed on the architecture of origin. This includes programs that modify themselves. Consequently, self-modification conditions need to be detected on the new target machine, at which time the affected code segments may have to be re-translated on-the-fly.

Due to its complexity, binary translation cannot really be seen as an answer to the portability problem, but only as an intermediate solution allowing us to keep using an existing software base while it is being rewritten for the new architecture. The technique that I am advocating in this paper is orders of magnitude simpler and concerned less with backward-compatibility than with forward-compatibility, with whatever may lie ahead in the future.

7.   Future Work

As a next step, a second CGLoader for a different architecture is planned that will operate on the identical SDE-file format. This would provide object-level portability between different processor architectures. In fact, it should be noted that traditional "binary" compatibility may not be good enough for future machines anyway. This is because different implementations of the same architecture begin to diverge by so much that it is becoming virtually impossible to generate native object-code that will perform well on all processors within a family. Among other features, different implementations of an architecture stand apart in the number of instructions that can be issued simultaneously to independent functional units, and in the depth of their instruction pipelines. On the other hand, the use of a CGLoader allows for the scheduling of instructions specifically for each particular target processor. It is thereby possible to deliver object code that is custom-tailored towards each processor's characteristics, along with a user convenience comparable to that of binary compatibility.

8.   Summary and Conclusion

This paper has described a new technique for representing programs abstractly, which yields a highly compact encoding and is able to provide a code generator with all the information available on the level of the source language, plus additional knowledge about the occurrence of common subexpressions in the source text. It facilitates simple and efficient on-the-fly code generation on relatively slow processors without precluding the use of highly optimizing code generation methods on faster ones.

The new technique is able to provide separate compilation of program modules with type-safe interfaces, independent module distribution, and interchangeability of modules with the same interface. These properties give it an advantage over other approaches to portability, such as using the programming language C [KR78]. It is also more practical than "shrouded" C [Mac93], which might be seen as a sort of portable assembly-language that is yet difficult to reverse-engineer. While it is true that many of the cost-lowering benefits of portability can be gained by using obscured C, or any high-level language for that matter, code generation from the SDE representation is orders of magnitude faster than the compilation of a C program source.

The proposed technique offers the potential of providing a universal software distribution format that is practical to use. Hence, it might encourage software developers to share reusable software components or offer them commercially. It eliminates the need for separate development environments and can reduce the number of recompilations after changes in library modules. Other potential applications of object-level portability may lie in heterogeneous distributed computing environments, in which the compactness of the SDE representation would be of particular advantage when network transfer is required.

Acknowledgments

The author is indebted to Niklaus Wirth for his guidance of the project described here, and for his numerous suggestions and comments. He would like to thank Jürg Gutknecht, who commented thoroughly on this paper and provided valuable criticisms that helped to improve the presentation of this material considerably.

References

[App85]   Apple Computer, Inc.; *Inside Macintosh*; Addison-Wesley; 1985ff.

[ADH89]   R. Atkinson, A. Demers, C. Hauser, Ch. Jacobi, P. Kessler and M. Weiser; Experiences Creating a Portable Cedar; *Proc. Sigplan '89 Conf. Programming Language Design and Implementation*, published as *Sigplan Notices*, 24:7, 322–329; 1989.

[BCF92]   M. Brandis, R. Crelier, M. Franz and J. Templ; *The Oberon System Family*; Report #174, Departement Informatik, ETH Zurich; 1992.

[Bro76]   P. J. Brown; Throw-Away Compiling; *Software–Practice and Experience*, 6:3, 423–434; 1972.

[CU89]   C. Chambers and D. Ungar; Customization: Optimizing Compiler Technology for SELF, a Dynamically-Typed Object-Oriented Programming Language; *Proc. ACM Sigplan '89 Conf. Programming Language Design and Implementation*, published as *Sigplan Notices*, 24:7, 146–160; 1989.

[CUL89]   C. Chambers, D. Ungar and E. Lee; An Efficient Implementation of SELF, a Dynamically-Typed Object-Oriented Language Based on Prototypes; *OOPSLA '89 Conf. Proc.*, published as *Sigplan Notices*, 24:10, 49–70; 1989.

[Con58]   M. E. Conway; Proposal for an UNCOL; *Comm. ACM*, 1:10 5–8; 1958.

[DF84]   J. W. Davidson and C. W. Fraser; Code Selection through Object Code Optimization; *ACM Trans. Programming Languages and Systems*, 6:4, 505–526; 1984.

[DRA93a]   United Kingdom Defence Research Agency; *TDF Specification, Issue 2.1*; June 1993.

[DRA93b]   United Kingdom Defence Research Agency; *A Guide to the TDF Specification, Issue 2.1.0*; June 1993.

[DRA93c]   United Kingdom Defence Research Agency; *Frequently Asked Questions about ANDF, Issue 1.1*; June 1993.

[DS84]   L. P. Deutsch and A. M. Schiffmann; Efficient Implementation of the Smalltalk-80 System; *Conf. Record 11th Annual ACM Symp. Principles of Programming Languages*, Salt Lake City, Utah, 297–302; 1984.

[Fel79]   S. I. Feldman; Make: A Program for Maintaining Computer Programs; *Software–Practice and Experience*, 9:4, 255–265; 1979.

[Fra90]   M. Franz; *The Implementation of MacOberon*; Report #141, Departement Informatik, ETH Zürich; 1990.

[Fra93a]   M. Franz; Emulating an Operating System on Top of Another; *Software–Practice and Experience*, 23:6, 677–692; June 1993.

[Fra93b]   M. Franz; The Case for Universal Symbol Files; *Structured Programming*, 14:3, 136–147; October 1993.

[FL91]   M. Franz and S. Ludwig; Portability Redefined; *Proc. 2nd Int. Modula-2 Conf.*, Loughborough, England; 1991.

[GF84]   M. Ganapathi and C. N. Fischer; Attributed Linear Intermediate Representations for Retargetable Code Generators; *Software–Practice and Experience*, 14:4, 347–364; 1984.

[GR83]   A. Goldberg and D. Robson; *Smalltalk-80: The Language and its Implementation*; Addison-Wesley; 1983.

[Gri72]   R. E. Griswold; *The Macro Implementation of SNOBOL4: A Case Study in Machine-Independent Software Development*; Freeman, San Francisco; 1972.

[KR78]   B. W. Kernighan and D. M. Ritchie; *The C Programming Language*; Prentice-Hall; 1978.

[Mac93]   S. Macrakis; *Protecting Source Code with ANDF*; Open Software Foundation Research Institute; June 1993.

[McI68]   M. D. McIlroy; Mass Produced Software Components; in Naur, Randell, Buxton (eds.), *Software Engineering: Concepts and Techniques*, Proceedings of the NATO Conferences, New York, 88–98; 1976.

[Mor91]   W. G. Morris; CCG: A Prototype Coagulating Code Generator; *Proc. ACM Sigplan '91 Conf. Programming Language Design and Implementation*, published as *Sigplan Notices*, 26:6, 45–58; 1991.

[Mot87]   Motorola, Inc.; *M68030 Enhanced 32-bit Microprocessor User's Manual*; Motorola Customer Order No. MC68020UM/AD; 1987.

[Nel81]   B. J. Nelson; *Remote Procedure Call*; Report #CLS-81-9, Palo Alto Research Center, Xerox Corporation, Palo Alto, California; 1981.

[NAJ76]   K. V. Nori, U. Amman, K. Jensen, H. H. Nägeli and Ch. Jacobi; Pascal-P Implementation Notes; in D.W. Barron, editor; *Pascal: The Language and its Implementation*; Wiley, Chichester; 1981.

[OSF91]   Open Software Foundation; *OSF Architecture-Neutral Distribution Format Rationale*; 1991.

[PW69]   P. C. Poole and W. M. Waite; Machine Independent Software; *Proc. ACM 2nd Symp. Operating System Principles*, Princeton, New Jersey; 1969.

[Rau78]   B. R. Rau; Levels of Representation of Programs and the Architecture of Universal Host Machines; *Proc. 11th Annual Microprogramming Workshop*, Pacific Grove, California, 67–79; 1978.

[Rei89]   M. Reiser; Private Communication; 1989.

[Ric71]   M. Richards; The Portability of the BCPL Compiler; *Software–Practice and Experience*, 1:2, 135–146; 1971.

[SCK93]   R. L. Sites, A. Chernoff, M. B. Kirk, M. P. Marks and S. G. Robinson; Binary Translation; *Comm. ACM*, 36:2, 69–81; February 1993.

[SWT58]   J. Strong, J. Wegstein, A. Tritter, J. Olsztyn, O. Mock and T. B. Steel; The Problem of Programming Communication with Changing Machines: A Proposed Solution: Report of the Share Ad-Hoc Committee on Universal Languages; *Comm. ACM*, 1:8, 12–18, and 1:9, 9–15; 1958.

[Szy92]   C. A. Szyperski; Write-ing Applications: Designing an Extensible Text Editor as an Application Framework; *Proc. 7th Int. Conf. Technology of Object-Oriented Languages and Systems (TOOLS'92)*, Dortmund, Germany, 247–261; 1992.

[TR74]   K. Thompson and D. M. Ritchie; The UNIX Time-Sharing System; *Comm. ACM*, 17:2, 1931–1946; 1974.

[Tic86]   W. F. Tichy; Smart Recompilation; *ACM Trans. Programming Languages and Systems*, 8:3, 273–291; 1986.

[TB85]   W. F. Tichy and M. C. Baker; Smart Recompilation; *Conf. Record 12th Annual ACM Symp. Principles of Programming Languages*, New Orleans, Louisiana, 236–244; 1985.

[US87]   D. Ungar and R. B. Smith; Self: The Power of Simplicity; *OOPSLA '87 Conf. Proc.*, published as *Sigplan Notices*, 22:12, 227–242; 1987.

[Wal91]   D. W. Wall; Predicting Program Behavior Using Real or Estimated Profiles; *Proc. ACM Sigplan '91 Conf. Programming Language Design and Implementation*, published as *Sigplan Notices*, 26:6, 59–70; 1991.

[Wel84]   T. A. Welch; A Technique for High-Performance Data Compression; *IEEE Computer*, 17:6, 8–19; 1984.

[Wir88]   N. Wirth; The Programming Language Oberon; *Software–Practice and Experience*, 18:7, 671–690; 1988.

[WG89]   N. Wirth and J. Gutknecht; The Oberon System; *Software–Practice and Experience*, 19:9, 857–893; 1989.

[WG92]   N. Wirth and J. Gutknecht; *Project Oberon: The Design of an Operating System and Compiler*; Addison-Wesley; 1992.

Distributed High-Level Module Binding
for
Flexible Encapsulation
and
Fast Inter-Modular Optimization

Christian S. Collberg

Department of Computer Science,* University of Auckland,
Private Bag 92019, Auckland, New Zealand
c_collberg@cs.aukuni.ac.nz

Abstract. We present a new modular object-oriented language with orthogonal encapsulation facilities. The language provides full support for encapsulation and separate compilation which makes it difficult to compile using standard techniques. We present new distributed translating techniques which overcome these difficulties by allowing inter-modular information to be exchanged at link-time. The same techniques may also be used with other modular and object-oriented languages to facilitate fast inter-modular optimizations such as inline expansion.

1 Introduction

ZUSE is a new modular object-oriented language with orthogonal encapsulation facilities: *concrete*, *abstract*, and *semi-abstract* export of types, constants, and (inline) procedures are all supported. ZUSE's encapsulation and separate compilation facilities make the language more difficult to compile than other similar languages (such as Ada [21], Mesa [10], Modula-2 [23], and Modula-3 [17]): inter-modular information necessary for memory allocation, semantic analysis, and code generation is not always present at compile time. The ZUSE translating system therefore exchanges inter-modular information at link-time. The module binder (called a PASTER) binds modules together partially at the intermediate code level, performs inter-modular optimizations such as inline expansion, allocates memory, and generates code for procedures for which code generation could not be performed at compile-time.

In addition to the description of the ZUSE language itself, the main contribution of this paper is the design of a distributed module binder (the DPASTER) which achieves better performance than its sequential counterpart (the SPASTER) by distributing its actions over the sites of a workstation-server network. The design of the ZUSE compiler and DPASTER assures that the cost in terms of

---

* This work was carried out while the author was at the Department of Computer Science, Lund University, Sweden.

translation-time, execution-time, and storage of using an abstract or semi-abstract item will be no greater than if the same item had been concrete. Although the DPASTER is designed specifically to support the encapsulation facilities of the ZUSE language, it can also be used with other modular and object-oriented languages to facilitate fast inter-modular optimizations.

Considering that all other aspects of language translation have been subject to attempts at parallelization,[2] it is remarkable that this appears to be the first attempt at parallelizing the linking process. It is also the first time distributed techniques have been used to tackle the problems inherent in combining separate compilation with both data abstraction and inter-modular optimization.

2 Background and Motivation

A desirable property of a modular language is that it should allow a module interface to be *fully abstract*, i.e. to contain only the information necessary in order for a client to correctly use the module and for an implementer to correctly implement it. This principle, apart from being basic to sound software engineering practices, is also essential in order to minimize compilation cost. If an interface is not fully abstract, containing extraneous information related more to the module's *implementation* than to its *specification*, any change to this information will affect the interface and hence will trigger the recompilation of the module's clients.

There is, however, an inherent conflict between the two goals of permitting fully abstract module interfaces and separate compilation of interfaces and bodies. The conflict stems from the fact that in many cases an abstract interface will not contain enough information in order for a compiler to assure the static semantic correctness of the module's clients or to generate code for the clients. The conflict is evident in the design of many current modular languages. Ada, for example, requires the implementation of private types to be given in the module interface, in order for the compiler to know the size of the type when compiling a client module. Modula-2, for the same reason, restricts its opaque types to pointers, while Mesa and Oberon-1 [24] require the *size* of opaque types and *public projection* types, respectively, to be given in the interface. Modula-3 instead relies on run-time processing in order to support its opaque and partially revealed object types, as does Albericht [18] – an Oberon descendant – to support its public projection type. Milano-Pascal [5], finally, employs a specialized module linker to share information between separately compiled Pascal modules, specifically sizes of abstract types and values of abstract constants.

It is not only the implementation of abstraction which is hampered by separate compilation but also inter-modular optimizations such as inline expansion and inter-procedural register allocation. C++ [1] and Mesa, for example, require inline procedures to be declared in the module specification, while Ada allows

---

[2] See, for example, Junkin [13] (compilation), Katseff [14] (assembly), Baalbergen [2] (making).

compilation dependencies between implementation units containing exported in-line procedures. The MIPS compiler-suite [11] solves the inter-modular inlining problem by concatenating the intermediate code of separately compiled modules prior to code-generation and optimization. Similarly, Wall [22] employs a specialized linker in order to do inter-modular register-allocation. While these implementations have some obvious advantages (good global code quality, no unnecessary recompilations) they suffer from the problem of excessive linking times. In this paper we will show that techniques from the field of *distributed compilation* allow full data encapsulation, separate compilation, and cross-module optimization to be achieved without undue translation-time or execution-time overhead.

3   The Language ZUSE

ZUSE contains the array of basic concepts found in most modular procedural and object-oriented languages and shares with these languages a common view of data and execution. We will focus our attention on the kinds of items ZUSE modules may export and the facilities available to a programmer for protecting the integrity of such items.

ZUSE borrows its module system from Modula-2: a module is made up of two separately compiled units, a *specification* and an *implementation*. Compilation dependencies generally exist between specification units and their clients, but ZUSE (like Modula-2) does not allow dependencies between implementation units.

ZUSE shares most of its basic type system with Modula-2. Also supported is a Modula-3-style object type which forms the basis of object-oriented programming, Mesa-style type initializers, structured manifest constants, and inline procedures. Unique to ZUSE are its three *modes of export*: a *concrete* item reveals its entire realization in the specification unit, a *semi-abstract* item reveals part of its realization and hides the rest in the implementation unit, and an *abstract* item hides all of its realization in the implementation unit.

3.1   Abstract export

Figure 1 shows a simple complex number package using abstract export. A similar package in Ada would require the realization of the exported type to be revealed in the specification unit. This, apart from being a breach of encapsulation, would require the recompilation of all client modules in the event of a change to the realization. Even worse, since Complex`Create[3] is an inline procedure, most Ada implementations would insert compilation dependencies between the implementation unit of Complex and all of its clients, forcing all clients (and possibly their clients) to be recompiled whenever the implementation unit was changed.

---

[3] M`t is a reference to the item t exported from module M.

```
SPECIFICATION Complex;          IMPLEMENTATION Complex;
    TYPE T = ;                      TYPE T += RECORD [Re, Im : REAL ];
    CONSTANT Zero : T =;            CONSTANT Zero : T += RECORD [
    PROCEDURE Create : (                Re :== 0.0 ;
        Re, Im : REAL) RETURN T =;      Im :== 0.0 ];
END Complex.                        INLINE PROCEDURE Create : (
                                        Re, Im : REAL) RETURN T +=
                                    BEGIN ... END Create;
                                    END Complex.
```

Fig. 1. Examples of ZUSE abstract export. The fact that T is realized as a record and that Create is an inline procedure is hidden from importing modules.

A Modula-2 or Modula-3 implementation, on the other hand, while not revealing the realization of the abstract type[4] or introducing any unwarranted compilation dependencies, would be forced to use a pointer-based realization of Complex`T. This has several disadvantages: (1) Every time a complex number is needed a call has to be made to the dynamic allocation routine, which would be expensive for, say, large arrays of complex numbers; (2) when using the built-in assignment operator on variables of Complex`T, the effect is *reference assignment* rather than *value assignment*, the preferred mode for complex values; (3) in Modula-2, which does not have garbage collection, temporary complex variables may never be reclaimed (see Feldman [9, pp. 127-131]).

Hence, ZUSE makes it possible to combine the static allocation, value semantics, and inline expansion of Ada, with the full data encapsulation and compilation independence of Modula-2 and Modula-3.

3.2 Semi-Abstract Export

It may sometimes be desirable to make part of the realization of an exported item known to its clients while keeping the rest private to the implementation. Oberon-1 and Modula-3 provide limited forms of such *semi-abstract* types, but at a cost: In Oberon-1 it is necessary to reveal the size of the full type in the specification unit [24], and in Modula-3 the use of partially revealed object types entails extra run-time analysis. ZUSE, on the other hand, has full support for semi-abstract types, constants, and inline procedures, and the ZUSE translation system assures that this will not involve any extra translation-time or run-time cost. Figure 2 gives some simple examples of ZUSE's semi-abstract export. Note that the ZUSE object type comes in three parts: an optional supertype (preceded by the keyword **WITH**), a sequence of fields (the first set of brackets), and a sequence of methods (second set of brackets).

---

[4] Modula-2 and Modula-3 do not support ZUSE-style abstract constants and inline procedures.

```
SPECIFICATION SemiAbstract;
  TYPE
    EnumT   = ENUM [Red, Blue, Green];
    RecordT = RECORD [x : CHAR ];
    ObjT    = OBJECT;
    SubObjT = OBJECT
                    [v : INTEGER ]
                    [R : (x : CHAR)] ;
  CONSTANT
    C : RecordT = RECORD [x :== "R"];
    NOT INLINE PROCEDURE P : (a : EnumT) =;
END SemiAbstract.

IMPLEMENTATION SemiAbstract;
  TYPE
    EnumT   += ENUM [Yellow, Orange];
    RecordT += RECORD [z : CARDINAL ];
    ObjT    += OBJECT [a : INTEGER] [ ];
    SubObjT += WITH ObjT OBJECT
                    [r : RecordT ]
                    [U : (z : CHAR)];
  CONSTANT
    C : RecordT += RECORD [z :== 15];
    PROCEDURE P : (a : EnumT) +=
    BEGIN ··· END P;
END SemiAbstract.
```

Fig. 2. Examples of ZUSE semi-abstract export. The module SemiAbstract reveals that the type RecordT is a record with a character field x, but hides the fact that RecordT also has an integer field z. SemiAbstract furthermore reveals that SubObjT is an object type with a field v and a method R, but hides the fact that SubObjT is a subtype of ObjT and that it declares an additional field r and a method U.

4   The ZUSE Compiler

A consequence of ZUSE's scheme for separate compilation and flexible encapsulation is that the ZUSE compiler does not always have available all the intermodular information necessary in order to completely check the static semantic correctness of the input program and to generate high-quality code. When the compiler determines that this is the case for a particular procedure, code generation and (some of the) semantic checking are deferred to module binding time when all relevant information is available. We list some of the circumstances under which such deferral will be necessary:

– The compiler will not always know the size of a variable which is defined in terms of imported abstract or semi-abstract types or constants. Hence it will be unable to generate code for any routine which allocates or references such a variable.

- The compiler cannot generate code for a call to an imported abstract procedure since it may be inline, and in this case the code of the procedure is unavailable for expansion.
- The compiler cannot construct the method template for an object type which has a semi abstract ancestor.
- The compiler cannot always check whether the declaration of an abstract type, constant, or inline procedure which is defined in terms of imported abstract items is part of an illegal recursive declaration (see Figure 3).

```
SPECIFICATION M;          IMPLEMENTATION M;
  TYPE T = ;                IMPORT N;
  PROCEDURE P : () =;       TYPE T += N`T;
END M.                      INLINE PROCEDURE P : () += BEGIN N`P (); END P;
                          END M.


SPECIFICATION N;          IMPLEMENTATION N;
  TYPE T = ;                IMPORT M;
  PROCEDURE P : () =;       TYPE T += RECORD [x,y : M`T ];
END N.                      INLINE PROCEDURE P : () += BEGIN M`P (); END P;
                          END N.
```

Fig. 3. Examples of illegal recursive declarations in ZUSE.

Deferred procedures and context conditions produced by the compiler for a certain module are stored in the module's object code file together with exported inline procedures, binary code produced for non-deferred procedures, etc. The object code file for a module M consists of seven major sections. The *Module Table* is a list of the names of the modules directly imported by M. It is used to determine the modules which will be part of the resulting program. The *Expression Section* is a set of constant expressions (in a DAG format) which organizes all the inter-modular information needed during binding. It stores, among other things, sizes of types, possible static error conditions, the allocation of global variables, and the inline-status and call graphs of procedures. The *Binary Code and Data Sections* contain code and data generated during compilation, and the *Relocation Section* contains relocation information for this code. The *Inline* and *Deferred* sections contain the intermediate code (in an *Attributed Linear Code* format) of inline and deferred procedures.

4.1 The Expression Section

The Expression Section is the central object code structure. The general structure of an expression is $e_{m,i}.a = f(e_{m_1,i_1}.a_1, \cdots, e_{m_n,i_n}.a_n)$, where $e_{m,i}.a$ is the attribute a of the i:th expression in module m, f is a function with n parameters, and the $e_{m_k,i_k}.a_k$:s are references to the values of other expressions. The function f may be an arithmetic operation, an operation to construct structured constants or method templates from their subparts, etc. It is sometimes convenient to think of the Expression Section as a graph where each $e_{m,i}.a$ is a node labeled with the operation f with n outgoing edges to the nodes $e_{m_1,i_1}.a_1, \cdots, e_{m_n,i_n}.a_n$. In most cases the graph will be a forest of expression DAGs, but the following example (the concatenation of the Expression Sections of the modules M and N which were given earlier) shows that the graph may sometimes contain cycles:

$$
\begin{array}{lll}
e_{M,1}.size & = e_{N,2}.size & \text{-- M`T} \\
e_{M,2}.inline & = \textbf{TRUE} & \text{-- M`P} \\
\underline{e_{M,2}.calls} & = [e_{N,3}] & \text{-- M`P} \\
e_{N,1}.expr & = 2 & \\
e_{N,2}.size & = e_{N,1}.expr * e_{M,1}.size & \text{-- N`T} \\
e_{N,3}.inline & = \textbf{TRUE} & \text{-- N`P} \\
e_{N,3}.calls & = [e_{M,2}] & \text{-- N`P}
\end{array}
$$

The illegal declarations of M and N are evident from these expressions: the recursive definitions of M`T and N`T correspond to a cycle between the expressions which express the types' size, and the recursive inline calls can be discerned from the call-graph of the procedures M`P and N`P.

5 The ZUSE Sequential Module Binder

In many ways the ZUSE sequential binder resembles ordinary systems link editors. They both combine the code produced by a compiler for separately compiled modules into an executable file, and they both run in several phases, the first phase loading definitions and the last phase performing relocations. However, unlike link editors the sPASTER is equipped to perform code generation, intermodular optimization, and some static semantic checking. Our present sPASTER design runs in four major phases (see Collberg [6] for a more detailed description):

- Phase 1 determines which modules will make up the resulting program, loads the expressions and inline procedures of all modules, and copies the binary code and data sections to the resulting executable file.
- Phase 2 evaluates the expressions, allocates global variables, performs deferred semantic checks, and expands inline calls in inline procedures. Phase 2 also makes a conservative estimate of which deferred procedures may be referenced at run-time.
- Phase 3 reads the intermediate code of every referenced deferred procedure, updates the code with the expression values calculated during Phase 2, expands inline calls, optimizes the intermediate code, generates machine code, and writes the generated code to the executable file.
- Phase 4 performs final relocation.

6 The ZUSE Distributed Module Binder

The high-level binding required by the ZUSE language can be a considerably more expensive operation than conventional link-editing, but fortunately it is a process which is amenable to parallelization. This section will describe the design of the ZUSE distributed module binder which distributes its actions over a workstation-server network, or similar loosely-coupled architectures.

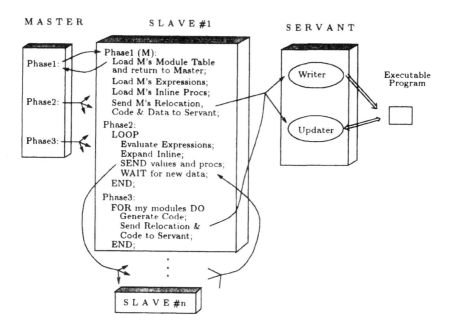

Fig. 4. An overview of the DPASTER.

The DPASTER can be said to be based on the *data decomposition*[5] model of compiler parallelization, in that each site taking part in the build process is responsible for a subset of the modules which make up the program. The DPASTER is made up of three kinds of processes: a *Master*, a *Servant*, and a set of *Slaves* (see the figure above). The Master is the main conductor of processing and is the program invoked by the user. It decides which modules each Slave should process, initiates the various processing phases, and relays any errors back to the user. The Slaves (there is at most one per site) are dormant until awakened by the Master, who assigns them a set of modules to process. The Slaves exchange information about the modules they have been assigned, optimize and generate

---

[5] See Khanna [15] for a taxonomy of approaches to parallel compilation.

code for their modules, and send the generated machine code and relocation information to the Servant, who (concurrently with the Slave's code generation) performs relocation. Once finished, the Slaves and the Servant return to their dormant state until next contacted by a Master.

The DPASTER runs in four main phases (see Figure 4), where the beginning of each phase serves as the primary *point of synchronization*; no Slave starts a phase until all other Slaves have finished the previous one. The rest of this section will present a detailed description of the DPASTER, one phase at a time. The algorithms presented assume that inter-process communication is realized by a Birrell and Nelson-type [3] remote procedure call (RPC) protocol, extended with facilities for multicast calls.

6.1 Phase 0

The initial phase awakens the Slaves and the Servant and relays to them global data pertinent to the session, including command line switches, environment variables, object file search paths, etc.

6.2 Phase 1

Finding the modules which are going to be part of the final program entails finding the closure of the module import graph with respect to the main module. During Phase 1 the Master repeatedly selects a module which has yet to be processed (starting with the main program module) and assigns it to a Slave. The assignment is based on an estimated processing time of each module, in an attempt to give the Slaves similar workloads. The Slave loads the module's Module Table, Expression Section, and Inline Section, and passes on the Binary Code, Data, and Relocation sections to the Servant. Finally, the Module Table is sent back to the Master, who uses it to find new unprocessed modules. The process terminates when all modules reachable from the main module have been processed.

6.3 Phase 2

During Phase 2 the Slaves cooperate in order to accomplish three tasks:

1. Evaluate all expressions and replicate the values at the Slaves who will need them during Phase 3.
2. Check deferred context conditions and report any violations back to the Master.
3. Expand inline calls in inline procedures and replicate the resulting code at the Slaves who will need it during Phase 3.

In the following we will concentrate on algorithms for expression evaluation (point 1 above), since the algorithms for point 2 and 3 are simple variants.

At the start of Phase 2 the expression graph has been partitioned (as a result of Phase 1) into s parts, one part for each Slave. The edges of the graph are either *internal* to the Slaves, connecting nodes residing on the same site, or *external*, connecting nodes on different sites. While the graph will most often be acyclic, illegal recursive declarations do introduce cycles.

Distributed expression evaluation can be approached in several different ways. The most intuitive method may be for each Slave to evaluate local subgraphs depth-first, and to retrieve the value of an external node by querying the Slave which owns the node. This method has three problems: (1) Each Slave has to know the owner of every node in the graph; (2) Cycles in the graph can lead to deadlock situations which either have to be avoided or detected; (3) The message complexity can be prohibitive (the number of messages is proportional to the number of external edges in the graph, which in turn grows with the number of Slaves). To reduce the message complexity we use a different algorithm which makes heavy use of multicast calls. It is a conceptually simple iterative method where each Slave evaluates as much as it can, distributes the new values to the other Slaves, and waits for the arrival of new values. The number of messages for this algorithm becomes proportional to the *height* of the graph formed by the external edges in the expression graph, rather than to the number of edges themselves, as in the previous algorithm. Our algorithm assumes the existence of a set of auxiliary boolean attributes of Table 1.

ATTRIBUTE	SEMANTICS	INITIAL VALUE
$e_{m,i}.known$	**TRUE** on Slave S_k if $e_{m,i}$ has a known value.	**TRUE** if $e_{m,i}$ was computed at compile-time.
$e_{m,i}.global$	**TRUE** if $e_{m,i}$ represents an exported entity.	Set at compile-time.
$e_{m,i}.referenced$	**TRUE** if $e_{m,i}$ represents a procedure reachable from the main module.	**TRUE** for main module bodies.

Table 1. Auxiliary attributes used by the expression evaluation algorithm.

The $e_{m,i}.referenced$ attribute is true for procedures which may actually be called at run-time. The $e_{m,i}.global$ attribute is used to avoid having to replicate expression values which are local to a module. During Phase 2 each Slave repeatedly executes the following algorithm:

1. Evaluate as many expressions as possible, i.e. all expressions whose arguments are known, and let E_{out} be the subset of these expressions which are global.
2. Compute R_{out}, the set of all global procedures $e_{m,i}$ reachable from other reachable procedures. I.e., $e_{m,i}$ is in R_{out}, iff for some $e_{n,j}$, $e_{m,i} \in e_{n,j}.calls$ and $e_{n,j}.referenced=$**TRUE**.

3. Multicast (E_{out}, R_{out}) to all other Slaves.

4. Wait for new incoming values. When a tuple (E_{in}, R_{in}) arrives from another Slave, store the incoming values of E_{in}, set $e_{n,j}.referenced=$**TRUE** for all $e_{n,j}$ in R_{in}, and set $e_{n,j}.known=$**TRUE** for all $e_{m,i}$ in E_{in} and R_{in}.

The algorithm terminates when no Slave can perform any more evaluations. There are two possible reasons for this: either all expressions have been evaluated, or there exists a cycle among the expressions. To differentiate between these situations, the Master employs a distributed termination detection algorithm which alerts it when all Slaves are idle. If at that point some Slaves still hold unevaluated expressions, the Master will know that the expressions contain a cycle and can issue the appropriate error message. The termination detection algorithm can be made particularly simple since all communication is in the form of multicast remote procedure calls:

1. A Slave is either *active* or *passive*. A Slave may enter a passive state only after having determined that all remote calls it has initiated have returned, all incoming calls have been returned, and that no further local processing is possible.

2. Let each Slave S_i maintain a count C_i of all multicast messages sent or received during the execution of the underlying algorithm.

3. Let the Master send a multicast remote procedure call **DETECT** to $S_1, \ldots, S_s$. On receipt of a **DETECT** call Slave S_i continues processing until a passive state is reached. At this point a reply is sent to the Master including the current value of C_i.

4. If all Slaves return the same value C, the Master can conclude that the underlying algorithm has terminated. Otherwise, the process is repeated with a new **DETECT** call.

Theorem 1. *The algorithm above detects termination iff the underlying computation has terminated.*

Proof. See [6, pp. 184–186]

The inline expansion part of Phase 2 uses an algorithm similar to the one just described for expression evaluation. One difference is that inline procedures (which may be large and expensive to transmit and store) are only replicated at the Slaves which actually need them, whereas all global expression values are always replicated at all Slaves.

The problem of distributed expression evaluation arises in two other areas related to language implementation, namely parallel attribute evaluation [16] and parallel evaluation of functional programs [12]. Since in general the expressions to be evaluated are known prior to evaluation, research in these areas emphasizes the development of algorithms which compute the optimal expression-to-site mapping. In our case, however, little is known about the structure of the expression graph before evaluation, and hence the expression-to-site mapping has to be done more or less at random. Our algorithm for expression evaluation is also reminiscent of *replicated blackboards* used in AI systems [7] and the parallel topological sorting algorithm of Er [8].

6.4 Phase 3

After Phase 2 all Slaves have local access to any inter-modular information they may need during Phase 3. This phase is therefore uncomplicated. Each Slave executes the following algorithm for each of the modules it owns:

1. Read the intermediate code of all referenced deferred procedures.
2. Update the code with the computed expression values and expand calls to inline procedures.
3. Optimize and generate machine code.
4. Send the code, relocation information, and new addresses of generated procedures to the Servant.

6.5 The Servant

It would be elegant if the Slaves could write their generated code directly to a shared global file. However, client file caching in network file systems makes concurrent access to global files problematic. Having a dedicated process handle all accesses to the executable file avoids this problem, and also makes it possible to perform relocations concurrently with code generation. In the current implementation the Servant process receives generated code and data, relocation information, and addresses of procedures and variables from the Slaves. Two lightweight processes (*Writer* and *Updater*) within the Servant operate concurrently on these data: *Writer* writes the code and data to the executable file, and *Updater* updates the file as relocation information and procedure addresses become available.

7 Evaluation

The current implementation of the ZUSE translation system runs on and produces code for Sun 3 workstations. For programs with a large amount of encapsulation and inlining the DPASTER achieves a speedup factor between 2.6 and 3.5.

Figure 5 shows the execution times (clock-time in number of seconds) for the sPASTER and the DPASTER, when binding five versions of the sPASTER itself. Table 2 gives the amount of encapsulation and inlining in each version. As a comparison the timing for the SUN Modula-2 linker *m2l* (a pre-linker which calls the UNIX systems linker *ld* as part of the linking process) for the same program is also given. Figure 5 also shows the results for four large artificially generated programs. Figure 6 shows how the speedup of the DPASTER varies with the amount of encapsulation and inlining.

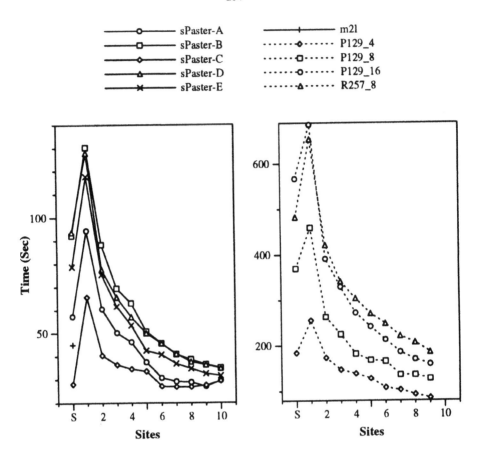

Fig. 5. On the left is shown the running time of the sPASTER (point S) and the DPASTER running on 1–10 sites when binding 5 versions of the sPASTER itself. The timing for the SUN Modula-2 linker m21 linking the sPASTER is also given. To the right is shown the total running time of the sPASTER and the DPASTER when binding 4 artificial programs.

8 Conclusions and Future Work

The main contributions of the ZUSE language and translating system is summarized by the following points:

Regular and flexible encapsulation ZUSE is more uniform than other similar languages in its ability to hide or reveal any part of any exported item.
Less dynamic allocation Since ZUSE's abstract types can be allocated statically there is less stress on the dynamic allocation system. Statically allocated abstract types also make it easy (compared to Modula-2 and Modula-3 which

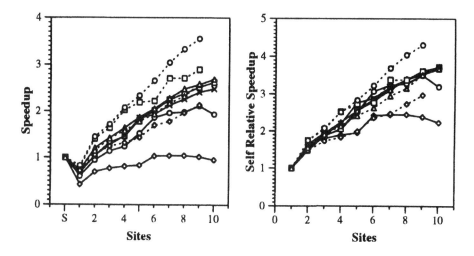

Fig. 6. Speedup of the DPASTER relative the SPASTER (left), and self-relative speedup of the DPASTER (right).

PROGRAM	M	LOC	S	P_t	T_a	C_a	I_a	I_2	I_3	I_c	P_d	S_d	E
sPaster-A	108	36	14	2024	28	34	0	0	0	0%	879	5	533
sPaster-B	108	36	14	2024	28	33	431	41	1325	55.6%	953	15	578
sPaster-C	108	36	14	2024	0	0	0	0	0	0%	0	0	55
sPaster-D	108	36	14	2024	0	0	431	41	1325	55.6%	760	10	99
sPaster-E	108	36	14	2024	16	27	202	10	724	34.8%	1024	15	497
P129_4	129	51	62	1905	471	448	653	133	1815	N/A	986	0	7542
P129_8	129	55	64	1905	471	448	653	268	3101	N/A	1129	0	8645
P129_16	129	56	64	1905	471	448	653	361	4040	N/A	1222	0	8360
P257_8	257	167	203	5158	203	800	775	647	2907	N/A	3889	0	13629

Table 2. Source code statistics regarding 5 versions of the SPASTER and 4 artificially generated programs. Abbreviations: M = Number of modules; LOC = Total size in thousands of lines of code. S = Total number of statements (in thousands). P_t = Total number of procedures. T_a, C_a, I_a = Number of abstract types, constants, and inline procedures; I_2, I_3 = Number of inline expansions performed during Phase 2 and 3; I_c = Percentage of dynamic calls expanded, for typical input. P_d = Number of referenced deferred procedures. S_d = Number of deferred context conditions. E = Number of constant expressions, excepting call-graph attributes.

only allow dynamically allocated abstract types) to implement abstract data types with value semantics.

No unnecessary recompilations Unlike Ada and C++, ZUSE's encapsulation and inter-modular inlining facilities do not create any unwarranted compilation dependencies, and hence will not generate any unnecessary recompilations.

Fast inter-modular optimization ZUSE's distributed module binder provides fast turn-around times even for large programs with much encapsulation and many inline calls.

No unwarranted run-time processing Unlike Modula-3 method lookup, object-type field accesses, and method template construction in ZUSE do not entail any extra run-time processing.

The inter-modular information discussed in this paper is of two different kinds: that which results from the use of encapsulation in modular and object-oriented programming languages and that which results from a desire to perform inter-modular optimizations. The distributed module binder presented here represents a framework in which both kinds of information can be computed at link-time, with little or no apparent extra translation-time cost compared to ·conventional link editing.

Our present work focuses on the use of better load balancing algorithms in the distributed binder. We are also investigating whether inter-procedural optimization techniques [20] other than inlining (such as inter-procedural register allocation [22] and constant propagation [4]) can be incorporated into our distributed binder. We are furthermore considering applying incremental techniques [19] to the distributed binder to further speed up binding when only small and local changes have been made between two consecutive binds.

References

1. Accredited Standards Committee X3, Information Processing Systems, The American National Standards Institute (ANSI), CBEMA, 311 First St. NW, Suite 500, Washington, DC 20001. *Draft Proposed American National Standard for Information Systems — Programming Language C++*, x3j16/91-0115 edition, September 1991.
2. Erik H. Baalbergen. Design and implementation of parallel make. *Computing Systems*, 1:135–158, 1988.
3. Andrew D. Birrell and Bruce Jay Nelson. Implementing remote procedure calls. *ACM Transactions on Computer Systems*, 2(1):39–59, February 1984.
4. David Callahan, Keith D. Cooper, Ken Kennedy, and Linda Torczon. Interprocedural constant propagation. In *Proceedings of the SIGPLAN '86 Symposium On Compiler Construction*. ACM, June 1986.
5. A. Celentano, P. Della Vigna, C. Ghezzi, and D. Mandrioli. Separate compilation and partial specification in Pascal. *IEEE Transactions on Software Engineering*, 6:320–328, July 1980.
6. Christian S. Collberg. *Flexible Encapsulation*. PhD thesis, Lund University, December 1992.

7. R. S. Engelmore and A. J. Morgan. *Blackboard Systems*, chapter 30. Addison-Wesley, 1988. ISBN 0-201-17431-6.

8. M. C. Er. A parallel computation approach to topological sorting. *The Computer Journal*, 26(4):293–295, 1983.

9. Michael B. Feldman. *Data Structures with Modula-2*. Prentice-Hall, 1988. ISBN 0-13-197666-4.

10. Charles M. Geschke, James H. Morris Jr., and Edwin H. Satterthwaite, Early experience with Mesa. *CACM*, 20(8):540–553, August 1977.

11. Mark Himelstein, Fred C. Chow, and Kevin Enderby. Cross-module optimizations: Its implementation and benefits. In *Proceedings of the Summer 1987 USENIX Conference*, pages 347–356, June 1987.

12. Simon L. Peyton Jones. *Parallel Graph Reduction*, chapter 24. Prentice-Hall, 1987. ISBN 0-13-453325-9.

13. Michael D. Junkin and David B. Wortman. The implementation of a concurrent compiler. Technical Report CSRI-235, Computer Systems Research Institute. University of Toronto, December 1990.

14. Howard P. Katseff. Using data partitioning to implement a parallel assembler. *SIGPLAN Notices*, 23(9):66–76, September 1988. ACM/SIGPLAN Parallel Programming: Experience with Applications, Languages, and Systems.

15. S. Khanna and A. Ghafoor. A data partitioning technique for parallel compilation. In *Proceedings of the Workshop on Parallel Compilation*, Kingston, Ontario, Canada, May 1990.

16. Eduard F. Klein. Attribute evaluation in parallel. In *Proceedings of the Workshop on Parallel Compilation*, Kingston, Ontario, Canada, May 1990.

17. Greg Nelson. *Systems Programming with Modula-3*. Prentice Hall, 1991. ISBN 0-13-590464-1.

18. Jukka Paakki, Anssi Karhinen, and Tomi Silander. Orthogonal type extensions and reductions. *SIGPLAN Notices*, 25(7):28–38, July 1990.

19. Russel W. Quong. *The Design and Implementation of an Incremental Linker*. PhD thesis, Stanford University, May 1989.

20. Stephen Richardson and Mahadevan Ganapathi. Code optimization across procedures. *Computer*, pages 42–49, February 1989.

21. M. W. Rogers, editor. *Ada: Language, Compilers and Bibliography*. Cambridge, 1984. ISBN 0-521-26464-2.

22. David W. Wall. Experience with a software-defined machine architecture. *ACM Transactions on Programming Languages and Systems*, 14(3):299–338, July 1992.

23. Niklaus Wirth. *Programming in Modula-2*. Springer Verlag, second edition, 1983.

24. Niklaus Wirth. Type extensions. *ACM Transactions on Programming Languages and Systems*, 10(2):204–214, April 1988.

Is Oberon as Simple as Possible? A Smaller Object-Oriented Language Based on the Concept of Module Type

Atanas Radenski

Department of Computer Science
Winston-Salem State University, P.O.Box 13027
Winston-Salem, North Carolina 27110, U.S.A.
E-mail: radenski@ecsvax.uncecs.edu

Abstract. The design of the programming language Oberon was led by the quote by Albert Einstein: 'make it as simple as possible, but not simpler'. The objective of this paper is to analyze some design solutions and propose alternatives which could both simplify and strengthen the language without making it simpler than possible.

The paper introduces one general concept, the module type, which can be used to represent records, modules, and eventually procedures. Type extension is redefined in terms of component nesting and incomplete designators. As a result, type extension supports multiple inheritance.

1 Introduction

The design of the programming language Oberon was led by the quote by Albert Einstein: 'make it as simple as possible, but not simpler'. The objective of this paper is to analyze some design solutions and propose alternatives which could both simplify and strengthen the language without making it simpler than possible.

The object orientation of Oberon is based on the concept of type extension. Section 2 of this paper outlines a problematic point in this concept as defined in Oberon: type extension applies to record and pointer types, but does not apply to procedure types. For this reason, procedures cannot be directly and conveniently redefined for extended types. As a consequence, method overriding may seem somewhat unnatural and tedious. This problematic point is eliminated with the concept of module type defined in Section 3. It is a generalization of record and procedure types and a single substitute for these types. As shown in Section 3, instances of module types can be used as record variables, or as procedures, or as Oberon modules. Overriding a method can be easily implemented by changing the module assigned to a field in an extension. Type extension itself is redefined in terms of component nesting and incomplete designators; as a result, it supports multiple inheritance.

Module types and type extension are integrated in an experimental object-oriented language that evolved from Oberon. The experimental language does not include record types, procedure types, procedures and modules, since all they are implemented by means of module types or module variables. The paper represents those features of the experimental language that are relevant to module types and type extension. The object orientation of this language is outlined in the end of Section 3.

2 The Need for Improvement

2.1 Type Extension as a Base of the Object Orientation of Oberon

Classes are implemented in Oberon as pointer types bound to record types with procedure variables. Objects are dynamic variables of such record types. For instance:

```
TYPE
  Class = POINTER TO ClassDesc;
  ClassDesc = RECORD
    x : INTEGER;
    method : PROCEDURE (self : Class; v : INTEGER);
  END;
VAR
  ptr : Class;
```

Note that *ptr.x* and *ptr.method* designate the fields *x* and *method* of the dynamic record variable *ptr^*.

Methods are implemented in Oberon as procedures. For example, a method may look like this:

```
PROCEDURE Method (self : Class; v : INTEGER);
BEGIN self.x := v END Method;
```

To create a new object, one has to assign specific procedures to all procedure variables:

```
NEW (ptr); ptr.method := Method;
```

Messages are calls of procedure variables, as, for instance:

```
ptr.method(ptr, 1);
```

Inheritance in Oberon is based on the concept of type extension [2, 3]. It permits the construction of new record types by adding fields to existing ones. For instance, type *SubclassDesc* extends type *ClassDesc* with the data field *y*:

```
TYPE
  Subclass = POINTER TO SubclassDesc;
  SubclassDesc = RECORD (ClassDesc)
    y : INTEGER
  END;
VAR
  subPtr : SubClass;
```

Type *SubclassDesc* is said to be a direct extension of type *ClassDesc*. Type *ClassDesc* is the direct base type of type *SubclassDesc*.

The fields of a record variable of an extended type can be referenced by usual field designators. For instance, *subPtr.x, subPtr.y, subPtr.method* are designators referencing the fields of the record variable *subPtr^*. A new object that belongs to *Subclass* can be created as follows:

NEW *(subPtr); subPtr.method := Method;*

An extended type is assignment compatible with its base type. For instance, the assignment *ptr^ := subPtr^* is legal and acts as a projection of record *subPtr^* onto record *ptr^*. The field *y* does not participate in the assignment. In contrary, the assignment *subPtr^ := ptr^* is illegal.

Type extension applies also to pointer types. By definition, the pointer type *Class* is extended by *Subclass* (see their declarations above), since the pointer base type *ClassDesc* of *Class* is extended by the pointer base type *SubclassDesc* of *Subclass*.

Since *subPtr* is an extension of *ptr*, the assignment *ptr := subPtr* is legal. After the assignment, *ptr* points to a dynamic variable of type *SubclassDesc*. After the assignment, *ptr* is said to be of dynamic type *Subclass*, while its declared (static) type continues to be *Class*. Thus, only *ptr.x* and *ptr.method* are accepted by the compiler as legal field designators. The field *y* can be referenced through *ptr* by means of a type guard, as illustrated by the following example:

ptr(Subclass).y := 0;

An attempt to execute the above statement when *ptr* does not actually point to a dynamic record of type *SubclassDesc* results in an abnormal halt. An abnormal halt can be prevented by a type test:

IF *ptr* IS *Subclass* THEN *ptr(Subclass).y := 0* END;

2.2 What is Problematic with Type Extension

Overriding a method in Oberon can be implemented by changing the procedure assigned to a field in an extension [1]. Unfortunately, procedures cannot be directly and conveniently redefined for extended types. For this reason, method overriding may seem somewhat unnatural and tedious. Consider, for example, the following procedure:

> **PROCEDURE** *OverridingMethod (self : Subclass; v : INTEGER);*
> **BEGIN**
>   *self.x := v; self.y := v*
> **END** *OverridingMethod;*

To override *Method* with *OverridingMethod*, one may wish to use the assignment *subPtr.method := OverridingMethod*. However, the definition of Oberon implies that *OverridingMethod* is not assignment compatible with *method*, and this assignment is not allowed.

More precisely, field method of *SubclassDesc* is inherited from *ClassDesc* and has the following procedure type:

> **PROCEDURE** *(self : Class; v : INTEGER)*

Besides, the heading of the newly created *OverridingMethod* is

> **PROCEDURE** *OverridingMethod (self : Subclass; v : INTEGER);*

The type of the formal parameter *self* of *OverridingMethod*, namely *Subclass*, is an extension of the type indicated in the declaration of *method*, namely *Class*. According to the definition of type extension, the type of *OverridingMethod* is not an extension of the type of *method*. Thus, *OverridingMethod* is not assignment compatible with *subPtr.method*.

The following implementation of *OverridingMethod* can be assigned to *subPtr.method*, since now *OverridingMethod* and *subPtr.method* have a single formal parameter of the same type:

> **PROCEDURE** *OverridingMethod (self : Class; v : INTEGER);*
> **BEGIN**
>   *self.x := v;*
>   **IF** *self* **IS** *SubClass* **THEN**
>   *self(SubClass).y := v;*
>   **END**
> **END** *OverridingMethod;*

Despite of the fact that the formal parameter of *OverridingMethod* is *Class*, it can and has to be called with actual parameters of type *SubClass*. By means of a type test and type guard, the overriding method treats the parameter as a variable of type *Subclass*. On the other end, the type of formal parameter of *OverridingMethod* is the same as that indicated for *subPtr.method*, and *OverridingMethod* can be assigned into *subPtr.method*. Such implementation of *OverridingMethod* seems somewhat unnatural and tedious.

3 Our Approach

A major problem with the object orientation of Oberon is that type extension applies to record and pointer types, but does not apply to procedure types. For this reason, methods cannot be directly and conveniently overridden for subclasses (see Section 2.2). The problem can be eliminated with the concept of module type defined in this section. Module types can be viewed as generalized record types. As shown in what follows, instances of module types can be used as record variables, or as procedures, or as Oberon modules. Overriding a method can be easily implemented by changing the module assigned to a field in an extension.

Module types and type extension are integrated in an experimental object-oriented language named K2 that evolved from Oberon. K2 does not include record types, procedure types, procedures and modules, since all they are implemented by means of module types or module variables. This section represents all features of the experimental language that are relevant to module types and type extension. The object orientation of this language is outlined in the end of the section.

3.1 Module Types

A *module type* consists of a definition, and optionally, a body. A *module definition* is a collection of declarations of constants, types, and variables. A *module body* is a collection of declarations, other bodies, and a sequence of statements. The statements are executed when the body is activated through a module call (Section 3.6). The definition of a global identifier and/or its body may include an import list (Section 3.7). A module type allows a body only if its definition contains a forward body declaration. Then a body can be declared within the same scope, or it can be left undefined (Section 3.4).

```
ModuleDefinition =
        "("[ImportList]
          DeclarationSequence
          [ForwardBodyDeclaration]
        ")"
DeclarationSequence = {declaration ";"}
declaration = ConstantDeclaration | TypeDeclaration | VariableDeclaration
ForwardBodyDeclaration = BODY
BodyDeclaration =
        BODY ident ";"
          [ImportList]
          DeclarationSequence
          BodySequence
        [BEGIN
          StatementSequence]
        END ident
BodySequence = {BodyDeclaration ";"}
```

Examples:

TYPE *Date* = *(day, month, year: INTEGER);*
TYPE *PersonalRecord* = *(*
  CONST *length* = *32;*
  TYPE *Name* = **ARRAY** *length* **OF** *CHAR;*
  name, firstName. Name,
  age: INTEGER
);

Constants, types and variables declared in a module definition are called *public components*, while those declared in the corresponding body are referred to as *local components*. Public components that are variables are also referred to as *parameters* (see Section also 3.6). Public types and constants are not parameters.

Example:

TYPE *Sample* = *(*
  publicVar: INTEGER;
  BODY
);
BODY *Sample;*
  localVar: INTEGER;
BEGIN *(* ... *)* **END** *Sample;*

The *scope* of an identifier which denotes a public component includes the module definition itself and the whole body, if any. Such an identifier is also visible within component designators. An identifier which declares a local component is not visible outside of the body that contains its declaration. Local variables keep their values between two successive calls of the body.

In addition to its public components and locally declared components, the entities declared in the environment of the body and its definition are also visible in the body. A local component hides non-local entities that have the same name. Hidden entities can still be referred to by component designators.

A variable declared in a module type definition can be followed by the read-only mark "-". Such a variable can be assigned values only from within the module body.

The identifier list of a variable declaration may contain the word *RESULT*. In this case, the type of the declared variable(s) can be neither a module type, nor an array type. Refer to Section 3.6 for the use of variables named *RESULT*.

Example:

TYPE *Log2* = *(*
  x: INTEGER;
  RESULT - : INTEGER;

BODY
);

3.2 Type Extension

A module type T_{ext} *directly extends* a module type T_{base} if T_{ext} has exactly one component of type T_{base}. T_{ext} *extends* a type T_{base} if it equals T_{base} or if it directly extends an extension of T_{base}.

Examples:

TYPE *Module1* = *(x : INTEGER);*
TYPE *Module2* = *(ancestor : Module1; y : INTEGER);*
TYPE *Module3* = *(ancestor : Module2; z : INTEGER);*

In the examples above, *Module3* directly extends *Module2* with component *z*. *Module3* is an indirect extension of *Module1*. *Module1* is a direct base type of *Module2* which is a direct base type of *Module3*. Nested components of *Module3* can be referenced by incomplete designators that do not contain the identifier *ancestor*, as explained below.

Components of module variables can be denoted by *incomplete designators* according to the following rules. It is said that *c* is a *nested component* of a module variable *m*, if *c* is a component of *m*, or *c* is a nested component of some component of *m*. Then, if the module variable *m* does not have a component *c*, then *m.c* designates a nested component of *m* determined by left-to-right level-order search among all nested components of *m*. If *p* designates a pointer, then *p.c* stands for $p\hat{}.c$ and *p[e]* stands for $p\hat{}[e]$ (that is, the dot and the opening bracket imply dereferencing).

Examples:

m3 : Module3;
m3.z
m3.y   (stands for *m3.ancestor.y*)
m3.x   (stands for *m3.ancestor.ancestor.x*)

3.3 Pointers

Variables of a *PointerType* assume as values pointers to variables of some *BaseType*. The *PointerType* is said to be bound to its *pointer BaseType*. Pointer types inherit the extension relation of their base types. A pointer type *P* bound to T_{base} is extended by any pointer type P_{ext} bound to an extension T_{ext} of T_{base}. For instance, type *Ptr3* extends type *Ptr1*, because *Module3* extends *Module1*:

```
TYPE Ptr1 = POINTER TO Module1;
TYPE Ptr3 = POINTER TO Module3;
p1 : Ptr1;        p3 : Ptr3;
```

The type with which a pointer variable is declared is called its *static type* (or simply its type). The type of the value assumed by a pointer variable at run time is called its *dynamic type*. The dynamic type of a pointer variable may be an extension of its static type (see examples in Section 3.5).

The *type guard PointerVariable(DynamicType)* asserts that the *PointerVariable* has the quoted *DynamicType*. If the assertion fails, the program execution is aborted, otherwise the *PointerVariable* is regarded as having the *DynamicType*. The guard is applicable only if the *DynamicType* is an extension of the static type of the *PointerVariable*.

The type test *v* IS *T* stands for "the dynamic type of *v* is *T*" and is called a *type test*. It is applicable if

(1) *T* is an extension of the declared type *T0* of *v*, and

(2) *v* is a pointer variable.

The monadic *address operator* "@" applies to an operand which is a variable of any type. The type of the result is a pointer to the operand's type. This operator is used to implement variable parameters (See an example in Section 3.6.)

Examples:

```
i        (INTEGER)              @i      (POINTER TO INTEGER)
```

3.4 Bodies for Module Variables

If a module type definition does not include a forward body declaration, variables of this type are not allowed to have bodies. If the definition does include a forward body declaration, two options exist.

First, let *T* be a module type for which a body *B* has been declared. The variable declaration *M ...: T* defines *B* as a body of *M*.

Second, let *T* be a module type which body has been left undefined. The declaration *M ...: T* does not define a body for *M*. An individual body B_m may be defined for *M* in the scope of *M*. In this way, module variables of the same type can have completely different bodies.

In all cases, a whole module assignment (Section 3.5) can be used to give a new value and a new body to a module variable.

Examples (refer to examples in Section 3.1):

```
log2: Log2;
BODY log2; (* assume x > 0 *)
BEGIN RESULT := 0;
  WHILE x > 1 DO x := x DIV 2; INC (RESULT) END;
END log2;
```

```
myLog2: Log2;
BODY myLog2; (* assume x > 0 *)
  y: INTEGER;
BEGIN RESULT := 0; y := 1;
  WHILE x > y DO ASH (y); INC (RESULT) END;
END myLog2;
```

3.5 Assignments

Assignments replace the current value of a variable by a new value specified by an expression. The expression must be assignment compatible with the variable. In particular, an expression e of type T_e is *assignment compatible* with a variable v of type T_v if:
- T_e and T_v are the same type, as specified below;
- T_e and T_v are pointer types and T_e is an extension of T_v;

Some less important cases of type compatibility (numeric types, strings, NIL and pointer types) need not to be discussed here.

T_a is the *same type* as T_b if:
- T_a and T_b are both denoted by the same type identifier, or
- T_a and T_b are denoted by type identifiers and T_a is declared to equal T_b in a declaration of the form TYPE $T_a = T_b$, or
- T_a and T_b are types of variables a and b which appear in the same identifier list in a variable declaration, provided T_a and T_b are not open arrays.

Note that module variables of the same type may have different bodies.

If an expression is assigned to a variable, the value of the variable becomes the same as the value of the expression. Besides:

(1) If the expression is of a module type, both its value and its body (if any) are assigned into the variable. If the body of the expression is undefined, the body of the variable becomes undefined.

(2) If the variable and the expression are of pointer types, the dynamic type of the variable becomes the same as the dynamic type of the expression.

Examples (refer to examples in Sections 3.3 and 3.4):

```
p1 := p3;      p1(Ptr3).z := 0;      log2 := myLog2;
```

Compared to Oberon, K2 offers a restricted form of assignment compatibility: In K2, an extended module type is not assignment compatible with its base type, while in Oberon an extended record type is assignment compatible with its base type.

3.6 Module calls

A *module call* consists of a module variable designator, followed by a (possibly empty) list of arguments. For the execution of the call, the arguments are assigned

(Section 3.5) to the parameters (Section 3.1), then the body of the module variable (if any) is executed. The association between the arguments and the parameters is positional, but the list of arguments may have less members than the total number of parameters. Module calls can appear as individual statements; they also can be used in expressions, as specified later in this section.

ModuleCall = designator "(" Arguments ")"

Examples:

Subroutine: (
  valuePar: INTEGER;
  variablePar: **POINTER TO** *INTEGER;*
  BODY
);
BODY *Subroutine;*
BEGIN *valuePar := valuePar + 1;*
  variablePar^ := variablePar^ + 1
END *Subroutine;*

i := 0; Subroutine(0, @i);   (* ... *)
Subroutine.valuePar := 0; Subroutine.variablePar := @i;
          Subroutine();   (* ... *)
Subroutine(); i := Subroutine.ValuePar + 1;

In an expression, a designator of a module variable which is not followed by an argument list refers to the current value of that variable. If it is followed by a (possibly empty) argument list, the designator implies the activation of the module body and stands for the value of the module variable resulting from the execution.
    A factor of the form

F(Arguments)

where *F* is a designator of a module variable which contains a component named *RESULT*, is evaluated as follows:
    (1) the module call *F(Arguments)* is executed first;
    (2) the value of *F.RESULT* is returned as value of *F(ARGUMENTS)*.

Example (refer to the examples in Section 3.4):

log2(k) + 1

If *designator* is a pointer variable with value NIL, the call *designator^(Arguments)* is executed as follows:
    (1) NEW(*designator*) allocates a dynamic module which is thereafter called

and executed;

(2) DISPOSE(*designator*) deallocates the dynamic module assigning NIL into designator.

An implementation may use a stack rather than a heap for such implicit module allocation/deallocation.

Example:

```
TYPE Factorial = (
  n: INTEGER; RESULT - : INTEGER;
  BODY
);
BODY Factorial;
  localFactorial: POINTER TO Factorial;
BEGIN
  IF n = 0 THEN RESULT := 1
  ELSE RESULT := n * localFactorial^(n - 1);
  END
END Factorial;
```

3.7 Compilation Units

A *compilation unit* is either a module type declaration eventually followed by a body, or a module variable declaration eventually followed by a body.

```
CompilationUnit =
        TypeDeclaration [";"BodyDeclaration]
      | VariableDeclaration [";"BodyDeclaration]
```

A compilation unit declares a single global identifier which is exported by the declaring unit. The exported identifier can be imported and used by other compilation units by means of an *import list* (see also Section 3.1).

```
ImportList = IMPORT ident [":="ident] {","ident [":="ident]}";"
```

Each identifier *I* from the import list of a module definition can be used in the definition itself, and in the type's body, if the type has a body. It the import list belongs to a module body, *I* can only be used in the body. If the form *I1* := *I* is used in the import list, then the imported entity is referred as *I1* rather than *I*.

A main program can be implemented as a compilation unit which consists of a module variable declaration and a body. A conventional module (or a package) is a also a compilation unit consisting of a module variable declaration plus eventually a body. A separately compiled class is a compilation unit which consists of a module type declaration and, in most cases, a body.

Examples:

```
TYPE ClassDesc = (
 TYPE Class = POINTER TO ClassDesc;
 x : INTEGER;
 method : (v : INTEGER; BODY);
 BODY
);

BODY ClassDesc;
 BODY method;
 BEGIN
  x := v;
 END method;
END ClassDesc;

MainProgram: (BODY);
BODY MainProgram;
 IMPORT ClassDesc;
 ptr: ClassDesc.Class;
 (* ... *)
BEGIN  (* MainProgram *)
 NEW (ptr);    ptr.method (1);
 (* ... *)
END MainProgram;
```

3.8 Module Types and Object Orientation

In K2, a pointer type bound to a module type represents a class (see *Class* and *ClassDesc* in Section 3.7). A variable (such as *ptr^*) of that module type is an object. A module component of that module type is a method. A call of a module component (such as *ptr.method(1)*) is a message.

Type extension implements inheritance in K2. For instance, *SubclassDesc* inherits field *x* from *ClassDesc* extending *ClassDesc* with a field *y*:

```
TYPE SubclassDesc = (
 IMPORT ClassDesc;
 superclass : ClassDesc;
 y : INTEGER;
 BODY
);
```

A module variable declared in the body of an extension can be used to override an inherited method:

```
BODY SubclassDesc;
  overridingMethod : (v : INTEGER; BODY);
  BODY overridingMethod;
  BEGIN x := v; y := v END;
BEGIN (* SubclassDesc *)
  superclass.method := overridingMethod;
END SubclassDesc;

subPtr : POINTER TO SubclassDesc;
NEW (subPtr); subPtr^();
```

The module call *subPtr^()* executes the assignment *superclass.method :=*
overridingMethod from the body of *SubclassDesc*. This assignment overrides (in
*subPtr^*) the method inherited from *ClassDesc*. Thus, overriding a method is simply
a module variable assignment. The difficulty with Oberon outlined in Section 2.2
does not exist in K2.

Note finally that the fields of an extension can be referred by incomplete
designators. For instance:

subPtr.x	(stands for *subPtr.superclass.x*)
subPtr.method	(stands for *subPtr.superclass.method*)

4 Conclusion

A problematic point in Oberon is that procedure fields of records cannot be directly
and conveniently redefined for extensions. From a standard object-oriented point of
view, method overriding in Oberon may seem unnatural and tedious (see Section
2.2). To cure this problem, Oberon-2 [4] extends Oberon with the new concept of
type bound procedures. Besides, Oberon-2 adds to Oberon open array variables, FOR
loops, and read-only export of data. (Object Oberon [5] is an experimental
predecessor of Oberon-2.) In fact, Oberon-2 implants the standard concept of method
in Oberon. The resulting language is not so simple and clean as Oberon was intended
to be. In particular, it supports too many different structures related to procedures:
type bound procedures, traditional constant procedures, procedure types, and
procedure variables.

K2 evolved from Oberon by introducing only one new feature, the module
type. Grace to the generality of the new concept, several features of Oberon were
eliminated. Namely, K2 does not contain record and procedure types (because they
are special kinds of module types), and does not need procedures and modules
(because they are modeled by module variables). While record extension is supported
by a specially designated language feature in Oberon, it is simply achieved by module
nesting and use of incomplete module component designators in K2.

The body of a K2 module that is a component of a larger module has access
to the components of the enclosing module. Thus, syntactical binding is as simple as

module nesting, and there is no need for a special concept such as the type-bound procedure of Oberon-2.

One more advantage of K2 compared to Oberon is that a module type that implements a class can be compiled separately and need not be enclosed in a package or Oberon module.

Most features of K2 have been tested by an experimental compiler implemented as a Turbo Pascal 6.0 program of about 3000 lines. A K2 compilation unit (a module type or variable declaration, eventually followed by a body) is translated into a Turbo Pascal unit; then this unit is compiled by the Turbo Pascal compiler. The K2 compiler extracts all constant and type declarations from module definitions and generates Turbo Pascal representations for those declarations. Module definitions are compiled into record types. Turbo Pascal objects are not used in the implementation. At present, type tests and type guards are not supported by the experimental compiler.

This paper describes an approach to the design of a small and simple, yet practically convincing object-oriented language. Our approach can be characterized as *simplicity through generality*. While we present a solution, we do not consider it as a final one. The absence of procedures as a special language feature and their implementation by means of module variables is a point that is widely open for criticism. Although a pointer variable of a module base type can be used as a conventional procedure (as illustrated in Section 3.6), programmers may wish to have procedures explicitly included in the language. Fortunately, our solution can be relatively easily modified to include procedures, while merging record types and modules in the same concept. A careful evaluation of this alternative is a subject of future work.

References

1. M. Reiser, N. Wirth: Programming in Oberon. Steps beyond Pascal and Modula. Wokingham: Addison-Wesley 1992

2. N. Wirth: The Programming Language Oberon. *Software - Practice and Experience* 18, 671-690 (1988)

3. N. Wirth: Type Extensions. *ACM Transactions on Programming Languages and Systems* 10, 204-214 (1987)

4. H. Moessenboeck, J. Templ: Object Oberon - A Modest Object-Oriented Language. *Structured Programming* 10, 44-46 (1989)

5. H. Moessenboeck: The Programming Language Oberon-2 Report. Computer Science Report 160, ETH Zurich 1991

Appendix: Syntax Description

declaration = ConstantDeclaration | TypeDeclaration | VariableDeclaration
ConstantDeclaration = CONST ident "=" ConstExpr
TypeDeclaration = TYPE ident "=" type "
type = ArrayDefinition | ModuleDefinition | PointerDefinition | TypeDesignator
TypeDesignator = qualident
qualident = {ident "."} ident
ArrayDefinition = ARRAY [ConstExpr {"," ConstExpr}] OF type
ModuleDefinition = "(" [ImportList] DeclarationSequence [BODY] ")"
ImportList = IMPORT ident [":=" ident] {","ident [":=" ident]}";"
DeclarationSequence = {declaration ";"}
BodyDeclaration =
      BODY ident ";" [ImportList] DeclarationSequence BodySequence
      [BEGIN StatementSequence] END ident
BodySequence = {BodyDeclaration ";"}
PointerDefinition = POINTER TO Type
VariableDeclaration = ident["-"] ["," ident["-"]] ":" type
expression = SimpleExpression [relation SimpleExpression]
relation = "=" | "#" | "<" | "<=" | ">" | ">=" | IN | IS
SimpleExpression = ["+" | "-"] term {AddOperator term}
AddOperator = "+" | "-" | "OR"
term = factor {MulOperator factor}
MulOperator = "*" | "/" | DIV | MOD | "&"
factor = number | CharConstant | string | NIL | set | "~" factor | "@" designator
      | designator ["("[ExprList]")"] | "("expression")"
designator = qualident {"."ident | "[" ExprList "]" | "("TypeDesignator")" | "^"}
set = "{" [element {"," element}] "}"
element = expression [".." expression]
statement = [assignment | ModuleCall | IfStatement | CaseStatement
      | WhileStatement | RepeatStatement LoopStatement | WithStatement
      | EXIT | RETURN]
assignment = designator ":=" expression
ModuleCall = designator "(" [ExprList] ")"
IfStatement = IF expression THEN StatementSequence {ELSIF expression
      THEN StatementSequence} [ELSE StatementSequence] END
CaseStatement =
      CASE expression OF case {"|"case} [ELSE StatementSequence] END
case = [CaseLabels {"," CaseLabels} ":" StatementSequence]
Caselabels = ConstExpr [".." ConstExpr]
WhileStatement = WHILE expression DO StatementSequence END
RepeatStatement = REPEAT StatementSequence UNTIL expression
LoopStatement = LOOP StatementSequence END
WithStatement = WITH qualident ":" typeDesignator DO StatementSequence END
CompilationUnit = TypeDeclaration [";"BodyDeclaration]
      | VariableDeclaration [";"BodyDeclaration]

On the Essence of Oberon

David A. Naumann

Southwestern University, Georgetown, TX 78626 U.S.A.

Abstract. Reynolds described the "essence of Algol" as the simple imperative language combined with the typed lambda calculus. We provide a similar description of Wirth's language Oberon as the simple imperative language combined with procedure types and record extension. Whereas the semantics of Algol has been given in terms of a (domain theoretic) model using an explicit representation of storage, our semantics uses predicate transformers; this is possible thanks to recent advances in the theory of predicate transformers. Predicate transformer semantics connects one of the most successful methods of rigorous program development with one of the most successful pragmatically-designed programming languages.

1 Introduction

In a seminal paper [18], Reynolds argued that the essence of Algol-60 is: (a) the simple imperative language as the basis for (b) call-by-name typed lambda calculus. Of Reynolds' other principles, the only one relevant here is: (c) storable values are distinguished from commands (and meanings of other "phrase types"). This analysis has been substantiated by elegant denotational models [18, 20, 15] and by the recent result [21] that computation in such a language can be separated into two phases: elimination of procedures (using the copy rule) and then execution of the procedureless program. It seems that the call-by-name reduction strategy is essential to both denotational and operational models, so they are not easily adapted to two essential features of Oberon [22, 16]: call-by-value (and result) parameter passing, and stored procedures (i.e. procedure variables). Moreover, implementations of languages like Oberon do not separate computations into two phases; the copy rule provides the correctness criterion but *not* the method for implementation. This paper is a preliminary report on an analysis of "theoretical Oberon" which justifies the design of Oberon from the point of view of systematic software development methods.

Our claim is that the essence of Oberon is characterized by the following.

1. The simple imperative language —assignments, local variables, and control constructs— is given a rich type system with procedure types and limited subtyping. In contrast with (c), procedures are a data type and can be stored as well as passed as parameters.
2. The call-by-name lambda calculus —the copy rule— accounts for the binding of names to constant values of all kinds, both "constant declarations" and procedure definitions.

3. Parameters are passed by value and result, not by name. Parameter passing is derived from the other imperative features, rather than from the lambda calculus.

These claims are substantiated by our semantics of a language we call "theoretical Oberon". The semantics interprets programs as predicate transformers; not only does this justify our claims, it also shows the close link between Oberon and standard methods of program development [4, 8]. In contrast, call-by-name is not well suited to these methods (without significant modifications, as for example in Reynolds' Specification Logic [17, 19]). The book *Programming in Oberon* [16] uses Hoare triples to present the formal semantics of assignment and control constructs; our purpose is to show that the entire language fits well with programming calculus. We do so by embedding Oberon in a language unencumbered by implementation-oriented restrictions. (Those restrictions are judiciously chosen, but not relevant to our analysis.)

Recent progress in the theory of predicate transformers is the technical foundation for our claims, which involve type structure including higher types [12, 9, 13]. Predicate transformers are well suited to modeling imperative languages, because the notion of state is explicated without recourse to an explicit operational model of the store as is used in the Algol models of Reynolds, et al [18, 20]. Using an explicit store, it is difficult to avoid making spurious distinctions between programs; abstract semantics for Algol-like languages remains an open problem [15].

The rest of the paper is organized as follows: Section 2 contrasts Wirth's language Oberon with theoretical Oberon (unqualified, "Oberon" always means the former). Section 3 discusses the specification and development of Oberon programs. Section 4 develops the predicate transformer semantics of theoretical Oberon. Section 5 explores the most significant new problem that arises in the predicate transformer semantics, namely program types.

2   Theoretical Oberon

Oberon differs from theoretical Oberon in several ways:

- the data types include pointer types
- expressions can have side effects
- it includes a notion of module for fine control on the scope of names, to encapsulate implementations of abstract objects
- it includes function procedures
- it restricts the use of some features

Pointers. We consider pointers to be a low-level feature which is usually used in very disciplined ways, e.g. to represent recursively defined types and unique identities of objects. We are not aware of programming methods with substantial general treatments of pointers; they are usually reduced to arrays, as they are in standard denotational semantics. Since we want to view Oberon as a subset

315

of a design language and programming calculus, we prefer to omit pointers. In particular, we view reference parameters as an implementation of value/result parameters; it is the latter that are treated here. For record types, the relation between type extension and assignment is treated the way Oberon treats pointers to records (see subsection 3.2).

Side effects. Reasoning about expressions with side effects is notationally complicated, and often best avoided. To simplify the formal semantics, we assume expressions are everywhere defined and have no side effects.

Modules. Oberon modules seem to be an orthogonal notion applicable to quite different kinds of languages. More to the point, we have nothing significant to contribute here on the topic of modules. They could be added, without essential change, to theoretical Oberon, although one might choose to decompose them into more primitive hiding and renaming operators. The programming method associated with modules is data refinement. Using predicate transformer semantics, the constructs of theoretical Oberon have all been shown to preserve data refinement [7, 5, 14], i.e. data refinement need only be done explicitly for those primitive subprograms that involve the data whose representation is changed (which are usually bundled together in a module). In contrast, call-by-name parameter passing does not preserve data refinement in general.

Function procedures. Function procedures in Oberon may have side effects, which we do not allow. Function procedures without side effects fit nicely with program derivation; since they pose no interesting difficulties, we omit them.

Restrictions. The term "restriction" should suggest that the language is a subset of a richer language suitable for program designs. For example, theoretical Oberon includes abstract types like exact reals and arbitrary sets that do not have efficient general implementations; it also has procedure-denoting expressions besides procedure identifiers. The restrictions of Oberon are a great engineering success: on the one hand facilitating systematic programming methods, and on the other hand facilitating implementation with efficiency significantly exceeding that of rival imperative languages like ML and call-by-name Algol.

In order to simplify the semantics presented here, we do adapt one restriction from Oberon: global variables of procedures assigned to variables must be declared in the global scope.

One way in which the restrictions of Oberon are relevant to program correctness is that they make compilation relatively simple, so writing a correct compiler is tractable. Indeed, Hoare, He, and Sampaio have shown how to derive a compiler for a related language [6].

Syntax For our expository purpose, it is appropriate to use notations more in the style of research literature than practical programming. In particular, scoped constructs such as local variable declarations are written using prefix keywords with scope delimited by parentheses rather than postfix keywords or special

brackets. The localization of variable x to program s, for example, is written
($\textbf{var}\,x\!:\!\textbf{int}\,\textbf{in}\,s$) rather than $\textbf{var}\,x\!:\!\textbf{int}$; $\textbf{begin}$ s $\textbf{end}$ as in Oberon or $\{\textbf{int}\ x\ ;\ s\}$
as in C. The call of a procedure variable or constant p is written "$\textbf{call}\,p$" instead
of just "p", for clarity.

Notations for derived programming idioms (e.g. iteration and Dijkstra's
guarded commands) are omitted, in spite of their importance; on the other hand,
the binding constructs (con, var, let, param, and rec) are not decomposed into
special constants applied to lambda abstractions (as they would be in a more
formal treatment). The syntax of theoretical Oberon is given in Table 1. Letters p,q,s range over program texts, w,x,y,r,pv over variables (in program
texts), P,Q over identifiers, and b,e over expressions (in which any and all
mathematical data types and operations are allowed).

Table 1. Syntax of programs in theoretical Oberon.

$\textbf{abort}$	divergence
$x := e$	assignment
$\textbf{call}\,m(e,w)$	call procedure m with actuals e,w
$[\varphi \,,\, \psi]$	prescription (specification)
($\textbf{con}\,X\,\textbf{in}\,p$)	specification constants
($\textbf{var}\,x\!:\!T\,\textbf{in}\,p$)	local variable of type T
($\textbf{let}\,Q\,\textbf{be}\,p\,\textbf{in}\,q$)	definition of constants
$p\ ;\ q$	sequential composition
($\textbf{if}\,b\,\textbf{then}\,p\,\textbf{else}\,q$)	alternatives
($\textbf{rec}\,Q\,\textbf{in}\,p$)	recursively defined program

Expressions m of type procedure are: variables, constant identifiers, and
parameterized programs of the form ($\textbf{param}\,(x\!:\!T,\textbf{var}\,y\!:\!U)\,p$). Parameter(s)
x are passed by value, y by value/result. An Oberon program like

$$\textbf{procedure}\ Q(x\!:\!T,\textbf{var}\,y\!:\!U);\textbf{begin}\ body\ \textbf{end}\ Q;\ldots Q(e,w)\ldots$$

is written

$$(\textbf{let}\,Q\,\textbf{be}\,(\textbf{param}\,(x\!:\!T,\textbf{var}\,y\!:\!U)\,body)\,\textbf{in}\,\ldots\textbf{call}\,Q(e,w)\ldots)\ .$$

3 Specification and Development of Sequential Programs

Programming begins with specifications —descriptions unfettered by concerns
of efficiency or even feasibility— and ends with code —efficient programs in a
language like Oberon that can be automatically translated to machine instructions. At intermediate steps, the programmer works with designs structured as
programs with specified but unimplemented parts. The constructs of Oberon are
well suited to programming because they serve well in the structuring of designs

but are also efficiently executable when their constituent parts are code rather than specifications.

This section reviews these ideas in some detail, to fix notation and discuss extension and procedure type features that distinguish Oberon from the notations hitherto used in programming methodology.

3.1 At the Predicate Level: Specifications and Assertions

The hallmark of imperative programming is *state*. Oberon is primarily for programming "state-transformation" programs, those that are specified by: the state space in which the program is to act, the precondition, and the postcondition. To specify relations between initial and final states, the state space is augmented with additional coordinates called *specification constants* (sometimes called "logical variables"). The assertion that program p meets such a specification is usually written as a Hoare triple

$$\{\varphi.X\}\, p \,\{\psi.X\} \tag{1}$$

where X is the specification constant, φ and ψ are state predicates. Formula (1) means: for any value of X, if p is started in a state satsifying $\varphi.X$ then it terminates, and the final state satisfies $\psi.X$.

The programming task is: given φ and ψ, find p satisfying (1). Thus the central objects of interest are state predicates and programs. For the programmer, a useful way to specify the semantics of the programming language is in the style of Hoare logic: rules for assertions of the form (1). In *Programming in Oberon*, for example, one finds the rule for sequential composition:

$$\frac{\{\varphi\}\, p \,\{\theta\} \qquad \{\theta\}\, q \,\{\psi\}}{\{\varphi\}\, p \,;\, q \,\{\psi\}}$$

This rule may be used from the bottom up: to meet the specification (φ,ψ) by the seqential composition of program p with q, it suffices to find an intermediate predicate θ and separately meet specification (φ,θ) with p and (θ,ψ) with q.

The axiom for assignment is striking in that it allows the determination of the precondition from the postcondition, by substitution:

$$\{\psi[x := e]\}\, x := e \,\{\psi\}$$

(Here $\psi[x := e]$ is the predicate ψ with every free occurrence of x replaced by e.) This suggests an alternate way to specify program meanings [4]. For each program p is given a function $[p]$ —usually written $wp.p$— from predicates to predicates, such that for each ψ, $[p]\psi$ is the weakest predicate φ such that (1) holds. Such a function is called a predicate transformer. Provided that the language of predicates is sufficiently expressive (e.g. [1]), Hoare triples and predicate transformers are equivalent: each can be defined from the other, by the correspondence

$$\{\varphi\}\, p \,\{\psi\} \qquad \text{iff} \qquad \varphi \Rightarrow [p]\psi$$

($\Rrightarrow$ means "implies in all states") or more precisely

$$\{\varphi.X\}\, p\, \{\psi.X\} \qquad \text{iff} \qquad (\forall X :: \varphi.X \Rrightarrow [\![p]\!](\psi.X))\ . \qquad (2)$$

Aside from mathematical elegance, predicate transformers have two significant advantages. The first is that they allow derivation of both state predicates (as in the intermediate θ needed above for sequential composition) and program expressions (in conditionals and assignments [8]). The second advantage is that predicate transformers can be used to give simple meanings to impractical program constructs such as unbounded nondeterminacy, angelic nondeterminacy, and even specifications themselves. This is the topic of subsection 3.4. The next two subsections consider assignments in light of the rich type system of Oberon.

3.2   Type Extension

A record type T' may be defined by extension from a record type T, i.e. type T' adds fields to the fields of T. We write $T' \Subset T$ for this direct extension, and $\Subset$ is also used for *type extension*: the reflexive transitive closure of direct extension. A record r' of type T' determines a projection of type T, which might be the intended value given to a variable r of type T in an assignment $r := r'$. Indeed, that is its meaning for records in Oberon. Projection is not, however the most useful interpretation, nor is it theoretical Oberon's (although projection is expressible). Instead, the effect of $r := r'$ is that r —though statically declared as type T— takes on a value of "dynamic type" T' that includes all of the fields of r' (just as for pointers to records and **var** parameters in Oberon). This implies the following interpretation of types: the values of a type T are all those entities of type T', for all extensions $T' \Subset T$. This principle applies to subtypes in general, not just to type extension for records. It can be formulated as a type subsumption rule: for all x of type T', if $T' \Subset T$ then x is also of type T. But subsumption does not fit well with Oberon's notation for static typing, in which each identifier has a single type. So instead of using a subsumption rule, the notion is expressed in terms of "assignment compatibility": an assignment $x := e$ is allowed if x is of type T and e is of type T' with $T' \Subset T$. Compatibility for parameters is derived from this rule (in subsection 4.2).

The term "dynamic type" refers to the fact that statically r may have type T, yet after the assignment $r := r'$ it has all the fields of type T'. A state predicate "r is T'" is needed both for program guards and for reasoning about dynamic types. A "type guard" is used in expression $r(T')$ to coerce the static type of r to be T'.

In addition to partially ordering types, extension also determines a partial ordering on values of type T, to wit: $r \preceq r'$ iff for each field f of r, f is also a field of r' and $r'.f = r.f$ (or rather $r'.f \preceq r.f$ if f is of record type). This in turn gives rise to an ordering on expressions, with respect to which all operators are monotonic because field selectors are. The relation $\preceq$ may be interpreted as an information ordering: if $r \preceq r'$ then r' has at least as much information as r. A predicate φ is called *monotonic in* r just if

$$r \preceq r' \Rrightarrow (\varphi.r \Rrightarrow \varphi.r')$$

for all r, r'. Given the interpretation of $\preceq$, sensible postconditions are monotonic: if certain information is desired, additional information is harmless and can be ignored. There are similar considerations for procedure types and the refinement ordering.

3.3 Procedure Types

In this subsection we consider parameterless procedures. The meaning of the call of a procedure constant is given by the copy rule, which says that the program

$$(\textbf{let } Q \textbf{ be } body \textbf{ in } \ldots \textbf{call } Q \ldots) \tag{3}$$

which calls declared procedure Q is the same as the program

$$\ldots body \ldots$$

in which calls of Q have been replaced by the body of Q. This justifies various rules for reasoning about procedure calls, along the following lines:

$$\frac{\{\varphi\}body\{\psi\} \qquad (Q \text{ is declared by (3)})}{\{\varphi\} \textbf{ call } Q \{\psi\}} \tag{4}$$

Rules of this form highlight an essential purpose of procedural abstraction: deriving one part of a program separately from its uses.

In contrast, there is no body for a procedure variable pv, because it gets its value by assignment. We are led to ask: what justifies an assertion

$$\{\varphi\} \textbf{ call } pv \{\psi\} \ ? \tag{5}$$

Operationally, **call** pv means to execute the program that is the current value of pv. Thus (5) holds just if that value satisfies (φ, ψ). Since the value of pv may change, assertions about it depend on the state. We are led to use the Hoare triple $\{\varphi\}pv\{\psi\}$ as a state predicate in a putative axiom for execution of procedure variables:

$$\{ (\{\varphi\}pv\{\psi\}) \wedge \varphi \} \textbf{ call } pv \{ \psi \} \tag{6}$$

There is a risk of confusion due to the implicit quantifications in the interpretation of Hoare triples (see (2)). We prefer the notations provided by refinement algebra. Before embarking in that direction, here is the companion axiom that is needed to establish preconditions like that in the preceding axiom:

$$\{ \{\varphi\}s\{\psi\} \} pv := s \{ \{\varphi\}pv\{\psi\} \} \tag{7}$$

Of course this is just an instance of the assignment axiom. Typically s in (7) is a procedure constant and one uses rule (4) to show that the precondition in (7) is everywhere true. Note, however, that we extend Oberon to allow procedure expressions s to be of the form $(\textbf{param}\,(x{:}T)\,q)$ where q is a program (like in Algol-W).

These simple axioms are the gist of what is needed to reason about procedure variables. In order to give correct formal rules, however, we must clarify the interpretations of state space coordinates apearing in s, φ, and ψ because these expressions play unusual roles in axioms (6) and (7).

3.4 At the Program Level: Refinement Algebra

Complications of nested Hoare triples can be dealt with by eliminating Hoare triples in favor of refinement. This has other benefits in program development that we do not explore here [2, 11].

Since the only relevant properties of programs are those specifiable by Hoare triples, it is natural to say that program q *refines* program p just if $\{\varphi\}p\{\psi\}$ implies $\{\varphi\}q\{\psi\}$, for all (φ, ψ). This is equivalent to

$$(\forall \psi :: [p]\psi \Rightarrow [q]\psi) ,$$

which we take to be the definition of $p \sqsubseteq q$.

Next we postulate an "imaginary program" $[\varphi , \psi]$, called a *prescription*, with the property that for any program p

$$\{\varphi\} p \{\psi\} \quad \text{iff} \quad [\varphi , \psi] \sqsubseteq p . \tag{8}$$

Given such a program —and henceforth we omit the qualification "imaginary"— we have no need for Hoare triples. The programming problem becomes: given (φ, ψ), find p such that $[\varphi , \psi] \sqsubseteq p$. Moreover, stepwise refinement can be precisely described: it boils down to the transitivity of $\sqsubseteq$. All of the Hoare logic rules [1] can be expressed in terms of refinement and prescriptions; for example the rule of consequence becomes

$$(\varphi' \Rightarrow \varphi) \wedge (\psi \Rightarrow \psi') \Rightarrow [\varphi' , \psi'] \sqsubseteq [\varphi , \psi] ,$$

which illustrates the way logical rules may be abandoned in favor of ordinary mathematical reasoning. This style may be called *refinement algebra*, because it emphasizes inequational reasoning with $\sqsubseteq$.

We write $(\mathbf{con}\, X \,\mathbf{in}\, p)$ to declare specification constant X with scope p. A typical case appears in this generalization (or clarification!) of (8):

$$\{\varphi.X\} p \{\psi.X\} \quad \text{iff} \quad (\mathbf{con}\, X \,\mathbf{in}\, [\varphi.X , \psi.X]) \sqsubseteq p . \tag{9}$$

Consider procedure variable pv. The formula

$$(\mathbf{con}\, X \,\mathbf{in}\, [\varphi.X , \psi.X]) \sqsubseteq pv \tag{10}$$

is like the right side of (9) except that pv is not a program but rather a program variable. Formulae like (10) can be used as state predicates (where φ, ψ have no specification constants besides X, and pv is not free on the left side): (10) depends only on the value of pv, and it is true in just the states where the value of pv satisfies the specification (φ, ψ). Ignoring specification constants, axiom (6) may be restated as:

$$\{ \,([\varphi , \psi] \sqsubseteq pv) \wedge \varphi \,\} \,\mathbf{call}\, pv \,\{ \,\psi \,\} . \tag{11}$$

The key to defining the semantics of $\mathbf{call}\, pv$ is to note that (11) holds for all predicates φ, ψ over the state space of pv. Quantifying over them gives

$$(\mathbf{con}\, \varphi, \psi \,\mathbf{in}\, [\,([\varphi , \psi] \sqsubseteq pv) \wedge \varphi , \psi \,]) \sqsubseteq \mathbf{call}\, pv .$$

Our semantics of **call** pv amounts to strengthening this to an equality (and treating specification constants properly; in fact section 4 uses a quite different formulation). Note that types are an essential ingredient: the quantified φ and ψ are predicates over the state space of pv, not over the state space of the program of which **call** pv is a constituent — which includes pv as one of its coordinates.

Axiom (7) can be written

$$\{\,[\varphi\,,\psi]\sqsubseteq s\,\}\,pv := s\,\{\,[\varphi\,,\psi]\sqsubseteq pv\,\}\ .$$

It is fortunate that the postcondition has this form. For stepwise refinement to be sound, program constructs should be monotonic with respect to refinement. That means if $p\sqsubseteq q$ and $C[-]$ is a context in which p can appear, then $C[p]\sqsubseteq C[q]$. The control constructs, recursion, and local variables are all monotonic (hence so are Oberon's parameter passing constructs; call-by-name is not monotonic, due to aliasing [10, 17][1]). An assignment $pv := s$ to procedure variable pv is a context with program s as constituent. Suppose we allow equality[2] as a predicate on program variables: the assignment axiom gives $\{true\}\,pv := s\,\{pv = s\}$, but if $s\sqsubseteq s'$ is a proper refinement (not equality) then it is not the case that $\{true\}\,pv := s'\,\{pv = s\}$; hence $pv := s$ does not refine to $pv := s'$. Thus assignment to procedure variables is not a monotonic program construct.

The monotonicity problem is solved as follows. In program development, prescriptions like $[true\,,\,s\sqsubseteq pv]$, briefly "$pv :\sqsupseteq s$", should be used in place of $pv := s$. Then the triple

$$\{true\}\,pv :\sqsupseteq s\,\{pv = s\}$$

fails to hold, because the program only establishes the postcondition $s\sqsubseteq pv$. Fortunately, the postconditions of interest have precisely this form (e.g. (7)). Now observe that for postconditions that are monotonic in pv, the programs $pv := s$ and $pv :\sqsupseteq s$ are equivalent. And $s\sqsubseteq pv$ is monotonic in pv, unlike $s = pv$. So in practice the prescription may always be refined to the assignment.

The same issues arise with other ordered data types, if one interprets the order relation as allowing improvement of the program involving expression e by replacing e with e' such that $e\preceq e'$. The solution is the same.

4 Predicate Transformer Semantics of Theoretical Oberon

It remains to interpret the language in a model that validates the reasoning of section 3.

---

[1] In theoretical Oberon, aliasing only arises with explicit multiple assignments like $x, x := e, e'$ (and analogous actual value/result parameters), because we treat assignable data structures as functions [4, 17, 8].

[2] Oberon proper has equality tests for both procedure and record types, but it is equality of pointers rather than equality of values (which is of limited use).

4.1 Basic Constructs

For the simple imperative language —even with local variable declarations—, predicate transformer semantics takes a very simple form which appears many times in the literature. In a fixed state space, each program p is interpreted as a function $[p]$ from predicates to predicates (over that state space). For a given program, it suffices to fix a state space that includes all variables mentioned in the program (and all specification constants of interest). In order to interpret prescriptions and **con**, the classical healthiness conditions [4] are dropped. We define *predicate transformer* to mean monotonic function from predicates to predicates. The definition of $[-]$ for most of the language is given in Table 2. Note:

- The clause for local variables is only correct for ψ that does not depend on x; renaming can be used to circumvent this limitation.
- The semantics of Oberon's regional type guard (**with** $r : T$ **do** p) can be derived directly from its translation into **if** r **is** T **then** p **else abort** (which is semantically but not syntactically type correct). The semantics of a program involving a type-guarded expression $r(T')$ can be derived from its translation into a regionally guarded program without the guard (T').
- μ denotes least fixed point; in **rec** , p usually depends on Q.

Table 2. Simple predicate transformer semantics, for predicates ψ over a fixed global state space.

$$
\begin{aligned}
[\text{abort}]\psi &= \textit{false} \\
[x := e]\psi &= \psi[x := e] \\
[[\varphi \text{ , } \theta]]\psi &= \varphi \wedge (\theta \Rightarrow \psi) \\
[\text{con } X \text{ in } p]\psi &= (\exists X :: [p]\psi) \\
[p \text{ ; } q]\psi &= [p]([q]\psi) \\
[\text{if } b \text{ then } p \text{ else } q]\psi &= (b \Rightarrow [p]\psi) \wedge (\neg b \Rightarrow [q]\psi) \\
[\text{rec } Q \text{ is } p]\psi &= (\mu Q :: [p]).\psi \\
[\text{var } x \text{ in } p]\psi &= (\forall x :: [p]\psi)
\end{aligned}
$$

This presentation of the semantics is informal in that semantics for basic data types is not given. A formal treatment would involve an interpretation of each data type T as a set $[T]$ of values of that type, and predicate φ would be interpreted as a subset $[\varphi]$ of the state space. For example, $[x \text{ is } T]$ would be the set of states where the value of x is a member of $[T]$. Except at the end of subsection 4.2 where it is needed for clarity, we also elide the interpretation $[e]$ of expressions.

4.2 Parameters and Procedure Constants

Call-by-value parameters (in imperative languages) have three features of note: actuals are evaluated before the procedure body is executed; formal parameters are the same as local variables in the procedure body; and stack allocation is used. All three features are transparent in the well known program transformation that replaces a call $P(e)$ of a procedure named P with the program

$$(\text{var } x \text{ in } x := e \ ; \ P') \tag{12}$$

where x is the formal parameter of P and P' is the parameterless procedure with the same body as P. Evaluation of e takes place before execution of P' thanks to the sequential composition of $x := e$ before P'. The formal x is declared to be a local variable. And stack allocation for the parameter has been reduced to stack allocation for local block, which in turn is evident from the predicate transformer semantics which uses a scoped quantifier.

Even more important than the fact that (12) accounts for the key features of value parameters is the fact that it provides for their elimination. What is left is the call of a parameterless procedure, which can be eliminated using the copy rule. Result and value/result parameters can be treated similarly. This is the effect of our semantics, but it is formalized using the **param** construct (analogous to lambda abstraction), for two reasons: first, the scope of formal parameters should be the procedure body — naively applying the above translation can result in scope conflicts; and second, typing problems arise with type extensions.

The type compatibility of actual parameter $e : T'$ with formal $x : T$ when $T' \in T$ follows directly from (12) and the compatibility rule for assignment. But if P has a value/result parameter r of type T, then a call $P(r')$ that passes record r' of type T' (with $T' \in T$) would be translated to the program

$$(\text{var } r : T \text{ in } r := r' \ ; \ P \ ; \ r' := r) \tag{13}$$

in which the second assignment is ill-typed. Oberon and theoretical Oberon allow the call, giving it a useful semantics: the extra fields of r' are retained unchanged. This effect can be obtained by using the type-guarded assignment $r' := r(T')$ in (13), but we choose not to do so. Instead of eliminating value/result parameters using (13), their semantics is given directly in the interpretation below of **call**. This in turn depends on the interpretation of **param** expressions as pairs consisting of the parameters and the meaning of the body:

$$[\text{param } (x : T, \text{var } y : U) \, p] = \langle \, (x : T, \text{var } y : U), \ [p] \, \rangle \ .$$

For definitions of procedure constants (and constants of all kinds) the **let** construct is used. In order to define the semantics of the language including **let**, we use *environments* (e.g. [20]) which associate identifiers with their definitions. All of the clauses in Table 2 are changed uniformly by adding the environment η, e.g.

$$[p \ ; \ q]_\eta . \varphi = [p]_\eta . ([q]_\eta . \varphi) \ .$$

The semantics of a complete program —one with no undeclared identifiers— is given as a predicate transformer by taking the environment to be empty.

The clause for **let** is the embodiment of the copy rule:

$$[\text{let } P \text{ be } e \text{ in } s]_\eta \cdot \varphi = [s]_{\eta[P \mapsto [e]]} \cdot \varphi \tag{14}$$

where $\eta[P \mapsto [q]]$ is η except that it maps P to $[q]$ (e may be a data or procedure expression). The environment is included in the "closure" representing a parameterized command:

$$[\text{param } (x{:}T, \text{var } y{:}U) \, p]_\eta = \langle (x{:}T, \text{var } y{:}U), \eta, [p]_\eta \rangle \; .$$

This leaves the calls of procedure values to be interpreted by

$$[\text{call } \langle (x{:}T, \text{var } y{:}U), \eta, [p]_\eta \rangle (e, w)]_{\eta'} \varphi = ([p]_\eta (\varphi[w := y]))[x, y := [e]_{\eta'}, w]$$

which can be obtained from (13) using the semantics of local variables, assignment, and sequential composition.

4.3 Procedure Variables

Unfortunately, the simple semantics based on a single state space is, in general, incompatible with procedure variables. In that semantics, to interpret a program p we need a single state space W whose coordinates include all those coordinates that appear in p. Suppose p has a global procedure variable pv. To reason about pv, we want to interpret its values as ranging over predicate transformers over some state space X. Write $PT.X$ for the set of predicate transformers on space X, i.e. the set of monotonic functions $\mathbb{P}.X \to \mathbb{P}.X$. If pv acts in state space X then W projects onto at least $PT.X$. Hence for reasons of cardinality it is impossible that X is W.[3] Equations like $X = (X \to X)$ have been solved, of course, using Scott domains (e.g. [20]). That approach is unsatisfactory for us, however, not just because it introduces a substantial complication but because it precludes prescriptions and such.

The time-honored alternative is careful use of types. For the rest of this section, we assume that procedures assigned to procedure variables have no global variables. Then the values of a procedure variable can be taken to be predicate transformers in a specific state space, namely the one determined by its parameter types. We return to this assumption in section 5. The values of a procedure variable are predicate transformers paired with parameters, but for simplicity we ignore parameters in the following. Assignment to procedure variables, and its generalization $pv :\sqsupseteq s$, is handled by the definitions in Table 2. It remains to define $[\text{call } pv]$. The environment can be elided because it plays no special role here. The weakest condition under which **call** pv establishes ψ

---

[3] The obvious way to tackle this is to impose healthiness conditions on $PT.X$. But even if it is restricted to all universally junctive predicate transformers (already too few), the result is isomorphic to the set $X \to X$ of ordinary functions, which is larger than X.

is that it's value establishes ψ under some precondition φ, and φ holds. This suggests the definition:

$$[\text{call } pv].\psi = pv.\psi \ . \tag{15}$$

This looks like a joke but it is not. Recall that in each state, the value of pv is a predicate transformer; so we can interpret $pv.\varphi$ as the state predicate true in states where the predicate obtained by applying the value of pv to φ is true. This would be less confusing if we had formalized the interpretation of predicates (writing $[\psi]$ in Table 2, and $[pv.\psi]$ here). In fact this is quite like the semantics for calls of procedure constants: for procedure constant P, (14) says to look up the value of P in the environment. For procedure variables, the value is in the store, that is all.

Space does not allow a complete formalization. For clarification it may be helpful to formulate (15) using an explicit parameter w for the state; the dependence of pv and ψ on the state is now explicit: $[\text{call } pv].(\psi_w) = pv_w.(\psi_w)$. But an important benefit of predicate transformer semantics is that the state space is implicit in the interpretation of predicates.

In program derivation, one does not calculate the weakest precondition of a procedure call, but rather uses the procedure specification to ensure that the precondition of the call is adequate (e.g. [3]). Similarly, for procedure variables one does not calculate the weakest precondition of the call, but rather asks that the precondition of the call is adequate for the call to establish its postcondition. That is the essence of axiom (6), which is validated by our semantics (as is (7)).

5 Program Types

Whereas variables and procedure parameters are typed in Oberon (and Algol-like languages) according to syntax-directed recursive rules, there is no such type system for commands. Instead, there are two global rules:

- when combining commands with control constructs, identifiers that occur in different constituents must have the same type (and are semantically identified), e.g. in $x := 0 ; x := 1$ the two occurrences of x refer to the same entity; and
- in a complete program, all identifiers must be declared.

The effect is that the "type" of $x := 0$ (i.e. the coordinates of its state space, some of which happen not to be modified by its execution) depends on its context. Since denotational semantics is defined by structural recursion, this contextual dependence is usually avoided by interpreting programs with respect to a fixed global state space, as in Table 2 (but see Reynolds et al [18, 20]).

The approach sketched in section 4, where we assume procedure variables have a completely specified type (because they have no globals), suggests using a type system for commands and defining the semantics by recursion on typings. The problem with this suggestion is that although it has sometimes been proposed that all globals be explicitly specified (e.g. [4]), that is certainly not done in

Oberon: parameters of procedure constants and variables are fully specified, but their globals are not specified at all. In a sense, commands are polymorphic,[4] but Oberon is not otherwise polymorphic — unless you count subtypes or ad hoc polymorphism for basic types. The single state space approach can be salvaged for procedures with global variables if those globals are required to be in the global scope (and cannot include procedure variables to which the procedure itself is assigned): the fixed collection of globals can be taken to be part of the single state space used to interpret procedures (in addition to the coordinates for their parameters and local variables). This includes all interesting Oberon programs, because Oberon has the stronger restriction that procedures assigned to variables must be defined in the global scope. In future work we will study less restricted languages using richer type systems for procedures — especially systems using a form of type extension for procedure types.

6 Conclusion

The remarkable point is not that we give semantics to theoretical Oberon but that we do so without the mathematical *tours de force* that seem to be needed for Algol-like languages (e.g. [15]). More importantly, the suitability of Oberon for systematic program development has been shown by taking it to be a subset of a rich calculus. Other subsets and other typing disciplines may also be of interest. To quote Reynolds [18]: "... the essence of Algol is not a straitjacket. It is a conceptual universe for language design... ." Control over global variables is needed in various theories for reasoning with procedure calls, suggesting the need for a more prominent role for globals in the typing of procedures. The analysis sketched in this paper and developed elsewhere shows that the author's program-level model of higher types [12] does provide an appropriate semantics, but practical experience with derivation of higher order programs is needed to guide the design of useful type systems.

Acknowledgements

Without the pioneering efforts of Niklaus Wirth and Jürg Gutknecht, this work would have had neither subject matter nor inspiration. John Reynolds' papers, lectures, and personal communication have been very helpful, as have discussions with Carroll Morgan, Peter O'Hearn, Bob Tennant, and Markus Kaltenbach. Development of predicate transformer semantics for higher types was instigated by Tony Hoare. Thanks also to the referees, whose helpful advice I was not entirely able to follow. This research was partly supported by a Cullen Foundation grant for travel to Oxford.

---

[4] This perspective has led to recent advances in semantics for Algol-like languages, using an explicit but "parametrically polymorphic" store [15].

References

1. K.R. Apt. Ten years of hoare's logic, a survey, part I. *ACM Transactions on Programming Languages and Systems*, 3, 1981.
2. R. J. R. Back. Correctness preserving program refinements: Proof theory and applications. Technical Report Tract 131, CWI, 1980.
3. A. Bijlsma. Calculating with procedure calls. *Information Processing Letters*, 46:211–217, 1993.
4. Edsger W. Dijkstra. *A Discipline of Programming*. Prentice-Hall, 1976.
5. Paul Gardiner and Carroll Morgan. Data refinement of predicate transformers. *Theoretical Computer Science*, 87:143–162, 1991.
6. C.A.R. Hoare, J. He, and A. Sampaio. Normal form approach to compiler design. to appear in *Acta Informatica*, 1993.
7. C.A.R. Hoare and Jifeng He. Data refinement in a categorical setting. Technical Monograph PRG-PRG-90, November 1990.
8. Anne Kaldewaij. *Programming: the Derivation of Algorithms*. Prentice-Hall, 1990.
9. Clare E. Martin. Preordered categories and predicate transformers. Dissertation, Oxford University, 1991.
10. Carroll Morgan. Procedures, parameters, and abstraction: Separate concerns. *Science of Computer Programming*, 11(1), 1988.
11. Carroll Morgan. *Programming from Specifications*. Prentice Hall, 1990.
12. David A. Naumann. Predicate transformers and higher order programs. Submitted to *Theoretical Computer Science*, 1992.
13. David A. Naumann. Two-categories and program structure: Data types, refinement calculi, and predicate transformers. Dissertation, University of Texas at Austin, 1992.
14. David A. Naumann. Data refinement, call by value, and higher types. Submitted to *Science of Computer Programming*, 1993.
15. P. W. O'Hearn and R. D. Tennant. Relational parametricity and local variables (preliminary report). In *Proceedings, Twentieth POPL*, pages 171–184, 1993.
16. Martin Reiser and Niklaus Wirth. *Programming in Oberon*. Addison-Wesley, 1992.
17. John C. Reynolds. *The Craft of Programming*. Prentice-Hall, 1981.
18. John C. Reynolds. The essence of Algol. In J. W. de Bakker and J. C. van Vliet, editors, *Algorithmic Languages*. North-Holland, 1981.
19. R. D. Tennant. Semantical analysis of specification logic. *Information and Computation*, 85, 1990.
20. R.D. Tennant. *Semantics of Programming Languages*. Prentice Hall, 1991.
21. Stephen Weeks and Matthias Felleisen. On the orthogonality of assignments and procedures in Algol. In *Proceedings, Twentieth POPL*, 1993.
22. N. Wirth. The programming language Oberon. *Software – Practice and Experience*, 18(7), 1988.

Adding Concurrency to the Oberon System

Spiros Lalis* and Beverly A. Sanders

Institut für Computersysteme
Swiss Federal Institute of Technology (ETH Zürich)
ETH Zentrum
CH-8092 Zürich
Switzerland

Abstract. The Oberon system, developed by Niklaus Wirth and Jürg Gutknecht is unusual in that, although it has a "single process multi-tasking" window user interface, it does not support concurrent execution of programs. This approach yields a simple design and works surprisingly well, but it does have limitations of practical importance. In this report we describe a project, Concurrent Oberon, which introduces concurrency into the Oberon system while maintaining the simplicity and spirit of the original system.

1 Introduction

Oberon is the name of both a programming language designed by Niklaus Wirth [RW92], and an operating system [WG92] for the Ceres workstation [Ebe87] designed by Wirth and Jürg Gutknecht.[2] Both the language, which supports object oriented programming, and the system are notable for the elegance of their designs and significant functionality with a very small use of resources.

In contrast to window systems that associate a process and a separate address space with each window, the Oberon system supports "single process multi-tasking" with a process called the Oberon loop. Associated with each window, or viewer object is a dynamically installed procedure referred to as a handler. Keyboard and mouse drivers are polled by the Oberon loop, and when an event is detected, a message is sent via an upcall to the handler of the affected viewer which processes it and returns control back to the Oberon loop.

The conventional method for executing a "program" is via special procedures called commands that are called from within the viewer handlers during event processing. Specifically, clicking the middle mouse button on the command name appearing in a viewer causes the handler for the viewer to call the corresponding procedure. Once the command execution is finished, the Oberon loop again polls for input events. As a consequence, all commands in the system are executed sequentially and the system does not respond to input events during command execution. This approach, with a single thread of control, simplifies the system

---

* Supported in part by Swiss National Science Foundation grant 21-25280-88.

[2] Oberon has been ported to other hardware platforms where it runs on top of a standard operating system.

significantly, and makes command execution efficient since there is no overhead due to context switching. Moreover, it works very well for interactive applications, since most event handling such as displaying a character is essentially instantaneous and the most commonly used commands require little time.

Nevertheless, there are cases where this model of control is less appropriate. For example, it is inconvenient to implement a long computation as a command because its execution will block the system until it is finished, perhaps for several minutes or even hours. Also, there are applications which react to external events (e.g. arrival of a packet over a network) and where polling via explicitly invoked commands is not an acceptable solution.

To accommodate such applications the Oberon system offers a mechanism called tasks. Tasks are special procedures that are invoked periodically from within the Oberon loop. In other words, tasks can be viewed as commands that are executed repeatedly in the background without an explicit user request.

Although this mechanism can be used to implement some simple applications, it does not adequately solve the above mentioned problems. Its most important deficiency is that task execution can be delayed arbitrarily long by input handling and long running commands. This makes it impossible to guarantee that a certain task will be executed within a reasonable amount of time, which is vital for applications with real time constraints such as connection-oriented network protocols.

Another problem is that tasks, like commands, run until they terminate, thus blocking the system during their execution. In order to use the task mechanism to implement background processing that still allows interactive use of the system, the computation must be broken down into a sequence of task invocations. Since tasks and the Oberon loop are executed on the same stack, this requires explicitly saving the intermediate state of the computation each time before returning control to the Oberon loop[Rei91]. This can be extremely cumbersome for programs that cannot be modeled as simple state machines, for example programs with recursion. In addition, since control transfer between a task and the Oberon loop is explicit, tasks must be implemented to execute "long enough" but not "too long". If this is not the case, either the task will block the system, or processor time will be wasted mostly in useless control transfers. However, thinking about how much time is needed to execute a particular piece of code has little to do with the problem the programmer is actually trying to solve. Such calculations are annoying, machine dependent, and even impossible to make if computation time is a function of parameters that are not known at implementation time.

Our experience with tasks showed that these limitations are indeed of practical importance, and therefore we decided to develop a special version of the Oberon system that would be free of these problems. An additional motivation was the desire to employ Oberon, which is currently used in all lower division computer science courses at ETH, for programming exercises in concurrent programming. Our objective was to offer basic facilities for straightforward background processing, timely response to events, and low level synchronization primitives that would allow various higher level techniques used in teaching and research in concurrency to be easily implemented. An important constraint was that the simplicity and spirit of the original system should be maintained as much as possible.

2   Introducing Concurrency

To achieve the goals stated above, we implemented a new version of the Oberon system called Concurrent Oberon. Concurrent Oberon provides threads (or light-weight processes) along with a simple priority scheduler. Threads encapsulate the state of a computation and allow control of the processor to be passed to different threads without requiring the programmer to explicitly save the state. The scheduler recognizes three priority levels. Round robin scheduling is used to allocate processor time among threads at a given level, provided there are no waiting threads with higher priority.

Since the "single process multi-tasking" approach has been demonstrated to work very well for user-driven applications, we maintained this aspect of the original Oberon system and also use a single thread to handle all keyboard and mouse events. By convention, the Oberon thread, which handles input from the mouse and keyboard, runs with normal priority, background threads have low priority, and threads which execute rarely, but must react quickly to events are assigned high priority. This convention provides fast response time where this is critical and makes the presence of background threads essentially invisible to the user. For reasons of compatibility with the standard system, the task mechanism is still available.

2.1   System Structure

Integrating concurrency in an elegant way into a system that was designed to be sequential posed an interesting engineering problem. Complicated globally shared data structures are pervasive in the Oberon system and careless use of concurrency could cause serious inconsistencies. There are several shared data structures which are not entirely encapsulated in an abstract data type and hence are accessed directly by clients, rather than via procedures. In addition, many Oberon applications are implemented using the implicit assumption that a sequence of operations is executed atomically, i.e, it is often assumed that the state of an abstract data type will not change between two successive operations. Thus the obvious approach of adding synchronization to the operations of an abstract data type would not be sufficient.

In Concurrent Oberon, the Oberon thread is a special thread which, by convention, is responsible for handling all keyboard and mouse events, controlling the screen, and accessing high level data structures related to input and display handling. The decision to delegate this work to a single thread means that most explicit synchronization can be avoided. New threads may be created to perform background activities and access such data structures only by communicating in a controlled way with the Oberon thread. This design provides the desired functionality without adding significant complexity, and as an important practical matter, remains compatible with the standard Oberon system. Existing applications can be run on Concurrent Oberon without changes.

An important consequence of this approach is that concurrency is not completely transparent. Programs which are to be run as threads must be programmed to do so, in particular, they must synchronize with the Oberon thread when accessing shared I/O related data structures. We do not feel that this is a significant restriction, because typical background threads do not interact with the display often and programming the required synchronization is not difficult.[3]

All threads in Concurrent Oberon occupy a single global address space. For each module, only a single image is loaded in memory. This means that global variables of the module are shared by all threads running procedures in the module or modules who import the variables. This fact affects the structure of a module whose procedures will be executed as threads, or whose data will be accessed concurrently.

The lack of memory protection between unrelated threads would be fraught with difficulties if we were using an unsafe language because small programming errors, especially with pointers and arrays, could easily cause system crashes by unintentionally modifying memory belonging to a different thread. In our case, the strong type checking provided by the Oberon programming language essentially guarantees that such errors will be detected by the compiler, or cause a trap at run time. Specifically, the compiler and run time system guarantee that all pointers always point to an object of a given type and which is in the scope of the code being executed, thus eliminating the sort of errors that cause system crashes by accidentally modifying memory belonging to unrelated threads. Our system also detects stack overflow and terminates the offending thread.

The main source of vulnerability to programming errors which can cause serious problems in Concurrent Oberon is failure to properly synchronize with the Oberon thread. According to our experience so far, this is not a serious problem because the critical shared data structures are well known and, as will be described in section 3.6, synchronization can be implemented in a straightforward way.

3 Threads

In this section, we describe the module Threads, which provides the new capabilities in Concurrent Oberon. This module provides procedures for creating and destroying threads, a set of operations that can be used to implement arbitrary synchronization tools (e.g. semaphores, signals, monitors, conditions), methods for synchronizing and communicating with the Oberon thread, and a few procedures used in system programs.

---

[3] An analogous lack of transparency exists in the standard system as well. Procedures that are to be executed as commands must follow a particular protocol to acquire input parameters, and are therefore implemented differently.

3.1 Thread Definition, Create and Destroy

A thread descriptor is a record (object) containing state information about the
thread needed by the scheduler. The definition is given below.

```
TYPE
  Thread = POINTER TO ThreadDesc;
  ThreadProc = PROCEDURE();
  ThreadDesc = RECORD
    state, priority: SHORTINT; (*read-only*)
    incNo: LONGINT; (*read-only*)
  END;

PROCEDURE Create (this: Thread; proc, trapproc: ThreadProc;
  wsp: LONGINT);
PROCEDURE Destroy (this: Thread);
```

To create a thread, the programmer must first allocate a thread descriptor using NEW and call the procedure Create with parameters body, trapbody
and wsp. wsp indicates the desired size of the workspace (stack) while proc and
trapproc are procedures to be executed when the thread is started or experiences a runtime error, respectively. The latter procedure can be used to perform
cleanup operations or to restart the thread. If trapbody is set to NIL, no actions
will be taken. Create allocates a workspace and initializes the state and priority
of the thread to the default values: suspended and low, respectively. If a newly
allocated thread descriptor is used as a parameter, the field incNo is set to 1, else,
if an old descriptor is used to host the execution, then the incarnation number
is incremented by 1. A descriptor may not be reused unless the corresponding
execution has terminated. Procedure Destroy destroys the specified thread and
releases its resources.

The Threads module exports two read-only variables:

```
VAR cur, oberon: Thread;
```

The variable oberon indicates the Oberon thread, and variable cur indicates
the thread that is currently executing, and is updated by the scheduler each time
control is transferred.

Parameters can be passed to a thread by using the type extension facility
of the Oberon language [RW92]. New descriptor types can be defined by augmenting the base thread descriptor with new fields, and instances thereof may
be used to invoke the Create operation. Thread procedures can access the additional fields of the descriptor of the currently executing thread using type guards
on the variable cur. An example is given in section 3.5.

3.2 Priorities

The scheduler recognizes three priority levels and considers a thread for execution only when there are no ready threads of higher priority. Undesired starvation is avoided by giving threads appropriate priorities and by changing their priority when this is required. The priority of a thread is given in the **priority** field of the thread descriptor and can be changed by the procedure SetPriority.

```
CONST
  low = 0; norm = 1; high = 2;

PROCEDURE SetPriority (this: Thread; prio: SHORTINT);
```

As an example of priority adjustment, the Oberon thread performs input handling at a normal priority and decreases its priority when no more input events are detected to allow execution of low priority threads. Conversely, as soon as keyboard and mouse activity is sensed the priority of the Oberon thread is reset to normal. High priority should be reserved for threads that are suspended most of the time waiting for events that must be handled very quickly.

3.3 Control Operations

The following exported constants and procedures control the control state of threads as indicated by the **state** field of the thread descriptor.

```
CONST
  ready = 0; asleep = 1; suspended = 2; destroyed = 3; trapped = 4;

PROCEDURE Suspend;
PROCEDURE Resume (this: Thread);
PROCEDURE Pass;
PROCEDURE Sleep (msecs: LONGINT);
```

A thread whose state is **ready** is either running or waiting for the scheduler to allocate the processor. The state of a thread is changed to **suspended** by executing **Suspend**. A thread which is suspended will become ready when another process executes **Resume**. Immediately after the call of **Create**, a thread has a state of **suspended** and **Resume** must be called in order to change its state to **ready** and allow its execution. **Resume** does not specify the next thread to be called by the scheduler, but rather changes the state of the specified thread so that it may be chosen by the scheduler. **Sleep** changes the state of the thread to **asleep**. A thread with state **asleep** will become **ready** when resumed by another thread or after the specified amount of time has expired. States **destroyed** and **trapped** indicate that the thread has been destroyed or has experienced a run-time error, respectively. **Pass** calls the scheduler in order to pass control to another thread without changing the state of the originating thread.

In order to implement higher level synchronization primitives, we offer primitives that can be used to define atomic segments, i.e. sequences of instructions which should be executed atomically.

```
PROCEDURE BeginAtomic;
PROCEDURE EndAtomic;
```

BeginAtomic marks the current thread so that it will not be preempted by the scheduler. EndAtomic unmarks the thread. Thus a sequence of statements between a BeginAtomic, EndAtomic pair will be executed atomically.[4] Properly nested BeginAtomic-EndAtomic pairs are also allowed. Within an atomic section, control can be explicitly passed to other threads using Suspend, Sleep or Pass. When the thread resumes execution, the remaining segment will be atomic. Atomic segments are intended to be used primarily for the implementation of higher level synchronization primitives, not as an all purpose synchronization device. We have introduced atomicity as a high level concept. Thus the implementation of synchronization primitives using the Threads module does not depend on how preemption is implemented. This is in contrast to the usual approach of requiring the programmer to turn off interrupts to achieve atomicity. This approach requires the programmer to turn off all possible sources of interrupts that might cause preemption, which may introduce unnecessary delays and requires modification of synchronization primitives if new interrupt sources are added later. It might also be the case that "preemption" is not implementeded via interrupts. For example one could modify the compiler to insert instructions (which are not under the programmers control) for releasing the processor.

3.4 Semantics of Operations

In this section, we give formally defined abstract implementations of the above operations. In particular the definitions of the control operations serve to clarify the semantics of the primitives as well as allow formal reasoning about programs using them. In order to specify the abstract implementations, we will use the programming notation given in [And91]. This notation is an extension of Dijkstra's notation including an await statement for synchronization, and using angle brackets "<>" to enclose atomic commands, or sequences of commands. The proof rules described in [And91], can be used for formal correctness proofs.

```
PROCEDURE Create (this: Thread; proc, trapproc: ThreadProc;
  wsp: LONGINT);
    <stack of size wsp allocated ∧ this.state := suspended
    ∧ this.priority := low>


PROCEDURE Destroy (this: Thread);
  <if
```

---

[4] Hardware interrupts are still serviced within an atomic segment.

```
    this.state < destroyed → this.state := destroyed
  []
    this.state >= destroyed → skip
  fi>
```

```
PROCEDURE SetPriority (this: Thread; prio: SHORTINT);
  <this.priority := prio>
```

```
PROCEDURE Suspend;
  <th.state := suspended>;<await(th.state = ready)>
```

```
PROCEDURE Resume (this: Thread);
  <if
    this.state < destroyed → this.state := ready
  []
    this.state >= destroyed → skip
  fi>
```

```
PROCEDURE Pass;
  skip
```

```
PROCEDURE Sleep (msecs: LONGINT);
  <t := Time; th.state := asleep>;
  <await(Time ≥ t+msecs ∨ th.state = ready) → th.state := ready>
```

In the above, t is a fresh variable local to the thread and th is the (fixed) name of the current thread. The variable Time represents the value of the system clock. Time is modified only by a clock process which repeatedly increments its value. As a result, the only assertions containing Time which are allowed are those which will not be falsified by increasing its value , for example $Time \geq n$. Typically, one does not formally reason with real time but uses Sleep to detect failures or indicate "wait only a finite amount of time", with the parameter serving as a hint to the scheduler about how long to wait. The value of the parameter then influences performance rather than correctness.

The procedures BeginAtomic and EndAtomic are equivalent to angle brackets. Calls to Pass and Suspend and Sleep appearing between BeginAtomic and EndAtomic explicitly release atomicity, which is continued after the call until the closing EndAtomic statement. In other words, in the context of a BeginAtomic, EndAtomic pair, the definition of Suspend and Sleep are modified by removing the initial " <" and the final " >". The definition of Pass becomes "> skip <". For example: BeginAtomic S1; S2; Suspend; S3; S4 EndAtomic is equivalent to < S2;S2 ; th.state := suspended>;<await(th.state = ready) ; S3;S4>. The statement sequence BeginAtomic S1; S2; Pass; S3; S4 EndAtomic is equivalent to < S2;S2 >;< S3;S4>.

3.5 Example

To demonstrate how threads can be programmed using these primitives we give an example of a thread that increments a variable approximately every T *msecs*. The value of the increment and T are passed to the thread as parameters. In addition, to guarantee that the variable is incremented during input handling, the priority of the thread is set to normal:

```
TYPE
  MyThread = POINTER TO MyThreadDesc;
     (*MyThreadDesc is extension of base type Threads.ThreadDesc*)
  MyThreadDesc = RECORD (Threads.ThreadDesc)
    i, inc, T: LONGINT
  END;

PROCEDURE Body;
BEGIN
  (*Type guard to ensure that Threads.cur has type MyThread*)
  WITH Threads.cur: MyThread DO
    LOOP Threads.Sleep(Threads.cur.T);
      INC(Threads.cur.i, Threads.cur.inc)
    END;
  END;
END Body;

PROCEDURE Start;
  VAR t: Thread;
BEGIN
    (* Allocate thread descriptor and initialize added fields*)
   NEW(t); t.inc := 100; t.i := 0; t.T := 50;
    (* Allocate workspace and initialize base fields *)
   Threads.Create(t,Body,NIL,128);
    (* Increase priority, default was low*)
   Threads.SetPriority(t,Threads.norm);
    (* Change thread state to ready to allow execution *)
   Threads.Resume(t)
  END Start;
```

3.6 Synchronization with the Oberon thread

In order for a thread to interact with Oberon, for example to display output, it must modify data structures which are also modified by the Oberon thread. In this section we describe two sets of primitives which have been provided in order to allow threads to synchronize with the Oberon thread without causing inconsistencies in shared data structures, and also without requiring explicit synchronization in programs which will be executed solely by the Oberon thread. The first approach ensures mutually exclusive access to shared data structures, while the second is similar to asynchronous message passing to the Oberon thread, which performs the requested command. Explicit synchronization is not necessary for file operations; in this case serializability is automatically guaranteed by the system.

Locking Shared Data Structures The easiest way to synchronize with the Oberon thread is using a special lock maintained by the Threads module. The lock is acquired and released with the following two procedures:

```
PROCEDURE LockOberon;
PROCEDURE UnlockOberon;
```

The lock is held initially by the Oberon thread and is released and reacquired between command and task execution. If a critical data structure is to be modified from within a thread, `LockOberon` must be invoked prior execution of the corresponding operations, and `UnlockOberon` must be called afterwards to release the lock. `LockOberon` suspends the current thread if another thread already holds the lock. Blocking is very likely to happen, since the lock is held by the Oberon thread during command and task execution. Nesting of these commands is allowed so that a command containing them may be executed from within the Oberon loop as well as by a background thread.

For example, if we wanted to modify the thread implemented in section 3.4 to print a message on the system log-viewer each time the variable is incremented, we could do this as follows:

```
PROCEDURE Body1;
  VAR W: Texts.Writer;
BEGIN
  WITH Threads.cur: MyThread DO
    Texts.OpenWriter(W); (*initialize writer*)
    LOOP
      Threads.Sleep(Threads.cur.T);
      INC(Threads.cur.i, Threads.cur.inc);
        (*write value to buffer*)
      Texts.WriteString(W, "i = ");
      Texts.WriteInt(W, Threads.cur.i, 0);
        (*display buffer contents in log viewer*)
      Threads.LockOberon;
      Texts.Append(Oberon.Log, W.buf);
      Threads.UnlockOberon
    END
  END
END Body1;
```

In the above example, W of type `Texts.Writer` is local to the thread and contains a buffer into which the desired output is written. The command `Texts.Append(Oberon.Log,W.buf)` appends the contents of the buffer to the Oberon Log, which is a shared text usually displayed in a viewer on the screen.

Asynchronous Communication An alternative approach to displaying output or otherwise accessing data structures shared with the Oberon thread is via "asynchronous message passing". This technique requires more effort to implement than using `LockOberon` and `UnlockOberon`, but is more appropriate when it is important that a thread produce output without blocking (e.g. monitoring a network or within a trap handler).

```
TYPE
  OberonAction = POINTER TO OberonActionDesc;
  OberonActionProc = PROCEDURE (this: OberonAction);
  OberonActionDesc = RECORD
    body: OberonActionProc
  END;

PROCEDURE QueueOberonAction (this: OberonAction);
PROCEDURE DoOberonActions;
```

The main idea of this approach is to put operations which access data structures shared with the Oberon thread into a queue which is processed from within the Oberon thread itself. `DoOberonActions` is not intended to be used by applications programmers but is called from within the Oberon thread. When it is called, it processes all `OberonActions` in the queue by executing the procedure which has been installed in the body field of the `OberonActionDesc` with the corresponding `OberonAction` pointer as a parameter. As with thread descriptors, arbitrary parameters can be passed to the body procedure by extending `OberonActionDesc` with additional fields.

We show how the previous example can be modified to use actions instead of locking.

```
TYPE
  MyAction = POINTER TO MyActionDesc;
  MyActionDesc = RECORD (Threads.OberonActionDesc)
    i: LONGINT
  END;

VAR W: Texts.Writer;

PROCEDURE ActionBody (this: Threads.OberonAction);
BEGIN
  WITH this: MyAction DO
    Texts.WriteString(W, "i = "); Texts.WriteInt(W, this.i, 0);
```

```
      Texts.WriteLn(W); Texts.Append(Oberon.Log, W.buf)
    END
  END ActionBody;

  PROCEDURE Body2;
    VAR a: Action;
  BEGIN
    WITH Threads.cur: MyThread DO
      LOOP
        Threads.Sleep(Threads.cur.T); INC(Threads.cur.i);
          (*Create and initialize Action*)
        NEW(a); a.i := Threads.cur.i; a.body := ActionBody;
          (*queue action*)
        Threads.QueueOberonAction(a)
      END
    END
  END Body2;
```

3.7 Systems Programming

Module Threads exports several procedures which are used by other system modules (Input and System) to implement preemption and trap handling. Although not intended to be used by application programmers, they appear in the interface and are briefly described here for completeness.

```
PROCEDURE Tick (ticks: LONGINT);

PROCEDURE Enumerate (proc: ThreadProc);
PROCEDURE StackState (this: Thread; VAR fp, pc: LONGINT);
```

Tick is used by the timer interrupt handler to advance the internal time of the scheduler. Enumerate applies an operation to all threads. StackState returns the last saved state of a thread, i.e. the frame pointer of the last procedure activation record and the address of the instruction last executed. This data is invalid for the currently executing thread. These procedures are used by the commands System.ShowThreads, System.DumpThread, System.DestroyThread and System.DestroyAllThreads which provide thread status information and control to the user.

The following procedures of module Threads may be called from withing interrupt handlers: Resume, SetPriority, Pass, and Destroy.[5]

---

[5] If destroy is called while the specified thread is within an atomic segment, the thread is marked, and its destruction actually carried out when the thread leaves the atomic segment. This allows, for example, the Oberon thread to be interrupted from the keyboard (with cntl-shift-del) safely.

4 Thread Status and Control

Module System has been extended with three new commands which allow the user to monitor and destroy threads via interactive commands.

```
PROCEDURE ShowThreads;
PROCEDURE DumpThread;
PROCEDURE DestroyThread;
PROCEDURE DestroyAllThreads;
```

Command ShowThreads opens a viewer that contains all running threads. For each thread, its identification number, body procedure and state are displayed. DumpThread opens a viewer with the stack dump of the selected thread and DestroyThread destroys the selected thread. In both cases, the target thread is identified by selecting its number on the screen. DestroyAllThreads destroys all threads except the Oberon thread. This should be done before turning the machine off to avoid potential inconsistencies in the file system.

5 Implementation

5.1 Module Structure

The module structure of the new system is given in figure 1. Modules belonging to the standard system which are not shown have been used without changes in Concurrent Oberon.

5.2 Scheduling

The scheduling policy has been designed to provide processor sharing between threads of the same priority, to make the presence of background threads invisible to the user, and to guarantee fast response time for specific events. The mechanism for preemption is a call to the scheduler from within an interrupt handler using the procedure Threads.Pass. The timer interrupt handler in module Input calls this procedure periodically in order to allow round robin scheduling of ready threads.

The priority of the Oberon thread is set to normal and the scheduler called whenever keyboard and mouse events are present. Mouse events are detected by polling the mouse device from within the timer interrupt handler. Keyboard events automatically result in an interrupt, which is also handled by module Input. After all input events have been processed the Oberon thread sets its priority to low, giving again low priority threads a chance to execute. With this scheme, background threads can utilize cycles not needed for interactive execution without affecting the response time of the interactive system. The structure of the Oberon thread is as follows:

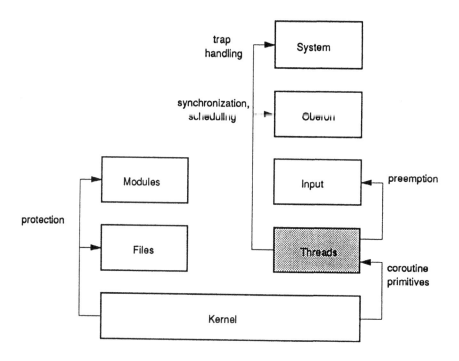

Fig. 1. Module Structure of Concurrent Oberon

```
Oberon-lock is already acquired
LOOP
  WHILE keyboard or mouse events present
  DO
    handle event;
    execute next task
    Threads.DoOberonActions;
  END;
  Threads.SetPriority(Threads.oberon, low);
  Threads.UnlockOberon ;
  Threads.Pass;
  Threads.LockOberon;
END;
```

5.3   Trap Handling

In the standard Oberon system, whenever a run time error occurs, a viewer
containing the stack dump of the failed execution is opened from within the

trap handler. In Concurrent Oberon, the trap handler is implemented in module Threads and uses the facilities for asynchronous communication with the Oberon thread to display trap information. An action which opens a viewer and prints the stack state is defined as a special `OberonAction`. On the occurrence of a runtime error, a new action is created, its fields are initialized with the state information provided by the system's kernel, the action is queued using `QueueOberonAction`, and control is switched to an available thread. A viewer displaying the corresponding trap information is opened at a later point in time, when the Oberon thread finally processes the queued `OberonAction`.

5.4 Stack Management and Garbage Collection

Module Kernel was extended to provide a set of primitives supporting simple coroutines [Wir83], thus providing the basic mechanism for implementing concurrency in the system. Module Threads uses these primitives to implement the desired scheduling policy.

```
TYPE
  Stack = POINTER TO StackDesc;
  StackDesc = RECORD
    status: SHORTINT;
    inSVC: BOOLEAN;
  END;

PROCEDURE NewStack (VAR stack: Stack; wsp: LONGINT);
PROCEDURE InitStack (stk: Stack; proc: PROCEDURE);
PROCEDURE Transfer (stk: Stack);
PROCEDURE Current (VAR cur: Stack);

PROCEDURE StackState (this: Stack; VAR fp, pc: LONGINT);
```

`NewStack` allocates stack space of size `wsp` and returns a pointer to the corresponding stack descriptor. Stacks are collected by the garbage collector as soon as they are no longer referenced and the corresponding memory is returned to the system. To initialize an execution on a stack, `InitStack` must be called with the `stack` and the procedure `proc` as parameters. The execution is started when control is transferred to the initialized stack for the first time. Control transfer is implemented by procedure `Transfer` that implements the appropriate actions depending on the status of the current execution. This information is recorded in the fields `status` and `inSVC`. The `status` field indicates whether the execution is in normal mode, trap mode, or interrupt mode. In the latter case, `inSVC` indicates if a supervisor call has been interrupted. The stack of the current execution is retrieved with calls to procedure `Current`. Finally, procedure `StackState` can be used to retrieve the state of a stack's execution. Data returned by `StackState` is invalid for the current stack. In the current version of the system, these procedures have been implemented as supervisor calls which allows the interface of module Kernel to be extended without requiring the symbol file to be changed. This is desirable since changing a symbol file of a module results in the need to recompile all clients of the module.[6]

To implement synchronization within module Files, two procedure variables are defined:

```
VAR lock, unlock: PROCEDURE,
```

These variables initially contain empty procedures, and at a later point in time, are overwritten by module Threads with real synchronization primitives (i.e. BeginAtomic and EndAtomic). This indirection allowed module Threads to be placed at a higher level in the module hierarchy of the Oberon system, namely outside the "inner core".

Additional changes to the Kernel module were modifications of the garbage collector to mark pointers on all stacks. In the standard system, there is only one stack belonging to the Oberon loop and the garbage collector is called only between commands when the stack is guaranteed to be empty. Obviously, in Concurrent Oberon, there is never a time when all stacks can be guaranteed to be empty, thus the more elaborate collector is required.

In older versions of the Concurrent Oberon system, stack memory was mapped on a separate address region and the memory management hardware of the processor was used to detect instances of stack overflow. In the current implementation stacks are allocated on the heap and like any other object, are automatically collected by the garbage collector. A modified Oberon compiler is used to make checks for stack overflow on each procedure call. This allows a safe implementation of Concurrent Oberon on machines which lack memory management hardware, such as the Ceres-3 workstation at the expense of slightly more expensive procedure calls.

6 Conclusion

Concurrency was not supported in the original Oberon system. We have argued that, although the Oberon experience has demonstrated that concurrency is not needed to support interactive applications in a system with a multiple window user interface, concurrency is nonetheless desirable in a single user system. We have demonstrated that it can be introduced without a significant increase in size or complexity by developing the Concurrent Oberon system. Since all interfaces remained the same or were extended, the semantics of the Oberon loop were preserved, Concurrent Oberon is completely compatible with the standard system and all applications run without change. So far, Concurrent Oberon has been successfully used as a basis for a research project on distributed programming [Lal94], several senior thesis projects (including the implementation of standard communication protocols TCP/IP [Ste92] which significantly increases

---

[6] If we had been willing to change the symbol file of the Kernel, the interface could have been simplified by replacing some of the procedures with exported variables.

the ability of the Oberon system to communicate with the outside world), and for student exercises in distributed systems and concurrent programming.

Acknowledgments: We would like to thank Professor Niklaus Wirth for providing the sources of the Oberon system, thereby making this project possible, and also for modifying the Oberon compiler to detect stack overflow. Professor Hanspeter Mössenböck, Martin Gitsels, and many others in the Institut für Computersysteme provided helpful comments on an earlier version of this paper. Finally, we would like to thank Philipp Heuberger and Martin Gitsels, who used Concurrent Oberon and provided valuable feedback during the development of the system.

References

[And91] Gregory R. Andrews. *Concurrent Programming: Principles and Practice.* Benjamin/Cummings Publishing Company, 1991.

[Ebe87] Hans Eberle. *Development and Analysis of a Workstation Computer.* PhD thesis, Swiss Federal Institute of Technology (ETH Zürich), 1987. Number 8431.

[Lal94] Spiros Lalis. *Distributed Object-Oriented Programming in a Network of Personal Workstations.* PhD thesis, Swiss Federal Institute of Technology (ETH Zürich), 1994. in preparation.

[Rei91] Martin Reiser. *The Oberon System: User Guide and Programmer's Manual.* ACM Press, Addison-Wesley, 1991.

[RW92] Martin Reiser and Niklaus Wirth. *Programming in Oberon: Steps Beyond Pascal and Modula.* ACM Press, Addison-Wesley, 1992.

[Ste92] Michael Steiner. TCP/IP für Ceres. ETH Informatik Diplomarbeit (Senior Thesis), 1992.

[WG92] Niklaus Wirth and Jürg Gutknecht. *Project Oberon: The Design of an Operating System and Compiler.* ACM Press, Addison-Wesley, 1992.

[Wir83] Niklaus Wirth. *Programming in Modula 2.* Springer Verlag, 1983.

Lecture Notes in Computer Science

For information about Vols. 1–704
please contact your bookseller or Springer-Verlag

Vol. 740: E. F. Brickell (Ed.), Advances in Cryptology – CRYPTO '92. Proceedings, 1992. X, 593 pages. 1993.

Vol. 741: B. Preneel, R. Govaerts, J. Vandewalle (Eds.), Computer Security and Industrial Cryptography. Proceedings, 1991. VIII, 275 pages. 1993.

Vol. 742: S. Nishio, A. Yonezawa (Eds.), Object Technologies for Advanced Software. Proceedings, 1993. X, 543 pages. 1993.

Vol. 743: S. Doshita, K. Furukawa, K. P. Jantke, T. Nishida (Eds.), Algorithmic Learning Theory. Proceedings, 1992. X, 260 pages. 1993. (Subseries LNAI)

Vol. 744: K. P. Jantke, T. Yokomori, S. Kobayashi, E. Tomita (Eds.), Algorithmic Learning Theory. Proceedings, 1993. XI, 423 pages. 1993. (Subseries LNAI)

Vol. 745: V. Roberto (Ed.), Intelligent Perceptual Systems. VIII, 378 pages. 1993. (Subseries LNAI)

Vol. 746: A. S. Tanguiane, Artificial Perception and Music Recognition. XV, 210 pages. 1993. (Subseries LNAI).

Vol. 747: M. Clarke, R. Kruse, S. Moral (Eds.), Symbolic and Quantitative Approaches to Reasoning and Uncertainty. Proceedings, 1993. X, 390 pages. 1993.

Vol. 748: R. H. Halstead Jr., T. Ito (Eds.), Parallel Symbolic Computing: Languages, Systems, and Applications. Proceedings, 1992. X, 419 pages. 1993.

Vol. 749: P. A. Fritzson (Ed.), Automated and Algorithmic Debugging. Proceedings, 1993. VIII, 369 pages. 1993.

Vol. 750: J. L. Díaz-Herrera (Ed.), Software Engineering Education. Proceedings, 1994. XII, 601 pages. 1994.

Vol. 751: B. Jähne, Spatio-Temporal Image Processing. XII, 208 pages. 1993.

Vol. 752: T. W. Finin, C. K. Nicholas, Y. Yesha (Eds.), Information and Knowledge Management. Proceedings, 1992. VII, 142 pages. 1993.

Vol. 753: L. J. Bass, J. Gornostaev, C. Unger (Eds.), Human-Computer Interaction. Proceedings, 1993. X, 388 pages. 1993.

Vol. 754: H. D. Pfeiffer, T. E. Nagle (Eds.), Conceptual Structures: Theory and Implementation. Proceedings, 1992. IX, 327 pages. 1993. (Subseries LNAI).

Vol. 755: B. Möller, H. Partsch, S. Schuman (Eds.), Formal Program Development. Proceedings. VII, 371 pages. 1993.

Vol. 756: J. Pieprzyk, B. Sadeghiyan, Design of Hashing Algorithms. XV, 194 pages. 1993.

Vol. 757: U. Banerjee, D. Gelernter, A. Nicolau, D. Padua (Eds.), Languages and Compilers for Parallel Computing. Proceedings, 1992. X, 576 pages. 1993.

Vol. 758: M. Teillaud, Towards Dynamic Randomized Algorithms in Computational Geometry. IX, 157 pages. 1993.

Vol. 759: N. R. Adam, B. K. Bhargava (Eds.), Advanced Database Systems. XV, 451 pages. 1993.

Vol. 760: S. Ceri, K. Tanaka, S. Tsur (Eds.), Deductive and Object-Oriented Databases. Proceedings, 1993. XII, 488 pages. 1993.

Vol. 761: R. K. Shyamasundar (Ed.), Foundations of Software Technology and Theoretical Computer Science. Proceedings, 1993. XIV, 456 pages. 1993.

Vol. 762: K. W. Ng, P. Raghavan, N. V. Balasubramanian, F. Y. L. Chin (Eds.), Algorithms and Computation. Proceedings, 1993. XIII, 542 pages. 1993.

Vol. 763: F. Pichler, R. Moreno Díaz (Eds.), Computer Aided Systems Theory – EUROCAST '93. Proceedings, 1993. IX, 451 pages. 1994.

Vol. 764: G. Wagner, Vivid Logic. XII, 148 pages. 1994. (Subseries LNAI).

Vol. 765: T. Helleseth (Ed.), Advances in Cryptology – EUROCRYPT '93. Proceedings, 1993. X, 467 pages. 1994.

Vol. 766: P. R. Van Loocke, The Dynamics of Concepts. XI, 340 pages. 1994. (Subseries LNAI).

Vol. 767: M. Gogolla, An Extended Entity-Relationship Model. X, 136 pages. 1994.

Vol. 768: U. Banerjee, D. Gelernter, A. Nicolau, D. Padua (Eds.), Languages and Compilers for Parallel Computing. Proceedings, 1993. XI, 655 pages. 1994.

Vol. 769: J. L. Nazareth, The Newton-Cauchy Framework. XII, 101 pages. 1994.

Vol. 770: P. Haddawy (Representing Plans Under Uncertainty. X, 129 pages. 1994. (Subseries LNAI).

Vol. 771: G. Tomas, C. W. Ueberhuber, Visualization of Scientific Parallel Programs. XI, 310 pages. 1994.

Vol. 772: B. C. Warboys (Ed.),Software Process Technology. Proceedings, 1994. IX, 275 pages. 1994.

Vol. 773: D. R. Stinson (Ed.), Advances in Cryptology – CRYPTO '93. Proceedings, 1993. X, 492 pages. 1994.

Vol. 774: M. Banâtre, P. A. Lee (Eds.), Hardware and Software Architectures for Fault Tolerance. XIII, 311 pages. 1994.

Vol. 775: P. Enjalbert, E. W. Mayr, K. W. Wagner (Eds.), STACS 94. Proceedings, 1994. XIV, 782 pages. 1994.

Vol. 776: H. J. Schneider, H. Ehrig (Eds.), Graph Transformations in Computer Science. Proceedings, 1993. VIII, 395 pages. 1994.

Vol. 777: K. von Luck, H. Marburger (Eds.), Management and Processing of Complex Data Structures. Proceedings, 1994. VII, 220 pages. 1994.

Vol. 778: M. Bonuccelli, P. Crescenzi, R. Petreschi (Eds.), Algorithms and Complexity. Proceedings, 1994. VIII, 222 pages. 1994.

Vol. 779: M. Jarke, J. Bubenko, K. Jeffery (Eds.), Advances in Database Technology — EDBT '94. Proceedings, 1994. XII, 406 pages. 1994.

Vol. 780: J. J. Joyce, C.-J. H. Seger (Eds.), Higher Order Logic Theorem Proving and Its Applications. Proceedings, 1993. X, 518 pages. 1994.

Vol. 782: J. Gutknecht (Ed.), Programming Languages and System Architectures. Proceedings, 1994. X, 344 pages. 1994.

Vol. 783: C. G. Günther (Ed.), Mobile Communications. Proceedings, 1994. XVI, 564 pages. 1994.